ITALIAN PHYSICAL SOCIETY

PROCEEDINGS

OF THE

INTERNATIONAL SCHOOL OF PHYSICS « ENRICO FERMI »

Course XXV

edited by M. N. Rosenbluth
Director of the Course

VARENNA ON LAKE COMO
VILLA MONASTERO
JULY 9-21 1962

Advanced Plasma Theory

1964

ACADEMIC PRESS • NEW YORK AND LONDON

SOCIETÀ ITALIANA DI FISICA

RENDICONTI
DELLA
SCUOLA INTERNAZIONALE DI FISICA «ENRICO FERMI»

XXV Corso

a cura di M. N. Rosenbluth
Direttore del Corso

VARENNA SUL LAGO DI COMO
VILLA MONASTERO
9-21 LUGLIO 1962

Teoria dei plasmi

1964

ACADEMIC PRESS • NEW YORK AND LONDON

ACADEMIC PRESS INC.
111 Fifth Avenue
New York 3, N. Y.

United Kingdom Edition
Published by
ACADEMIC PRESS INC. (London) Ltd.
Berkeley Square House, London W. 1

Library of Congress Catalog Card Number: 63-20572

PRINTED IN ITALY

INDICE

M. N. Rosenbluth – Introduction . . . . pag. IX

Gruppo fotografico dei partecipanti al Corso . . . . fuori testo

Lezioni

W. B. Thompson – Kinetic theory of plasma . . . . pag. 1

Russel Kulsrud – General stability theory in plasma physics . . . . » 54

G. Ecker – Gas discharge theory . . . . » 97

M. N. Rosenbluth – Topics in microinstabilities . . . . » 137

H. P. Furth – Instabilities due to finite resistivity or finite current-carrier mass . . . . » 159

P. A. Sturrock – Nonlinear theory of electrostatic waves in plasmas . » 180

C. Mercier – Equilibre et stabilité des systèmes toroïdaux en magnétohydrodynamique au voisinage d'un axe magnétique , ... » 214

B. Bertotti – Boundary layer problems in plasma physics . . . . . . » 234

Seminari sull'invarianza adiabatica

R. M. Kulsrud – Introduction for papers on adiabatic invariance . . » 252

R. M. Kulsrud – Adiabatic invariant of the harmonic oscillator . . . » 254

M. Kruskal – The gyration of a charged particle . . . . » 260

Introduction.

M. N. Rosenbluth

General Atomic, Division of General Dynamics Corporation - San Diego, Cal.

Unhappily, I am finally writing this introduction in the days following the awful events of President Kennedy's assassination which is dominating my thoughts now. Perhaps I will be excused then for saying some words about a strong nontechnical recollection I have of the Varenna summer school. I think everyone who has been there will recall it as a vital experience in international co-operation where new friends from many countries are made and ideas and feelings—technical, political and personal—freely exchanged. Even after making allowances for the idyllic setting and the common technical interest of the participants, one is bound to receive a strong impression of the common humanity of us all and a hope that the obstacles to the late President's vision of a world of peace and progress may indeed be overcome.

The field of plasma physics is especially suitable for an international school since so many of the participants have been involved in the world-wide quest for thermonuclear power. Fortunately, a happy tradition of total co-operation between all countries exists in this area, perhaps because it is almost unique in being a well-defined and challenging scientific problem from which great economic, but not military, benefits may be foreseen as the result of scientific progress.

My own interest in the thermonuclear problem has undoubtedly led to a certain bias in the selection of the topics to be covered in this course. It has always seemed to me that the best starting point for the study of plasma physics is to try to understand the simplest possible, most idealized, situation. This implies a high-temperature plasma in which collective effects predominate and a quiescent plasma in a simple equilibrium so that linear wave dynamics can be considered. The experimental attainment of this theoreticians' dream is of course precisely the thermonuclear problem. After this understanding has been reached it may be possible to comprehend more fully complex astrophysical and geophysical plasmas In the meantime, considerable technological progress may be made through intuition and invention, but I believe that the ultimate exploitation of plasmas will depend on systematic scientific theory and experiment.

The aim of this series of lectures then has been to present the elements of the theory of high-temperature plasmas. I believe that most of the important topics have been at least sketchily covered, with the exception of the study of the various complex types of high frequency plasma waves—plasma oscillation, cyclotron waves, whistlers, etc. All of the students had considerable knowledge of the field already so that the lectures are on a fairly advanced level. Some preliminary study of the field would probably be advisable before attempting to read these notes, although I do not feel that they should be useful only to experts. For example, the lecture notes from the Riso course in 1960 (Riso Report No. 18 Danish AEK) are somewhat more elementary. It must be pointed out that these lecture notes occupy a no-man's land between original research papers and a coherently organized and polished book of a single author. Their virtue perhaps is that they are broader and more up to date than a monograph could be.

The first step in understanding a plasma is the reduction of this many-body system to a 6-dimensional problem described by the Vlasov and Fokker-Planck equations. This topic is covered in the Kinetic Theory lectures of the noted Shakespearian actor, Dr. W. BE. THOMPSON. A further reduction to the modified magnetohydrodynamic fluid-type equations is possible in the limit of small gyro-radius and low frequency. This development culminates in the energy-principle for determining hydrodynamic stability and is discussed by one of its originators, Dr. RUSSEL KULSRUD. The energy principle has been exploited to great effect in the study of complex plasma confinement geometries. Some of these very general results are presented here by Dr. CLAUDE MERCIER. These sets of lectures may be considered to be on basic and already well-understood aspects of plasma physics.

The other lectures concern topics still imperfectly understood and under development.

An important area in which much work continues to be done is the exploitation of the rich variety of phenomena contained within the Vlasov equations but not the simple fluid equations. A superficial treatment of some of these phenomena is contained in my lectures on Microinstabilities. It should be noted that a slight extension of these techniques to include variations along the magnetic field leads to the « universal » instability.

It has become apparent that much of the content of the usual MHD equations is contained in the infinite conductivity constraint that particles remain « tied » to field lines in any motion. Thus when this constraint is relaxed even slightly new types of motion become possible as is discussed in the lectures on Resistive Instabilities by Dr. HAROLD FURTH, who also enlivened the proceedings notably in his role of Court Jester.

As in other types of fluid dynamics a great many of the answers we seek are obscured by our inability to treat nonlinear problems effectively. Some important beginnings have been made and are discussed in the lectures of Professor PETER STURROCK. Another topic, of both mathematical and physical importance, concerns the proper ordering scheme for the approximate treatment of boundary layers and is discussed in the lectures of Dr. BRUNO BERTOTTI. Finally, while the course was in general directed toward the study of high-temperature plasma a very fascinating sidelight was given to us in the lively lectures of Dr. GUNTER ECKER on the low-temperature gas-discharge regime. It is interesting to note the differences and similarities in two such closely-related topics and sobering to realize how rare and difficult communication between them has been.

In addition to these formal lectures the theory of adiabatic invariants was discussed by R. KULSRUD following two important papers by KULSRUD himself and M. KRUSKAL. They are reprinted «in extenso». Further seminars on specialized and current research were presented by G. SANDRI on The New Foundations of Statistical Dynamics, D. PFIRSCH on Microinstabilities of the Mirror Type in Inhomogeneous Plasmas, G. LAVAL and R. PELLAT on the Boundary Layer between a Plasma and a Magnetic Field, G. KNORR on Nonlinear phenomena in Microscopic Wave Propagation, S. CUPERMAN, F. ENGELMAN and J. OXENIUS on Nonthermal Impurity Radiation from a Hot Spherical Plasma, K. VON HAGENOW and H. KOPPE on The Partition Function of a Completely Ionized Gas, F. ENGELMAN on Quantum-Mechanical Treatment of Electric Microfield Problems in Plasmas, J. B. TAYLOR on Finite Larmor Radius Effect and the Rotational Instability of Plasma in Fast B_z Compression Experiments (Theta-Pinch), D. VOSLAMBER and D. K. CALLEBAUT on Stability of Force-Free Magnetic Fields, C. F. WANDEL and O. KOFOED-HANSEN on the Eulerian-Lagrangian Transformation in the Statistical Theory of Turbulence, N. A. KRALL on Oscillations in Nonuniform Plasmas, P. J. KELLOG on The Earth's Environment: Observations on a Very Large Mirror Machine, E. T. KARLSON on the Motion of a Plasma in an Inhomogeneous Magnetic Field.

Unfortunately it is only possible to list these by title and author here as we dispose only of their summaries. These seminars helped a great deal to give some experimental back-up for these theoretical lectures.

Finally, I would like to give my deep thanks to all those who have participated so whole-heartedly in the course—to the knowledgeable attentive and long-suffering student body, to those who presented seminars; and to the lecturers. I hope the reader will concur with my judgement of the excellence of their efforts.

The chief thanks of course go to our hosts, the Italian Physical Society, who have provided the magnificent villa on Lake Como, with its excellent facilities and amenities, making a physical background most conducive to relaxed and harmonious discussion. I would like to mention especially those representatives on the scene who have been so helpful and made us feel so welcome; Dr. Bruno Bertotti, the scientific secretary, who has done most of the director's work, the secretarial staff—Misses LaFarge, Navone, and Variola—who not only coped with these mountains of manuscript but also went far out of their way to take care of all the troublesome details which invariably arise, and Professor G. Germanà who ran the whole operation so skillfully.

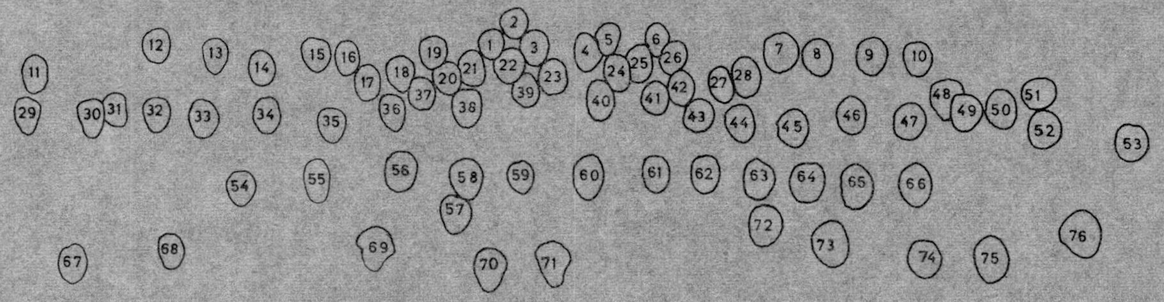

1. D. Callebaut
2. A. Kildal
3. I. Spalding
4. T. E. Stringer
5. F. Engelmann
6. M. A. Hellberg
7. A. Tozzi
8. R. Giovanelli
9. E. Tränkle
10. G. Brifford
11. K. Von Hagenow
12. U. Felderhof
13. J. B. Taylor
14. A. Cavaliere
15. E. Witalis
16. A. Lemaire
17. H. Skullerud
18. P. Bertrand
19. D. Voslamber
20. S. Tjøtta
21. N. Warmoltz
22. Ch. Wu
23. P. Kellogg
24. R. D. Dietz
25. M. G. Rusbridge
26. V. Jensen
27. L. Dupas
28. A. C. Levi
29. C. F. Wandel
30. G. Nozza
31. R. T. P. Whipple
32. N. Krall
33. C. Bartoli
34. D. Lepechinsky
35. T. Consoli
36. F. Gratton
37. E. Karlson
38. S. Cuperman
39. J. Monaghan
40. D. Pfirsch
41. B. Liley
42. H. Repp
43. A. Skorupski
44. P. Lauscher
45. S. Lafleur
46. G. Laval
47. Mrs. Laval
48. G. Sandri
49. I. P. Plantevin
50. M. Fournier
51. G. Knorr
52. Mrs. Engelmann
53. H. Kulsrud
54. R. Kulsrud
55. G. Ecker
56. W. Thompson
57. P. Kulsrud
58. B. Bertotti
59. M. N. Rosenbluth
60. H. Furth
61. P. S. Sturrock
62. R. B. Hall
63. Ch. Lafleur
64. I. Archambault
65. Mrs. Archambault
66. J. Naze
67. P. Schram
68. M. Weenink
69. A. Lafarge
70. E. Variola
71. L. Navone
72. G. Szamosi
73. Szamosi (son)
74. Mrs. Szamosi
75. Mrs. Karlson
76. B. Robinson

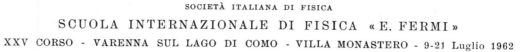

SOCIETÀ ITALIANA DI FISICA

SCUOLA INTERNAZIONALE DI FISICA « E. FERMI »

XXV CORSO - VARENNA SUL LAGO DI COMO - VILLA MONASTERO - 9-21 Luglio 1962

LEZIONI

Kinetic Theory of Plasma.

W. B. Thompson

Atomic Energy Research Establishment - Culham

Part I

1. – Introduction.

The object of kinetic theory is to extract from an exact and detailed description of a complex physical system that information needed to describe its gross behaviour.

For example, consider a system composed of many charged particles having mass m_i, charge e_i and velocity $\boldsymbol{v}_i$; then in a volume V large enough to contain many particles, we may define a macroscopic density ϱ, velocity $\boldsymbol{V}$, charge Q, and current $\boldsymbol{J}$, by

$$\frac{1}{V}\sum m_i = \varrho\,; \qquad \frac{1}{V}\sum_i m_i \boldsymbol{v}_i = \varrho \boldsymbol{V}\,;$$

$$\frac{1}{V}\sum e_i = Q\,; \qquad \frac{1}{V}\sum_i e_i \boldsymbol{v}_i = \boldsymbol{J}\,.$$

It is often convenient to write

$$\boldsymbol{v}_i = \boldsymbol{V} + \boldsymbol{c}_i\,; \qquad \boldsymbol{J} = Q\boldsymbol{V} + \boldsymbol{j}\,.$$

The macroscopic variables ϱ, $\boldsymbol{V}$ etc. satisfy equations of motion that may be deduced from the laws of motion for the individual particle, *i.e.* from

$$m\dot{\boldsymbol{v}}_i = \boldsymbol{F}_i = e_i(\boldsymbol{E} + \boldsymbol{v}_i \times \boldsymbol{B}) + \boldsymbol{F}_{\text{int}}\,,$$

where

$$\sum_i \boldsymbol{F}_{\text{int}} = 0$$

from conservation of momentum. Then $\partial\varrho/\partial t$ is determined by equating the rate of change of density to the flux into a volume; *i.e.*

$$\frac{\partial\varrho}{\partial t} + \nabla\cdot(\varrho\boldsymbol{V}) = 0\,. \tag{I.1.1}$$

The rate of change of momentum is similarly

$$\frac{\partial}{\partial t}(\varrho\boldsymbol{V}) + \frac{\partial}{\partial\boldsymbol{x}}\cdot\sum_i m\boldsymbol{v}_i\boldsymbol{v}_i = \frac{\partial}{\partial t}(\varrho\boldsymbol{V}) + \frac{\partial}{\partial\boldsymbol{x}}\cdot(\varrho\boldsymbol{V}\boldsymbol{V}) + \frac{\partial}{\partial\boldsymbol{x}}\cdot\sum m\,\boldsymbol{c}_i\,\boldsymbol{c}_i =$$
$$= \sum_i \boldsymbol{F}_i = \sum_i e_i(\boldsymbol{E} + \boldsymbol{v}_i\times\boldsymbol{B}) = Q(\boldsymbol{E} + \boldsymbol{V}\times\boldsymbol{B}) + \boldsymbol{j}\times\boldsymbol{B}\,,$$

and, defining the stress tensor $p_{rs} = \sum_i mc_{ir}c_{is}$ and using the continuity eq. (I.1.1),

$$\varrho\left(\frac{\partial\boldsymbol{V}}{\partial t} + \boldsymbol{V}\cdot\nabla\boldsymbol{V}\right) + \nabla\cdot\boldsymbol{p} = Q(\boldsymbol{E} + \boldsymbol{V}\times\boldsymbol{B}) + \boldsymbol{j}\times\boldsymbol{B}\,. \tag{I.1.2}$$

In a similar way, we may equate the rate of gain of energy to the rate of doing work, *i.e.*

$$\frac{\partial}{\partial t}\left(\sum_i\frac{1}{2}mv_i^2\right) + \nabla\cdot\sum_i\boldsymbol{v}_i\frac{1}{2}mv_i^2 = \sum_i\boldsymbol{F}_i\cdot\boldsymbol{v}_i = \sum_i e\boldsymbol{v}_i\cdot\boldsymbol{E} = \boldsymbol{J}\cdot\boldsymbol{E}\,,$$

or

$$\boldsymbol{V}\cdot\left(\varrho\frac{\mathrm{D}\boldsymbol{V}}{\mathrm{D}t} + \nabla\cdot\boldsymbol{p}\right) + \frac{\mathrm{D}}{\mathrm{D}t}\sum_i\frac{1}{2}mc_i^2 + \sum_i\frac{1}{2}mc_i^2\nabla\cdot\boldsymbol{V} + (\nabla\cdot\boldsymbol{p})\cdot\boldsymbol{v} + \nabla\cdot\sum\boldsymbol{c}\frac{1}{2}mc^2 =$$
$$= Q\boldsymbol{V}\cdot\boldsymbol{E} + \boldsymbol{j}\cdot\boldsymbol{E}\,.$$

and defining the internal, energy

$$U = \sum_i\tfrac{1}{2}mc_i^2 = \tfrac{3}{2}nkT\,,$$

and the heat flux

$$\boldsymbol{q} = \sum\boldsymbol{c}\,\tfrac{1}{2}mc^2\,,$$

$$\frac{\mathrm{D}U}{\mathrm{D}t} + U\nabla\cdot\boldsymbol{V} + (\boldsymbol{p}\cdot\nabla\boldsymbol{V}) + \nabla\cdot\boldsymbol{q} = \boldsymbol{j}\cdot\boldsymbol{E}\,. \tag{I.1.3}$$

For an ionized gas, the charged particles are of two sorts; electrons and ions, and the interdiffusion of these particles gives rise to a current; the electron and ions moving with speeds $\boldsymbol{V}+\Delta\boldsymbol{V}_-$, $\boldsymbol{V}+\Delta\boldsymbol{V}_+$. By considering the two gases separately the quantity $\Delta\boldsymbol{V}_-$ is found to satisfy

$$\frac{\mathrm{D}\boldsymbol{V}}{\mathrm{D}t} + \frac{\mathrm{D}\Delta\boldsymbol{V}}{\mathrm{D}t} + \Delta\boldsymbol{V}\cdot\nabla\Delta\boldsymbol{V} + \Delta\boldsymbol{V}\cdot\boldsymbol{V} + \frac{1}{nm}\nabla\cdot\boldsymbol{p}_- = \frac{e_-}{m_-}(\boldsymbol{E} + \boldsymbol{V}\times\boldsymbol{B} + \Delta\boldsymbol{V}\times\boldsymbol{B}) + \frac{\boldsymbol{F}_-}{m_-}\,.$$

If this and a similar equation for ΔV_+ are multiplied by a product of density and charge there results

$$\frac{\mathrm{D}\boldsymbol{j}}{\mathrm{D}t} + \frac{e_-}{m_-}\nabla\cdot\boldsymbol{p}_- + \frac{e_+}{M_+}\nabla\cdot\boldsymbol{p}_+ - \frac{Q}{\varrho}\nabla\cdot\boldsymbol{p}_+ + \left[\frac{Q^2}{\varrho} - \left[n_-\left(\frac{e^2}{m_-}\right) + n_+\left(\frac{e}{m_+}\right)^2\right]\cdot\right.$$

$$\left.\cdot[\boldsymbol{E} + \boldsymbol{V}\times\boldsymbol{B}] + \frac{Q}{\varrho}(\boldsymbol{j}\times\boldsymbol{B}) - \left[n_-\left(\frac{e_-^2}{m_-}\right)\Delta\boldsymbol{V}_- + n_+\left(\frac{e}{M_+}\right)^2\Delta\boldsymbol{V}_+\right]\right]\times\boldsymbol{B} = 0\,,$$

$$\text{(I.1.4)}\qquad \frac{m_-}{ne^2}\frac{\mathrm{D}\boldsymbol{j}}{\mathrm{D}t} + \frac{m_-}{ne_-}[\nabla\cdot\boldsymbol{p}_- - \boldsymbol{j}\times\boldsymbol{B}] + [\boldsymbol{E} + \boldsymbol{V}\times\boldsymbol{B}] - \eta\boldsymbol{j} = 0\,,$$

where

$$\eta\boldsymbol{j} = \frac{m_-}{ne^2}\left[\frac{ne_-}{m_-}\boldsymbol{F}_- + \frac{ne_+}{m_+}\boldsymbol{F}_+\right].$$

A more formal treatment of the relation between the microscopic and the macroscopic can be effected by employing a distribution function f; a quantity which describes the statistical evolution of the system, in which case the underlying microscopic dynamics of the system is embraced in an equation of motion for f, the equation of transport.

There are several sorts of distribution function f, ranging from the Liouville function $F(x_1, x_2, \ldots, x_N, v_1, \ldots, v_N)$ to the Boltzmann single-particle function $f(x, v)$. The Liouville function is a function of the complete set of micro coordinates, and satisfies the equation

$$\text{(I.1.5)}\qquad \frac{\partial F}{\partial t} + [H(x_1 \ldots x_N,\ v_1 \ldots v_N), F] = 0\,,$$

which is completely equivalent to the microscopic dynamics, H being the complete Hamiltonian. This can be written, introducing the acceleration field $\boldsymbol{A}_i(x_1 \ldots x_N)$,

$$\frac{\partial F}{\partial t} + \sum_i\left(\boldsymbol{v}_i\cdot\frac{\partial F}{\partial\boldsymbol{x}_i} + \boldsymbol{A}_i(x_1 \ldots x_N)\cdot\frac{\partial F}{\partial\boldsymbol{v}}\right) = 0\,.$$

The equivalence of Liouville's equation and the equations of motion is established by observing that if the system is given as in the state specified by

$$\boldsymbol{x}_1(0),\ \boldsymbol{x}_2(0) \ldots \boldsymbol{x}_N(0),\ \boldsymbol{v}_1(0),\ \boldsymbol{v}_2(0) \ldots \boldsymbol{v}_N(0)\,,$$

so that

$$F(0) = \Pi\,\delta(\boldsymbol{x}_i - \boldsymbol{x}_i(0))\,\delta(\boldsymbol{v}_i - \boldsymbol{v}_i(0))\,,$$

its subsequent evolution is described by

$$F(t) = \Pi\, \delta(\boldsymbol{x}_i - \boldsymbol{X}_i(t))\, \delta(\boldsymbol{v}_i - \boldsymbol{V}_i(t))\,, \tag{I.1.6}$$

where $\boldsymbol{X}_i(t)$ and $\boldsymbol{V}_i(t)$ are the relevant solutions of the equations of motion, *i.e.*

$$\boldsymbol{X}_i(t) = \boldsymbol{x}_i(0) + \int_0^t \mathrm{d}t'\, \boldsymbol{V}_i(t')\,,$$

$$\boldsymbol{V}_i(t) = \boldsymbol{v}_i(0) + \int_0^t \mathrm{d}t'\, \boldsymbol{A}_i(\boldsymbol{X}_i(t'), \boldsymbol{V}_i(t');\, \sum \boldsymbol{X}_{\mathrm{not}\,i}(t'), \boldsymbol{V}_{\mathrm{not}\,i}(t'))\,. \tag{I.1.7}$$

The Boltzmann function on the other hand satisfies an equation of the form

$$\frac{\partial f}{\partial t} + \boldsymbol{v}\cdot\frac{\partial f}{\partial \boldsymbol{x}} + \boldsymbol{A}_0(x, t)\cdot\frac{\partial f}{\partial \boldsymbol{v}} = \left.\frac{\partial f}{\partial t}\right|_{\mathrm{int}}. \tag{I.1.8}$$

The transport equation for the Boltzmann function may be obtained by repeated integration of the Liouville equation, for f itself is defined by

$$f(x_1, v_1) = V\int F(x_1 \dots x_N, v_1 \dots v_N)\, \mathrm{d}^3x_2 \dots \mathrm{d}^3x_N\, \mathrm{d}^3v_2 \dots \mathrm{d}^3v_N\,.$$

If there were no interaction between particles so that the acceleration field $\boldsymbol{A}_i$ could be factored as $\boldsymbol{A}_1(x_1, v_1)\boldsymbol{A}_2(x_2, v_2)$ ... etc., then a closed equation for f could be obtained by integrating over the Liouville equation as

$$\frac{\partial f}{\partial t} + \boldsymbol{v}\cdot\frac{\partial f}{\partial \boldsymbol{x}} + \boldsymbol{A}\cdot\frac{\partial f}{\partial \boldsymbol{v}} = 0\,, \tag{I.1.9}$$

the collisionless Boltzmann, or Vlasov equation.

If, however, there exists an inter-particle potential $\phi(x_i, x_j)$ the final integral cannot be evaluated in terms of $\boldsymbol{x}_1$, $\boldsymbol{v}_1$ and f alone, instead it becomes

$$-\frac{V}{m_1}\sum_j \frac{\partial}{\partial \boldsymbol{v}_1}\cdot\int \frac{\partial}{\partial \boldsymbol{x}_1}\phi(\boldsymbol{x}_1, \boldsymbol{x}_j)\, F(\boldsymbol{x}_1 \dots \boldsymbol{x}_N)\, \mathrm{d}^3v_2 \dots \mathrm{d}^3v_N\, \mathrm{d}^3x_2 \dots \mathrm{d}^3x_N\,,$$

or, introducing the two-particle function

$$f(1, 2) = V^2\int F(\boldsymbol{x}_1, \boldsymbol{x}_2 \dots \boldsymbol{x}_N)\, \mathrm{d}^3x_3 \dots \mathrm{d}^3x_N\, \mathrm{d}^3v_3 \dots \mathrm{d}^3v_N\,,$$

$$\left.\frac{\partial f}{\partial t}\right|_{\mathrm{int}} = \frac{N}{Vm_1}\frac{\partial}{\partial \boldsymbol{v}_1}\cdot\int \frac{\partial \phi(1, 2)}{\partial \boldsymbol{x}_1} f(1, 2)\, \mathrm{d}^3x_2\, \mathrm{d}^3v_2\,,$$

and the general equation of transport for f becomes

$$\frac{\partial f}{\partial t}+\boldsymbol{v}\cdot\frac{\partial f}{\partial \boldsymbol{x}}+\boldsymbol{A}_0\cdot\frac{\partial f}{\partial \boldsymbol{v}}-\frac{n}{m_1}\frac{\partial}{\partial \boldsymbol{v}}\cdot\int\frac{\partial}{\partial \boldsymbol{x}_1}\phi(x_1, x_2)\,f(1, 2)\,\mathrm{d}^3x_2\,\mathrm{d}^3v_2\,,$$

or

$$\frac{\partial f}{\partial t}+\boldsymbol{v}\cdot\frac{\partial f}{\partial \boldsymbol{x}}+\boldsymbol{A}_0\cdot\frac{\partial f}{\partial \boldsymbol{v}}+I(f)=0\,. \tag{I.1.10}$$

The first major problem of kinetic theory is to find an approximate form for $I(f)$; the second being that of solving (I.1.10) for a given form of I and deducing the moments required for a macroscopic description of phenomena.

For diffuse gases in which a *strong* but *localized interaction* occurs between the particles, a coarse-grained equation for f may be obtained in which the interaction term I is represented by the rate of change of f produced by impulsive collisions between particles; ($\boldsymbol{g}=\boldsymbol{v}_1-\boldsymbol{v}_2$), $\theta=$ scattering angle

$$\left.\frac{\partial f}{\partial t}\right|_{\text{int}}=\int\mathrm{d}\Omega\int\mathrm{d}^3v_2\,g\,\sigma(g, \theta)\,[f(\overline{\boldsymbol{v}}_1)\,f(\overline{\boldsymbol{v}}_2)-f(\boldsymbol{v}_1)\,f(\boldsymbol{v}_2)]\,,$$

where $\overline{v}_1$, $\overline{v}_2$ are related to v_1, v_2 and θ, being in fact the negatives of the velocities resulting when a collision between v_1 and v_2 occurs with a scattering angle θ. Our later work will be devoted to showing that, with certain corrections, this result is a valid approximation for an ionized gas where the forces of interaction are *weak*, but *long ranged*. At present we will concentrate on the second problem, that of solving Boltzmann's equation and determining the transport coefficients.

2. – Hydrodynamic equations from the transport equation.

As a preliminary to any attempt to solve the Boltzmann equation we will use it to form the hydrodynamic equations. To do this we use the definitions of Section **1**, which, expressed in terms of the distribution function f become the following moments of f

$$\varrho=\int fm\,\mathrm{d}^3v;\qquad \varrho\boldsymbol{V}=\int fm\boldsymbol{v}\,\mathrm{d}^3v;\qquad \boldsymbol{c}=\boldsymbol{v}-\boldsymbol{V};$$

$$\boldsymbol{p}=\int fm\boldsymbol{c}\boldsymbol{c}\,\mathrm{d}^3v;\qquad \frac{3}{2}kT=\int f\,\tfrac{1}{2}mc^2\,\mathrm{d}^3v;\qquad \boldsymbol{q}=\int f\,\tfrac{1}{2}mc^2\,\boldsymbol{c}\,\mathrm{d}^3v\,.$$

Since the B.E. forms a representation of the dynamics of the system, the macroscopic equations for the moments may be formed therefrom, *i.e.* from

$$\int m\left\{\frac{\partial f}{\partial t}+\boldsymbol{v}\cdot\frac{\partial f}{\partial \boldsymbol{x}}+\boldsymbol{A}\cdot\frac{\partial f}{\partial \boldsymbol{v}}-I(f)\right\}\mathrm{d}^3v=0\,, \tag{I.2.1}$$

$$\frac{\partial \varrho}{\partial t}+\operatorname{div}(\varrho\boldsymbol{V})=0\,,$$

since $\int mI(f)\,\mathrm{d}^3v=0$, from mass conservation.

From $\int \mathrm{d}^3m\, m\boldsymbol{v}\{\ \}=0$,

$$\varrho\frac{\mathrm{D}\boldsymbol{V}}{\mathrm{D}t}+\boldsymbol{\nabla}\cdot\boldsymbol{p}-\boldsymbol{F}=0 \qquad \text{where} \qquad \boldsymbol{F}=\int m\boldsymbol{A}f\,\mathrm{d}^3v\,, \tag{I.2.2}$$

and (I.2.1) has been used. Finally from $\int \mathrm{d}^3v\,\frac{1}{2}mv^2\{\ \}$,

$$\frac{\mathrm{D}U}{\mathrm{D}t}+U\operatorname{div}\boldsymbol{V}+\boldsymbol{p}:\boldsymbol{\nabla}\boldsymbol{V}+\operatorname{div}\boldsymbol{q}=0\,, \tag{I.2.3}$$

and

$$U=\tfrac{3}{2}nkT\,.$$

For an ionized gas, there are similar equations for each component, although now the interaction integral does not vanish, but leads to terms representing the transfer of energy and momentum between the two components. Alternatively these equations may be combined and as in Section **1**, the mean velocity may be defined as $\varrho\boldsymbol{V}=(\varrho_++\varrho_-)\boldsymbol{V}=\varrho_+\boldsymbol{V}_++\varrho_-\boldsymbol{V}_-$, and $\boldsymbol{p}$, T etc. defined relative to V, whereupon

$$\frac{\partial \varrho}{\partial t}+\operatorname{div}\varrho\boldsymbol{V}=0\,, \tag{I.2.4}$$

$$\varrho\frac{\mathrm{D}\boldsymbol{V}}{\mathrm{D}t}+\boldsymbol{\nabla}\cdot\boldsymbol{p}-Q(\boldsymbol{E}+\boldsymbol{V}\times\boldsymbol{B})-\boldsymbol{j}\times\boldsymbol{B}=0\,, \tag{I.2.5}$$

and

$$\frac{\mathrm{D}U}{\mathrm{D}t}+U\operatorname{div}\boldsymbol{V}+\boldsymbol{p}:\boldsymbol{\nabla}\boldsymbol{V}+\operatorname{div}\boldsymbol{q}-\boldsymbol{j}\cdot\boldsymbol{E}=0\,. \tag{I.2.6}$$

To illustrate important methods used in solving the B.E. we shall first consider some simple representations of a simple gas, and only gradually approach the complexities of the ionized gas in a magnetic field.

3. – The normal solution: Hilbert's procedure.

To derive meaningful hydrodynamic equations it is useful to restrict attention first to those situations in which the rate of change of the distribution function is slow so that the collision frequency is much greater than any hydrodynamic frequency, *i.e.* if we introduce a macroscopic time scale, T, length scale, L, and a characteristic velocity, $V = LT^{-1}$, then, if the external forces are small, so that $T^{-2} < 1$, the L.H.S. of the B.E. scales as T^{-1}. We can also define a collision time by $\tau^{-1} \simeq n\sigma_0 V$, where σ_0 is a mean cross-section, whereupon the condition, collision frequency is much greater than hydrodynamic frequency, becomes $\tau/T = \varepsilon \ll 1$, and the B.E. may be written

$$I(f, f) = \varepsilon \left\{\frac{\partial}{\partial t} + \boldsymbol{v}\cdot\frac{\partial}{\partial \boldsymbol{x}} + \boldsymbol{A}_0\cdot\frac{\partial}{\partial \boldsymbol{v}}\right\} f = \varepsilon \, \mathrm{D}f . \tag{I.3.1}$$

It now makes sense to seek a solution expanded in powers of ε:

$$f = f_0 + \varepsilon f_1 + \dots ,$$

which on being introduced into (I.3.1) reduces it to

$$I(f_0, f_0) = 0 , \tag{I.3.2}$$

$$I(f_0, f_1) = Df_0 . \tag{I.3.3}$$

The first equation here is satisfied by the locally Maxwellian distribution; *i.e.* by

$$f_0 = n(x, t) \left(\frac{m}{2\pi} kT\right)^{\frac{3}{2}} \exp\left[-\frac{m[\boldsymbol{v} - \boldsymbol{V}(x, t)]^2}{2kT(x, t)}\right] , \tag{I.3.4}$$

where n, T, V are undetermined functions of x, t.

Hilbert observed that by writing $f_1 = f_0 \phi$, (I.3.3) may be written

$$\int \mathrm{d}^3 v' f_0(\boldsymbol{v}) f_0(\boldsymbol{v}') |\boldsymbol{v} - \boldsymbol{v}'| \sigma(\boldsymbol{v} - \boldsymbol{v}', \theta) [\phi(\bar{\boldsymbol{v}}) + \phi(\bar{\boldsymbol{v}}') - \phi(\boldsymbol{v}) - \phi(\boldsymbol{v}')] = \mathrm{D}f_0 ,$$

i.e.

$$\int K(\boldsymbol{v}, \boldsymbol{v}') \phi \, \mathrm{d}^3 v' = \mathrm{D}f_0 ,$$

an integral equation in which conservation laws require that K should be symmetric in $\boldsymbol{v}$, $\boldsymbol{v}'$. This has the following interesting consequence: that a solution can be obtained only if Df_0 is orthogonal to the solution $h(\boldsymbol{v})$ of the homogeneous equation

$$\int K(\boldsymbol{v}, \boldsymbol{v}')\, h(\boldsymbol{v}')\, \mathrm{d}^3 v' = 0\,,$$

for

$$\int \mathrm{d}v \int \mathrm{d}v'\, h(\boldsymbol{v})\, K(\boldsymbol{v}, \boldsymbol{v}')\,\phi(\boldsymbol{v}') = 0 = \int \mathrm{d}v\, h(\boldsymbol{v})\, \mathrm{D}f_0(\boldsymbol{v})\,.$$

Since

$$\int K(v, v')\,\phi(v')\, \mathrm{d}v'$$

gives the rate of change of ϕ produced by collision, the solutions to the homogeneous equation are the collision invariants, m, $m\boldsymbol{v}$, $\frac{1}{2}mv^2$, and the constraints on Df_0 become the zero-order hydrodynamic equations. Since for a Maxwellian distribution $p_{ij} = nkT\delta_{ij}$, $\boldsymbol{q} = 0$, $U = \frac{3}{2}nkT$, these become

(I.3.5) $$\frac{\partial n}{\partial t} + \boldsymbol{V}\cdot\boldsymbol{\nabla} n + n\boldsymbol{\nabla}\cdot\boldsymbol{V} = 0\,,$$

(I.3.6) $$\frac{\partial \boldsymbol{V}}{\partial t} + \boldsymbol{V}\cdot\boldsymbol{\nabla}\boldsymbol{V} + \frac{1}{nm}\left[\boldsymbol{\nabla}(nkT) - \boldsymbol{F}\right] = 0\,,$$

(I.3.7) $$\frac{\partial}{\partial t}\left(\frac{3}{2}nkT\right) + \boldsymbol{V}\cdot\boldsymbol{\nabla}\left(\frac{3}{2}nkT\right) + \frac{5}{2}nkT\,\mathrm{div}\,\boldsymbol{V} = 0\,.$$

Furthermore, since f_0 depends on x, and t through n, T and $\boldsymbol{V}$, we may write, with $\boldsymbol{c} = \boldsymbol{v} - \boldsymbol{V}$

$$\mathrm{D}f_0 = \left\{\frac{1}{n}\left(\frac{\partial}{\partial t} + \boldsymbol{v}\cdot\boldsymbol{\nabla}\right) n - \left(\frac{3}{2} - \frac{1}{2}\frac{mc^2}{kT}\right)\frac{1}{T}\left(\frac{\partial}{\partial t} + \boldsymbol{v}\cdot\boldsymbol{\nabla}\right) T + \right.$$
$$\left. + \frac{m\boldsymbol{c}}{kT}\cdot\left(\frac{\partial}{\partial t} + \boldsymbol{v}\cdot\boldsymbol{\nabla}\right)\boldsymbol{V} - \frac{m\boldsymbol{c}}{kT}\cdot\boldsymbol{A}\right\} f_0\,.$$

The time derivatives may be eliminated with aid of (I.3.5)-(I.3.7) and

(I.3.8) $$\mathrm{D}f_0 = \left\{\left[\frac{m}{kT}\,\boldsymbol{c}\cdot(\boldsymbol{c}\cdot\boldsymbol{\nabla}))\boldsymbol{V} - \frac{1}{3}\frac{mc^2}{kT}\,\mathrm{div}\,\boldsymbol{V}\right] - \left[\frac{5}{2} - \frac{1}{2}\frac{mc^2}{kT}\right]\frac{1}{T}\,(\boldsymbol{c}\cdot\boldsymbol{\nabla})T\right\} f_0\,.$$

4. – Mean free time theory for a simple gas.

A good many of the features of the B.E. may be retained by replacing the collision integral by a simple relaxation term, *i.e.* retaining only the tendency for a distribution function f to relax back to the Maxwellian, and representing $I(f)$ by $-1/\tau\,(f-f_0)$, whereupon (I.3.3) becomes $1/\tau\,(f_0-f)=\mathrm{D}f_0$, or $f=f_0-\tau\mathrm{D}f_0$ and using (I.3.8)

$$f=f_0\left\{1-\tau\left[\frac{m}{kT}\,\boldsymbol{c}\cdot(\boldsymbol{c}\cdot\boldsymbol{\nabla})\boldsymbol{V}-\frac{1}{3}\frac{mc^2}{kT}\,\mathrm{div}\,\boldsymbol{V}\right]-\left(\frac{5}{2}-\frac{1}{2}\frac{mc^2}{kT}\right)\frac{1}{T}\,\boldsymbol{c}\cdot\boldsymbol{\nabla}T\right\}.$$

We need this to form the moments

$$p_{ij}=\int fmc_ic_j\,\mathrm{d}^3v\,,\qquad \boldsymbol{q}=\int f\,\frac{1}{2}\,mc^2\,\boldsymbol{c}\,\mathrm{d}^3v\,,$$

since $\langle\boldsymbol{c}\rangle$ and $\langle\frac{1}{2}c^2\boldsymbol{c}\rangle=0$. Carrying out the integrals yields

$$p_{ij}=\frac{\tau}{2}\,n(kT)\left[\frac{\partial V_i}{\partial x_j}+\frac{\partial V_j}{\partial x_i}-\frac{2}{3}\,\mathrm{div}\,\boldsymbol{V}\,\delta_{ij}\right]=-\frac{\tau}{2}\,p[\boldsymbol{\nabla V}]\,,$$

and

$$\boldsymbol{q}=-\frac{75}{32}\,\tau\,\frac{2kT}{m}\,k\boldsymbol{\nabla}T\,.$$

II. – Transport Processes in Fully Ionized Gases - Normal Solutions.

1. – The linearized equations.

When instead of a simple gas, an ionized gas in a magnetic field is considered several new complications arise, many of which may be clarified by a consideration of the simple m.f.t. theory. In the first place, instead of a simple gas, an ionized gas in a mixture of two gases, ions and electrons, which interdiffuse. As a result, even starting from a common Maxwell distribution the elimination of the time derivatives produces some new results, since neither component moves exactly as does the mixture. Next, for many plasmas, the gyro frequency $\omega=eB/m$ is comparable to or larger than the collision frequency, and the technique used for splitting up the terms in the B.E. becomes inappropriate. To handle the magnetic term it may itself be split into two parts, one involving the mean velocity $(e/m)\boldsymbol{V}\times\boldsymbol{B}$ which is small, and a second involving $\boldsymbol{c}\,(=\boldsymbol{v}-\boldsymbol{V})$ the « peculiar » velocity $(e/m)\boldsymbol{c}\times\boldsymbol{B}$, which is not necessarily small. The B.E. may then be written as

$$\text{(II.1.1)}\quad \underbrace{\left[\frac{\partial}{\partial t}+\boldsymbol{v}\cdot\frac{\partial}{\partial \boldsymbol{x}}+\frac{e}{m}(\boldsymbol{E}+\boldsymbol{V}\times\boldsymbol{B})\cdot\frac{\partial}{\partial \boldsymbol{v}}\right]f}_{A}+\underbrace{\frac{e}{m}(\boldsymbol{c}\times\boldsymbol{B})\cdot\frac{\partial f}{\partial \boldsymbol{v}}}_{B}-\underbrace{I(f)}_{C}=0\,,$$

$A:B:C::1:\omega T:T/\tau$. Different orderings are possible for these terms, *e.g.* ωT may be ~ 1 while T/τ is large, whereupon the ordering valid in the nonmagnetic problem becomes useful. ωT and T/τ may be of the same order, or finally, in a magnetic field, hydrodynamic behaviour is possible when $\omega T\gg 1\gg T/\tau$. In any case, it is possible to start with a Maxwellian distribution

$$f_0=n(x)\left(\frac{m}{2\pi kT}\right)^{\frac{3}{2}}\exp\left[-\tfrac{1}{2}m(\boldsymbol{v}-\boldsymbol{V})^2/kT\right],$$

where, as before, n, T, $\boldsymbol{V}$ are functions of position and time, and $\boldsymbol{V}$ is the velocity of the fluid as a whole. Now as in I, we may write the B.E. as

$$\text{(II.1.2)}\qquad \left[I(f)-\frac{e}{m}\boldsymbol{c}\times\boldsymbol{B}\cdot\frac{\partial}{\partial \boldsymbol{v}}f\right]_i=[\mathrm{D}f^0]_i\,,$$

and in forming the R.H.S. observe that space and time variation arises only through n, T, $\boldsymbol{V}$, and further, that the time derivatives may once more be eliminated by use of the zero-order equations of motion. Now, however, certain complications arise, for in the equation for each component there appear terms such as $\boldsymbol{\nabla}\log n_i+\boldsymbol{\nabla}\log T$, which no longer is equivalent to $\boldsymbol{\nabla}\log p$; moreover $(1/nm)\boldsymbol{F}_i$ is no longer eliminated by the equation of momentum conservation. Instead Df takes the form

$$\text{(II.1.3)}\qquad [\mathrm{D}f^0]_i=f_i^0\left\{\frac{m_i}{kT}\left[\boldsymbol{c}\cdot(\boldsymbol{c}\cdot\boldsymbol{\nabla})\boldsymbol{V}-\frac{1}{3}c^2\operatorname{div}\boldsymbol{V}\right]-\right.$$
$$\left.-\left[\frac{5}{2}-\frac{1}{2}\frac{m_ic^2}{kT}\right]\boldsymbol{c}\cdot\boldsymbol{\nabla}\log T+\frac{n}{n_i}\boldsymbol{c}\cdot\boldsymbol{d}_i\right\},$$

where

$$\text{(II.1.4)}\quad \boldsymbol{d}_i=\boldsymbol{\nabla}\frac{n_i}{n}+\frac{1}{p}\left[\frac{n_i}{n}-\frac{n_im_i}{\varrho}\right]\boldsymbol{\nabla}p-\frac{1}{p}\left[n_ie_i-\frac{n_im_i}{\varrho}Q\right][\boldsymbol{E}+\boldsymbol{V}\times\boldsymbol{B}]\,,$$

or more (anti!) symmetrically as

$$\text{(II.1.5)}\quad \boldsymbol{d}_1=-\boldsymbol{d}_2=\boldsymbol{\nabla}\frac{n_1}{n}+\frac{n_1n_2(m_2-m_1)}{n\varrho p}\boldsymbol{\nabla}p-\frac{n_1n_2}{p\varrho}(e_1m_2-e_2m_1)\,[\boldsymbol{E}+\boldsymbol{V}\times\boldsymbol{B}]\,.$$

The problem of finding the normal solutions to the B.E. then reduces to that of solving the linear integro-differential equation

$$\left[I(f)-\frac{e}{m}\boldsymbol{c}\times\boldsymbol{B}\cdot\frac{\partial f}{\partial \boldsymbol{v}}\right]_i = P\exp\left[-\frac{\frac{1}{2}m_i c^2}{kT}\right]. \tag{II.1.6}$$

2. – Mean free time model.

As for the simple gas, the general form of the transport coefficients may be obtained by studying a simple model — one in which the collision integral is replaced by a simple relaxation term; $I(f)=(1/\tau)(f_0-f)$. Furthermore, we may introduce axis $OX\|$ the magnetic field and represent the peculiar velocity c by

$$c_x = c_\parallel\,, \qquad c_y = c_\perp \cos\varphi\,, \qquad c_z = C_\perp \sin\varphi\,, \tag{II.2.1}$$

whereupon

$$\frac{e}{m}\boldsymbol{c}\times\boldsymbol{B}\cdot\frac{\partial f}{\partial \boldsymbol{v}} = -\omega\frac{\partial f}{\partial\varphi}\,, \tag{II.2.2}$$

and the equation to be solved becomes

$$f-\omega\tau\frac{\partial f}{\partial\varphi} = f^0[1-\tau P]\,, \tag{II.2.3}$$

and on introducing the abbreviation $\boldsymbol{w}=\sqrt{(m/2kT)}\,\boldsymbol{c}$

$$P = 2\left[\frac{\partial V_i}{\partial x_j}-\frac{1}{3}\operatorname{div}V\delta_{ij}\right]w_i w_j - \left[\left(\frac{5}{2}-w^2\right)\frac{1}{T}\nabla T-\frac{n}{n_i}\boldsymbol{d}_i\right]\cdot\boldsymbol{c}\,. \tag{II.2.4}$$

In terms of φ, the differential equation contains terms ~ 1, $\cos\varphi$, $\sin\varphi$, $\cos^2\varphi$, $\sin^2\varphi$, $\sin\varphi\cos\varphi$, and the *periodic* contributions due to these terms are

$$1:\frac{1}{1+\omega^2\tau^2}(\cos\varphi-\omega\tau\sin\varphi)\,, \qquad \frac{1}{1+\omega^2\tau^2}(\sin\varphi+\omega\tau\cos\varphi)\,,$$

$$-\frac{1}{2}+\frac{1}{2}\frac{1}{1+4\omega^2\tau^2}[(\cos^2\varphi-\sin^2\varphi)-4\omega\tau\cos\varphi\sin\varphi]\,,$$

$$-\frac{1}{2}-\frac{1}{2}\frac{1}{1+4\omega^2\tau^2}[(\cos^2\varphi-\sin^2\varphi)-4\omega\tau\cos\varphi\sin\varphi]\,,$$

and

$$\frac{1}{1+4\omega^2\tau^2}[\cos\varphi\sin\varphi+\omega\tau(\cos^2\varphi-\sin^2\varphi)]\,,$$

(II.2.5)
$$f=f_0\Bigg[1-\tau\Bigg\{2\left[(\operatorname{div}\boldsymbol{V})_\parallel w_\parallel^2+\frac{1}{2}(\operatorname{div}\boldsymbol{V})_\perp w_\perp^2-\frac{1}{3}\operatorname{div}\boldsymbol{V}w^2\right]-$$
$$-\left[\left(\frac{5}{2}-w^2\right)\frac{1}{T}\boldsymbol{\nabla}_\parallel\cdot T-\frac{n}{n_i}d_\parallel\right]c_\parallel+\frac{1}{1+\omega^2\tau^2}\left[\frac{kT}{m}\left(\frac{\partial\boldsymbol{V}_\perp}{\partial x_\parallel}+\frac{\partial V_\parallel}{\partial\boldsymbol{x}_\perp}\right)c_\parallel-\right.$$
$$\left.-\left(\frac{5}{2}-w^2\right)\frac{1}{T}\boldsymbol{\nabla}_\perp T-\frac{n}{n_i}\boldsymbol{d}_\perp\right]\cdot(\boldsymbol{c}_\perp+\boldsymbol{c}_\perp\times\boldsymbol{b}\omega\tau)+\frac{1}{2(1+4\omega^2\tau^2)}\cdot$$
$$\cdot[(\boldsymbol{\nabla}_\perp\boldsymbol{V}_\perp)+\boldsymbol{b}\times(\boldsymbol{\nabla}\boldsymbol{V})\times\boldsymbol{b}]:\boldsymbol{ww}+4\omega\tau[(\boldsymbol{\nabla}_\perp\boldsymbol{V}_\perp)\times\boldsymbol{b}-\boldsymbol{b}\times(\boldsymbol{\nabla}_\perp\boldsymbol{V}_\perp)]:\boldsymbol{ww}\Bigg\}\Bigg]\,,$$

(II.2.6)
$$\boldsymbol{j}=n_1e_1\boldsymbol{c}_1+n_2e_2\boldsymbol{c}_2=ne_1\sqrt{\frac{kT}{m_1}}\,\tau_1\left[\boldsymbol{d}_\parallel+\frac{\boldsymbol{d}_\perp+\boldsymbol{b}\times\boldsymbol{d}_\perp}{1+\omega^2\tau^2}\right]_1+$$
$$+ne_2\sqrt{\frac{kT}{m_2}}\,\tau_2\left[\boldsymbol{d}_\parallel+\frac{\boldsymbol{d}_\perp+\boldsymbol{b}\times\boldsymbol{d}_\perp}{1+\omega^2\tau^2}\right]_2,$$

(II.2.7)
$$q=-\frac{3}{2}\frac{k^2T}{m_1}n_1\tau_1\left[\boldsymbol{\nabla}_\parallel T+\frac{1}{1+\omega^2\tau^2}\boldsymbol{\nabla}_\perp T+\boldsymbol{b}\times\boldsymbol{\nabla}_\perp T\right]_1-$$
$$-\frac{3}{2}\frac{k^2T}{m^2}n_2\tau_2\left[\boldsymbol{\nabla}_\parallel T+\frac{1}{1+\omega^2\tau^2}\boldsymbol{\nabla}_\perp T+\boldsymbol{b}\times\boldsymbol{\nabla}_\perp T\right]_2+$$
$$+\frac{5}{2}n(kT)^2\left\{\frac{\tau_1}{m_1}\left[\boldsymbol{d}_\parallel+\frac{\boldsymbol{d}_\perp+\boldsymbol{b}\times\boldsymbol{d}_\perp}{1+\omega^2\tau^2}\right]_1+\frac{\tau_2}{m_2}\left[\boldsymbol{d}_\parallel+\frac{\boldsymbol{d}_\perp+\boldsymbol{b}\times\boldsymbol{d}_\perp}{1+\omega^2\tau^2}\right]_2\right\},$$

(II.2.8)
$$P_{ij}=p_0-[\tau_1p_1+\tau_2p_2][(\operatorname{div}\boldsymbol{V})_\parallel-\tfrac{1}{3}\operatorname{div}\boldsymbol{V}]\boldsymbol{bb}-$$
$$-[\tau_1p_1+\tau_2p_2][\tfrac{1}{2}(\operatorname{div}\boldsymbol{V})_\perp-\tfrac{1}{3}\operatorname{div}\boldsymbol{V}][\boldsymbol{1}-\boldsymbol{bb}]-$$
$$-\left\{p_1\frac{1}{8}\frac{1}{1+4\omega^2\tau^2}[\boldsymbol{\nabla}_\perp\boldsymbol{V}_\perp+\boldsymbol{b}\times\boldsymbol{V}\times\boldsymbol{b}]+2\omega\tau[\boldsymbol{\nabla}_\perp\boldsymbol{V}_\perp\times\boldsymbol{b}-\boldsymbol{b}\times\boldsymbol{\nabla}_\perp\boldsymbol{V}_\perp]\operatorname{diag}\right\}_{1+2}+$$
$$+\frac{1}{4}\frac{\tau p}{1+\omega^2\tau^2}\left\{\left(\frac{\partial\boldsymbol{V}_\perp}{\partial x_\parallel}+\frac{\partial\boldsymbol{V}_\parallel}{\partial x_\perp}\right)+\omega\tau\,\boldsymbol{b}\times\left(\frac{\partial\boldsymbol{V}_\perp}{\partial x_\parallel}+\frac{\partial\boldsymbol{V}_\parallel}{\partial x_\perp}\right)\right\}_{1+2}+$$
$$+\frac{1}{8}\frac{p\tau}{1+4\omega^2\tau^2}[\boldsymbol{\nabla}_\perp\boldsymbol{V}_\perp+\boldsymbol{b}\times\boldsymbol{\nabla}\boldsymbol{V}\times\boldsymbol{b}]+2\omega\tau[\boldsymbol{\nabla}_\perp\boldsymbol{V}_\perp\times\boldsymbol{b}-\boldsymbol{b}\times\boldsymbol{\nabla}_\perp\boldsymbol{V}_\perp]_{1+2}\,.$$

From these expressions it is clear that, for two reasons: (1) because the centre of mass of the total system does not coincide with the centre of mass of either component, and (2) because trajectories are curved by the magnetic field, the transport processes are considerably complicated; the heat flux, for example, depends on the vector $\boldsymbol{d}$, which could be eliminated in favour of the current, or the pressure gradient and electric field. The current itself depends on temperature and density gradients as well as on the electric fields — moreover, currents either of heat or electricity, are not in the direction of applied forces. These complications persist when attempts are made to solve the integral equations posed by the Boltzmann equation. Note that this model has proved incapable of describing the explicit dependence of $\boldsymbol{j}$ on $\boldsymbol{\nabla} T$.

3. – Methods used to obtain the normal solution.

We will now consider the procedures that have been developed for solving the integral eq. (II.1.2) which, on introducing $f_1 = f_0 \phi$ may be written as

$$I(f) = \int \mathrm{d}^3 v' \int \mathrm{d}\Omega \, |\boldsymbol{v} - \boldsymbol{v}'| \sigma(|\boldsymbol{v} - \boldsymbol{v}'|, \theta) \cdot$$
$$\cdot \{f_0(\boldsymbol{v})[1 + \phi(\boldsymbol{v})] f_0(\boldsymbol{v}')[1 + \phi(\boldsymbol{v}')] - f_0(\bar{\boldsymbol{v}})[1 + \phi(\bar{\boldsymbol{v}})] f_0(\bar{\boldsymbol{v}}')[1 + \phi(\bar{\boldsymbol{v}}')]\} ,$$

or

$$\text{(II.3.1)} \qquad \int \mathrm{d}^3 v' \int \mathrm{d}\Omega \, |\boldsymbol{v} - \boldsymbol{v}'| \sigma(|\boldsymbol{v} - \boldsymbol{v}'|, \theta) \cdot \{f_0(\boldsymbol{v}) f_0(\boldsymbol{v}')[\phi(\boldsymbol{v}) + \phi(\boldsymbol{v}')] -$$
$$- f_0(\bar{\boldsymbol{v}}) f_0(\bar{\boldsymbol{v}}')[\phi(\bar{\boldsymbol{v}})] + \phi(\bar{\boldsymbol{v}}')]\} .$$

We must now consult the dynamics of a collision in order to discover the value of $\bar{\boldsymbol{v}}, \bar{\boldsymbol{v}}'$. On a collision, the centre-of-mass motion is constant; *i.e.*

$$\text{(II.3.2)} \qquad \boldsymbol{V} = \frac{m_1 \boldsymbol{v}_1 + m_2 \boldsymbol{v}_2}{m_1 + m_2} = \frac{m_1 \bar{\boldsymbol{v}}_1 + m_2 \bar{\boldsymbol{v}}_2'}{m_1 + m_2} ,$$

and on removing this, the collision is described as the motion of a particle of the reduced mass $m_1 m_2/(m_1 + m_2)$, moving with the relative velocity $\boldsymbol{g} = \boldsymbol{v}' - \boldsymbol{v}$. Conservation of energy on collisions requires that only the direction of $\boldsymbol{g}$ be changed, *e.g.* that $\boldsymbol{g}'$ is

$$\text{(II.3.3)} \qquad O\boldsymbol{g} = (\bar{\boldsymbol{v}}' - \bar{\boldsymbol{v}}) ,$$

thus, the final velocities and the $\bar{\boldsymbol{v}}$ are given by

$$\text{(II.3.4)} \qquad -\bar{\boldsymbol{v}}_1 = \boldsymbol{v}_{1f} = \boldsymbol{v}_1 + \frac{m_2}{m_1 + m_2} (\boldsymbol{g} - O \cdot \boldsymbol{g}) ; \qquad \boldsymbol{v}_{2f} = \boldsymbol{v}_2 - \frac{m_1}{m_1 + m_2} (\boldsymbol{g} - O \cdot \boldsymbol{g}) .$$

From conservation of energy also

$$\tfrac{1}{2}m_1 v_1^2 + \tfrac{1}{2}m_2 v_2'^2 = \tfrac{1}{2}m_1 \bar{v}_1^2 + \tfrac{1}{2}m_2 \bar{v}_2'^2 ,$$

hence

$$f_0(\bar{\boldsymbol{v}})\, f_0(\bar{\boldsymbol{v}}') \sim \exp\left[-\frac{\tfrac{1}{2}m_1 \bar{v}_1^2 + \tfrac{1}{2}m_2 \bar{v}_2'^2}{kT}\right] = f_0(\boldsymbol{v})\, f_0(\boldsymbol{v}') ,$$

and

$$\text{(II.3.5)} \qquad I(f_0, \phi) = \int \mathrm{d}^3 v' \int \mathrm{d}\Omega\, |\boldsymbol{v} - \boldsymbol{v}'|\, \sigma(|\boldsymbol{v} - \boldsymbol{v}'|, \theta)\, f_0(\boldsymbol{v})\, f_0(\boldsymbol{v}') \cdot \\ \cdot [\phi(\boldsymbol{v}) + \phi(\boldsymbol{v}') - \phi(\bar{\boldsymbol{v}}) - \phi(\bar{\boldsymbol{v}}')] .$$

Observe that $I(\boldsymbol{v}, \boldsymbol{v}')$ is unchanged by interchanging $m_1\boldsymbol{v}$ and $m_2\boldsymbol{v}'$ for; since this changes the sign of g, (II.3.4) is unchanged, while all other terms are symmetric in these quantities, and the integral equation may be written

$$\text{(II.3.6)} \qquad \int \mathrm{d}^3 v' \int \mathrm{d}\Omega\, |v - v'|\, \sigma_{11}(|v - v'|, \Theta)\, f_1(v)\, f_1(v')[\phi_1(v) + \phi_1(v') - \phi_1(\bar{v}) - \phi_1(\bar{v}')] + \\ + \int \mathrm{d}^3 v' \int \mathrm{d}\Omega\, |v - v'|\, \sigma_{12}(|v - v'|, \Theta)\, f_1(v)\, f_2(v')[\phi_1(v) + \phi_2(v') - \phi_1(\bar{v}) - \phi_2(\bar{v}')] + \\ + \Omega f_0 \frac{\partial \phi_1}{\partial \varphi} = f_0^1 \left\{ 2\left(\frac{\partial V_i}{\partial x_j} - \frac{1}{3}\operatorname{div} V\, \delta_{ij}\right) w_i w_j - \left[\left(\frac{5}{2} - w^2\right)\frac{1}{T}\nabla T - \frac{n}{n_i}\cdot d_i\right]\cdot c\right\}_1 ,$$

with a similar equation for ϕ_2.

One technique for solving these equations is a development of that due to Chapman and Cowling, and exploited by Landshoff and Marshall. In it, the function ϕ, which must be proportional to a linear combination of the forces on the right is written as

$$\{V_{i,j}\}\{[w_i w_j - \tfrac{1}{3}w^2]\phi_1(v^2) + [(b \times w)_i w_j]\phi_2(v^2) + [(b \times w)_i (b \times w)_j - \tfrac{1}{3}\delta_{ij}(b \times w)^2]\phi_3\} + \\ + \{\phi_4(v^2)\boldsymbol{w} + \phi_5(v^2)\boldsymbol{b} \times \boldsymbol{w} + \phi_6(v^2)\boldsymbol{b}(\boldsymbol{b}\cdot\boldsymbol{w})\} \cdot (\boldsymbol{\nabla} \log T + \boldsymbol{d}) ,$$

and the scalar quantities $\phi_i(v')$ determined by a set of scalar equations.

Approximate solutions to these equations may be obtained by expanding in powers of certain polynomials: the Sonine polynomials, which are orthogonal with $\exp[-x^2]x^n$: *i.e.* such that

$$\int \exp[-x^2]\, x^{n+1} S_n^m(x) S_n^{m^1}(x)\, dx = \delta(m, m^1)\, \frac{\frac{1}{2}(n+m)!}{m!}.$$

If the magnetic term is absent, a simple variational principle exists, which aids in obtaining approximate solutions. In a magnetic field, Marshall treated the differential term by using a complex representation of ϕ; $\partial/\partial\varphi$ then interchanged real and imaginary parts: this, however, reduced the variational principle from a maximal to a simple stationary form. Simple rational expressions were obtained for the transport coefficient, the simplicity being enforced by the trial functions used.

An alternative method has been employed by Bernstein and Robertson. They first observe that if the limit $m/M \to 0$ is taken, conservation of energy permits a separation of the two equations for the perturbations ϕ_- in the electron distribution and ϕ_+ in the ion distribution; the equations becoming

$$f_-^0 \left\{ 2\nabla \boldsymbol{V} : \left(\boldsymbol{ww} - \frac{1}{3} w^2 \boldsymbol{I}\right) + \left(w^2 - \frac{5}{2}\right) \boldsymbol{c} \cdot \nabla \log T - \boldsymbol{c} \cdot \boldsymbol{d} - \frac{e}{m} \boldsymbol{B} \times \boldsymbol{c} \cdot \frac{\partial \phi_-}{\partial \boldsymbol{c}} \right\} =$$

$$= K_{--}\phi_- + K_{+-}\phi_-,$$

and

$$f_+^0 \left\{ 2\nabla \boldsymbol{V} : \left(\boldsymbol{ww} - \frac{1}{3} w^2 \boldsymbol{I}\right) + \left(w^2 - \frac{5}{2}\right) \boldsymbol{c} \cdot \nabla \log T - \frac{e_+}{m_+} \boldsymbol{B} \times \boldsymbol{c} \cdot \frac{\partial \phi_+}{\partial \boldsymbol{c}} \right\} - I_{++} = \tag{II.3.7}$$

$$= \left[I_{+-} - \frac{1}{nkT} \boldsymbol{j} \times \boldsymbol{B} \cdot \boldsymbol{c} f_+^0 - \frac{2_-}{kT} \boldsymbol{c} \cdot \boldsymbol{d} f_+^0 \right] = 0,$$

where from conservation of energy, the R.H.S. vanishes. Further, in their analysis, Bernstein and Robertson use a Fokker-Planck representation of the interaction term which introduces some slight simplification. They now represent the solutions ϕ as linear combinations of the forces as does Marshall, but instead of his simple complex representation, represent the angular dependence by a harmonic expansion, *i.e.*

$$\phi^- = \sum \phi_{l\pm n} \{T_{l\pm n}(\vartheta, \varphi) = N_{l,n} P_l^n(\cos\theta) \begin{Bmatrix} \sin n\varphi \\ \cos n\varphi, \end{Bmatrix}$$

where

$$N^2 = \frac{2l+1}{2\pi} \frac{(l-|n|)!}{(l+|n|)!} \frac{1}{1+\delta|n|},$$

is a normalizing factor.

If now the ϕ's are substituted in (II.3.7) and the contributions to the various sources are separated, integro-differential equations are derived for the separate contributions to ϕ — *e.g.* those arising from $\boldsymbol{d}$ become

$$f^0_-[d_z c T_{1,0} - I_-(\phi_{1,0}, T_{1,0}, N_{1,0})] = 0 , \tag{II.3.8}$$

$$f^0_-[d_y c T_{1,1} - I_-(\phi_{1,1}, T_{1,1}, N_{1,1}) + \omega\phi_{1,-1} N_{1,1} T_{1,1}] = 0 ,$$

and

$$f^0_-[d_x c T_{1,-1} - I_-(\phi_{1,-1} T_{1,-1} N_{1,-1}) - \omega\phi_{1,1} N_{1,1} T_{1,-1}] = 0 ,$$

ω being the gyro frequency.

Now since the integral operator I_- is invariant under rotation the spherical harmonics are eigen-functions thereof and

$$I(\phi TN) = K\phi TN = \lambda(c^2)\phi TN ,$$

where λ is a function (unknown!) of c^2; and K is an integral operator acting on the functions $\phi(c^2)$. If now we consider $d \| OY$, the second two equations take the form

$$\frac{d_y c}{N_{1,1}} - K\phi_{1,1} + \omega\cdot\phi_{1,-1} = 0 \tag{II.3.9}$$

$$- K\phi_{1,-1} - \omega\cdot\phi_{1,1} = 0$$

By setting $\omega = 0$, and replacing $N_{1,1}$ by $N_{1,0}$ and d_y by d_z, the first equation in (II.3.8) may be reduced to (II.3.9). The problem of solving this equation is however still sufficiently formidable that recourse is had to a variational procedure — which is a suitable modification of that introduced by HIRSCHFELDER *et al.* for normal gases, and used by Marshall.

4. – The variational procedure.

In the absence of a magnetic field, the equation to be solved takes on the form

$$\psi(c) = K\phi(c) , \tag{II.4.1}$$

where ψ is a known function and K the integral operator representing the change in ϕ produced by collisions. A variational principle may be derived by observing first that K is a symmetric operator, hence for any two functions of c, η and ξ,

$$(\eta, K\xi) = \int d^3c\,\eta(c)\,K\xi(c) = (\xi, K\eta)\,. \tag{II.4.2}$$

Further $(\xi, K\xi) \leqslant 0$.

To obtain approximate solutions to (II.4.1) consider

$$\lambda(\chi) = -(\chi, \psi)^2/(\chi, K\chi)\,. \tag{II.4.3}$$

If now, we insist that λ be stationary with respect to χ, then

$$\delta\lambda = -2\left\{[(\chi, K\chi)\psi - (\chi, \psi)K\chi]2\,\frac{(\chi, \psi)}{(\chi, K\psi)^2}\cdot\delta\chi\right\} \tag{II.4.4}$$

which vanishes if

$$\psi = \frac{(\chi, \psi)}{(\chi, K\chi)}\,\chi$$

satisfies (II.4.1). By carrying the variation out to 2nd order in $\delta\chi$, λ is shown to be a maximum at the solution.

In a magnetic field, we must consider a pair of equations

$$\psi(c) = \omega\,\frac{\partial}{\partial\varphi}\,\phi_2(c) + K\phi_1(c) \tag{II.4.5}$$

$$-\omega\,\frac{\partial}{\partial\varphi}\,\phi_1(c) = K\phi_2(c)\,.$$

A partial integration shows

$$\left(\chi_1, \omega\,\frac{\partial}{\partial\varphi}\,\chi_2\right) = -\left(\chi_2, \omega\,\frac{\partial}{\partial\varphi}\,\chi_1\right). \tag{II.4.6}$$

Now consider

$$\lambda_1 = -\frac{(\chi_1, \psi)^2}{[(\chi_1, K\chi_1) + (\chi_1, \omega\,\partial\chi_2/\partial\varphi)^2/(\chi_2, K\chi_2)}\,. \tag{II.4.7}$$

If χ_1 is fixed this is a minimum with respect to χ_2 when

$$-\left(\chi_2, \omega\,\frac{\partial}{\partial\varphi}\,\chi_1\right)^2\Big/(\chi_2, K\chi_2)$$

is a *maximum*, Re χ_2. This, however, by comparison with (II.4.3) holds if

$$-\omega\frac{\partial}{\partial\varphi}\chi_1 = K\chi_2\,.$$

If, on the other hand, χ_2 is taken as fixed, λ_1 is stationary when

$$-\frac{2(\chi_1,\psi)}{[(\chi_1,K\chi_1)-(\chi_1,\omega(\partial\chi_2/\partial\varphi))^2/(\chi_2,K\chi_2)]^2}\left\{\left[-\left\{(\chi_1,K\chi_1)+\frac{(\chi_1,\omega\partial\chi_2/\partial\varphi)^2}{(\chi_2,K\chi_2)}\right\}\cdot\psi-\right.\right.$$
$$\left.\left.-(\chi_1,\psi)^2\frac{(\chi_1,\omega\partial\chi_2/\partial\varphi)}{(\chi_2,K\chi_2)}\cdot w\frac{\partial\chi_2}{\partial\varphi}-(\chi_1,\psi)^2K\chi_1\right]\cdot\delta\chi_1\right\}=0\,,$$

i.e. when

$$\psi=\frac{(\chi_1,\psi)^2}{-[(\chi_1,K\chi_1)+(\chi_1,\omega\partial\chi_2/\partial\varphi)^2/(\chi_2,K\chi_2)]}\cdot\left[K\chi_1+\frac{(\chi_1,\omega\partial\chi_2/\partial\varphi)}{(\chi_2,K\chi^2)}\,\omega\frac{\partial\chi_2}{\partial\varphi}\right].\tag{II.4.8}$$

This second stationary point may be shown to be a maximum, hence

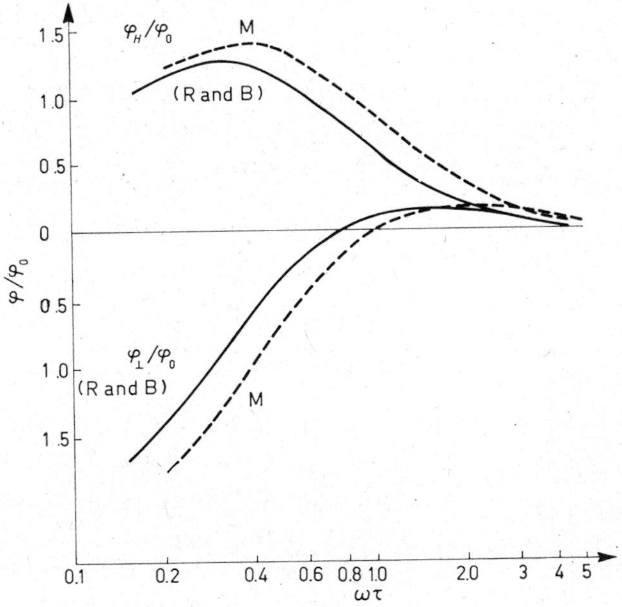

Fig. 1. – Comparison of results for the electrical conductivity of MARSHALL and of BERNSTEIN and ROBINSON.

constants may be chosen so that χ_1 and χ_2 satisfy (II.4.5) if λ is a maximum Re χ_1 and a minimum Re χ_2. At this stage trial functions may be introduced (usually

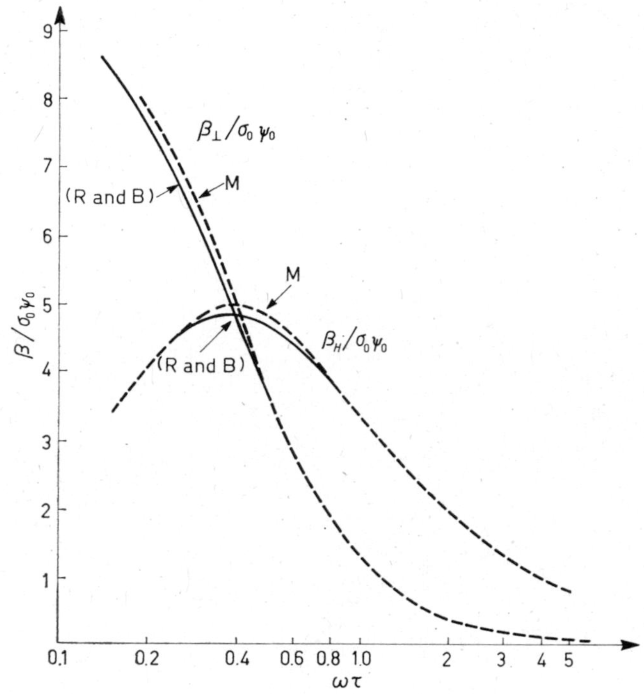

$\beta_\perp/\sigma_0\psi_0$
(R and B)
M
M
$\beta_H/\sigma_0\psi_0$
(R and B)
$\beta/\sigma_0\psi_0$
9
8
7
6
5
4
3
2
1
0.1
0.2
0.4
0.6 0.8 1.0
2
3
4
5
$\omega\tau$

Fig. 2.

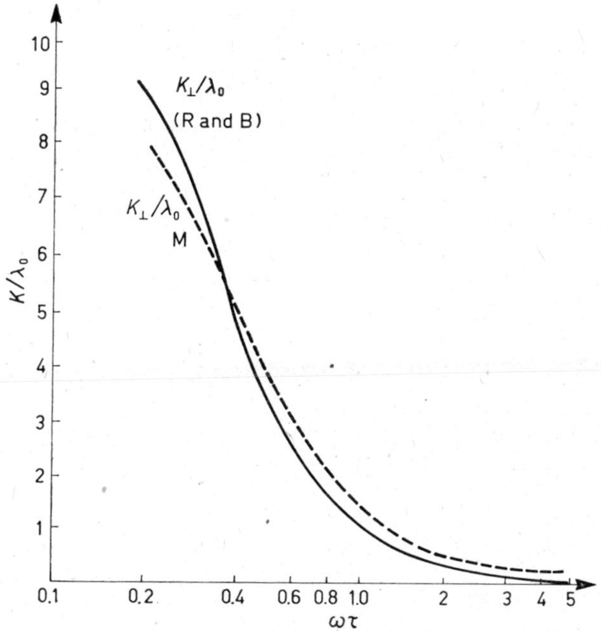

$K_\perp/\lambda_0$
(R and B)
$K_\perp/\lambda_0$
M
K/λ_0
10
9
8
7
6
5
4
3
2
1
0.1
0.2
0.4
0.6 0.8 1.0
2
3
4
5
$\omega\tau$

Fig 3.

polynomials!) and constants selected to determine the extrema. Bernstein and Robinson evaluated this result by considering a Lorentz gas (in which $I(\_\ \_) = 0$), which may be solved exactly, and claim 10% accuracy for all results.

The results of these rather elaborate calculations differ from the m.f. time

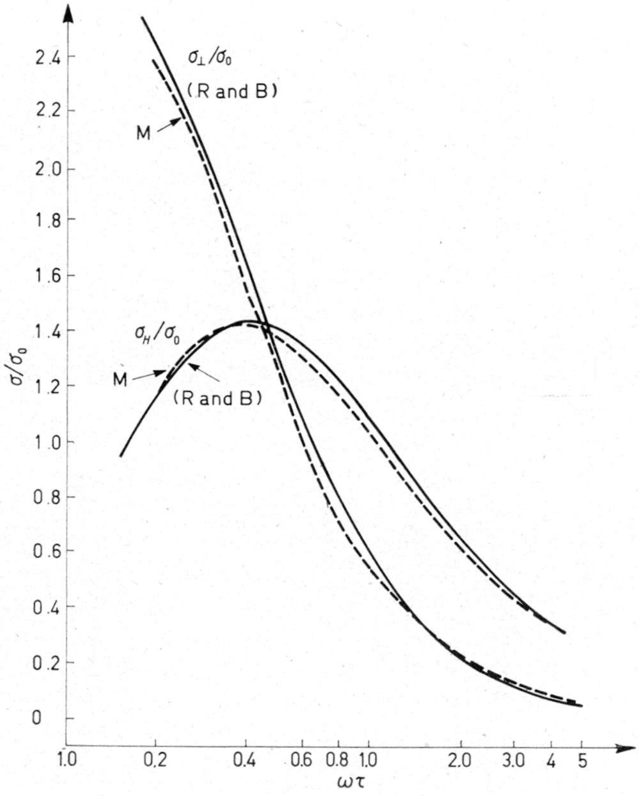

Fig. 4.

theory in two ways; a value is given for τ, which depends on the particular cross-section considered, and the simple rational functions $\tau/(1+\omega^2\tau^2)$, $\omega\tau^2/(1+\omega^2\tau^2)$ are replaced by more elaborate quantities of roughly the same form. In addition $\boldsymbol{j}$ contains a term $\sim \nabla T$, the thermal diffusion effect, which is missing from the m.f.t. theory.

Marshall's results are given in the accompanying figure; a comparison with the numerical results of Bernstein and Robinson is effected in the curves shown.

5. – Alternative approaches to solving the B.E.

In addition to the methods described above for obtaining the normal solution some other approaches have been used. One, due to Rosenbluth and Kaufmann, alters the ordering of the three terms in the B.E. The order selected is

$$B \gg A \gg C ,$$

an order which is valid for rapid motions in strong fields. By selecting as a zero-order distribution the Maxwellian, this is also valid for slow motions in strong fields. The distribution function is written as $f = f_0 + f_1 + f_2$, where $Df_0 = \omega \partial f_1/\partial\varphi$ so that

$$f_1 = \frac{1}{\omega} f_0 \left\{ \left(w^2 - \frac{5}{2} \right) \boldsymbol{c} \times \boldsymbol{b} \cdot \nabla \log T - \frac{m}{kT} (\boldsymbol{c} \times \boldsymbol{b}) \cdot \boldsymbol{d} - (\nabla_\perp \boldsymbol{V}_\perp \times \boldsymbol{b} - \boldsymbol{b} \times \nabla \boldsymbol{V}) : \boldsymbol{ww} \right\} ,$$

f_2 is now determined by

$$-\omega \frac{\partial}{\partial\varphi} f_2 = I(f^0, f^1) ,$$

where $I = I(f^0, \varphi)$ and $\varphi = f_1/f_0$, and the problem is then reduced to that of calculating the integrals appearing in I. There is a happy agreement between transport coefficients calculated this way and the strong field limits of those calculated by Marshall. (As corrected by Haas and Vaughan Williams.)

Yet another method is due to H. Grad. Here it is observed that what is desired from the B.E. is not the solution f, but the value of the moments $\boldsymbol{c}$, $\boldsymbol{cc}$, $\boldsymbol{c}c^2$ which have hydrodynamic significance, and further, that what is obtained is not the solution f, but some rather crude approximation to the normal solution. This being so, it might be appropriate to consider not the equation for f, but the equations for the moments of f. The moment equations do not, of course, close unless forced to: but they may be forced to close by choosing a trial distribution function of the correct form. The first five moments, l, $\boldsymbol{V}$, $\frac{1}{2}mc^2$, yield hydrodynamic quantities of immediate interest, while the hydrodynamic equations involve a further eight, the remaining five components of mc_ic_j and the three $\frac{1}{2}mc^2\boldsymbol{c}$. Grad's method involves writing the distribution function as a linear combination of these moments: 1, φ, T, P_{ij} and q, chosen to be consistent. The moment equations then are forced to close after the first 13, and differential equations are obtained for the time-dependence of the moments. This method has been applied to the ionized plasma by Kolodner, and by Lyley and Herdan; it is not noticeably simpler than the normal solution procedure.

6. – Transport coefficients for a fully ionized gas in a magnetic field.

(Marshall, corrected by Haas and Vaughan-Williams).

— *Current density*

$$\boldsymbol{j}=\sigma_{\parallel}\boldsymbol{D}_{\parallel}+\sigma_{\perp}\boldsymbol{D}_{\perp}+\sigma_{H}\boldsymbol{b}\times\boldsymbol{D}+\phi_{\parallel}\boldsymbol{\nabla}_{\parallel}T+\phi_{\perp}\boldsymbol{\nabla}_{\perp}T+\phi_{H}\boldsymbol{b}\times\boldsymbol{\nabla}T\,,$$

where for any vector $\boldsymbol{V}$

$$\boldsymbol{V}_{\parallel}=(\boldsymbol{V}\cdot\boldsymbol{b})\boldsymbol{b}\,,\quad \boldsymbol{V}_{\perp}=\boldsymbol{V}-\boldsymbol{V}_{\parallel}\,,$$

and

$$\boldsymbol{B}=B\boldsymbol{b}\,,$$

and

$$\boldsymbol{D}=\boldsymbol{E}+\boldsymbol{v}\times\boldsymbol{B}+\frac{1}{ne}\boldsymbol{\nabla}p\,.$$

Heat flux

$$\boldsymbol{q}=-\lambda_{\parallel}\boldsymbol{\nabla}_{\parallel}T-\lambda_{\perp}\boldsymbol{\nabla}_{\perp}T-\lambda_{H}\boldsymbol{b}\times\boldsymbol{\nabla}T+\psi_{\parallel}\boldsymbol{j}_{\parallel}+\psi_{\perp}\boldsymbol{j}_{\perp}+\psi_{H}\boldsymbol{b}\times\boldsymbol{j}\,,$$

or

$$\boldsymbol{q}=-K_{\parallel}\boldsymbol{\nabla}_{\parallel}T-K_{\perp}\boldsymbol{\nabla}_{\perp}T-K_{H}\boldsymbol{b}\times\boldsymbol{\nabla}T-\beta_{\parallel}\boldsymbol{D}_{\parallel}-\beta_{\perp}\boldsymbol{D}_{\perp}-\beta_{H}\boldsymbol{b}\times\boldsymbol{D}\,.$$

Stress tensor is most intelligible in component form. If $OX\parallel\boldsymbol{b}$ and

$$S_{ij}=\frac{1}{2}\left(\frac{\partial v_i}{\partial x_j}+\frac{\partial v_j}{\partial x_i}-\frac{1}{3}\,\delta_{ij}\,\mathrm{div}\,v\right),$$

$$p_{xx}=p-2\mu S_{xx}\,,$$

$$p_{yy}=p-\frac{2\mu}{1+r_+^2}\left\{S_{yy}+\frac{1}{2}\,r_+^2(S_{yy}+S_{zz})+r_+S_{yz}\right\},$$

$$p_{zz}=p-\frac{2\mu}{1+r_+^2}\left\{S_{zz}+\frac{1}{2}\,r_+^2(S_{yy}+S_{zz})-r_+S_{yz}\right\},$$

$$p_{xy}=p_{yx}=-\frac{2\mu}{1+\frac{1}{4}r_+^2}\left\{S_{xy}+\frac{1}{2}\,r_+S_{xz}\right\},$$

$$p_{xz}=p_{zx}=-\frac{2\mu}{1+\frac{1}{4}r_+^2}\left\{S_{xz}-\frac{1}{2}\,r_+S_{xy}\right\},$$

$$p_{yz}=p_{zy}=\frac{2\mu}{1+\frac{1}{4}r_+^2}\left\{S_{yx}-\frac{1}{2}\,r_+(S_{yy}-S_{zz})\right\}.$$

To define the coefficients in these expressions we introduce the following collision frequencies:

$$\nu_- = \frac{2}{3}\sqrt{\frac{2\pi}{m_-}}\, ne^4 (\log \Lambda)/(kT)^2 \qquad \text{with} \qquad \Lambda = 2(kT)^{\frac{3}{2}}/e^3(\pi n)^{\frac{1}{2}} ,$$

$$\nu_+ = \frac{2\sqrt{2}}{5}\sqrt{m_-/m_+}\cdot\nu_- ,$$

the gyro frequencies

$$\Omega_- = (eB/m)_- \qquad \Omega_+ = (eB/m)_+ ,$$

and the ratios

$$r = \Omega_-/\nu_- \qquad r_+ = \Omega_+/\nu_+ ,$$

the mass ratio

$$M_- = m_-/(m_- + m_+) ,$$

the unmodified transport coefficients,

$$\sigma_0 = ne^2/m_-\nu_- , \qquad \lambda_0 = \sigma_0 k^2 T/e^2 , \qquad \phi_0 = \sigma_0/e , \qquad \psi_0 = kT/e ,$$

and the r-dependent factors,

$$\Delta_1^{-1} = r^4 + 6.28r^2 + 0.93 , \qquad \Delta_2^{-1} = r^4 + 16.20r^2 + 44.3 ,$$

$$\Delta_3^{-1} = r^2 M_- + 9M_- + 3.39\sqrt{M_-} + 0.32 , \qquad \Delta_4^{-1} = r^2 + 3.48 , \qquad \Delta_5^{-1} = r^2 + 12.72$$

The transport coefficients for a magnetized plasma then become

$$\sigma_\parallel = \sigma_0 1.93/2 \qquad \phi_\parallel = 0.777\phi_0 ,$$

$$\sigma_\perp = \sigma_0 \tfrac{1}{2}(r^2 + 1.8)\Delta_1 \qquad \phi_\perp = -\phi_0 0.75(r^2 - 0.966)\Delta_1 ,$$

$$\sigma_H = -\sigma_0 \tfrac{1}{2} r(r^2 + 4.4)\Delta_1 \qquad \phi_H = -\phi_0 2.15 r\Delta_1 ,$$

$$\lambda_\parallel = 1.02\lambda_0 \qquad \psi_\parallel = 3.30\psi_0$$

$$\lambda_\perp = 1.25\lambda_0[(3M_- + 0.56\sqrt{M_1})\Delta_3 + (5.43r^2 + 36.1)\Delta_2 - 3.56\Delta_5] \qquad \psi_\perp = \psi_0[2.5r^2 + 11.5]\Delta_4 ,$$

$$\lambda_H = 1.25\lambda_0[rM_-\Delta_3 - 9.28r\Delta_2 - r\Delta_5] \qquad \psi_H = \psi_0 1.5r\Delta_4 ,$$

$$\mu = \tfrac{1}{4} nkT/\nu_+ .$$

The second form of the thermal conductivity may be written in terms of the first, thus

$$K_{\parallel} = \lambda_{\parallel} - \phi_{\parallel}\psi_{\parallel} \qquad \beta_{\parallel} = -\sigma_{\parallel}\psi_{\parallel} ,$$

$$K_{\perp} = \lambda_{\perp} + \phi_H\psi_H - \phi_{\perp}\psi_{\perp} \qquad \beta_{\perp} = \sigma_H\psi_H - \sigma_{\perp}\psi_{\perp} ,$$

$$K_H = \lambda_H - \phi_{\perp}\psi_H - \phi_H\psi_{\perp} \qquad \beta_H = -\sigma_H\psi_{\perp} - \sigma_{\perp}\psi_H .$$

In the limit of strong magnetic fields $r, r_+ \to \infty$

$$\sigma_{\parallel} = 0.8\sigma_0 \qquad \phi_{\parallel} = 0.68\phi_0 ,$$

$$\sigma_{\perp} = 0.5\sigma_0 r^{-2} \qquad \phi_{\perp} = 0.75\phi_0 r^{-2} ,$$

$$\sigma_H = r\sigma_{\perp} \qquad \phi_H = 2.13\phi_0 r^{-3} ,$$

$$\lambda_{\parallel} = \lambda_0 \qquad \psi_{\parallel} = 3.3\psi_0$$

$$\lambda_{\perp} = 0.7\sqrt{\left(\frac{m_+}{m_-}\right)}\,\lambda_0 r^{-2} \qquad \psi_{\perp} = 2.5\psi_0 ,$$

$$\lambda_H = -0.5\lambda_0(0.4 + r^2 m_-/m_+)r^{-1} \qquad \psi_H = 1.5\psi_0 r^{-1} ,$$

$$p_{xx} = p_0 - 2\mu S_{xx} , \qquad p_{yy} = p_0 - \mu(S_{yy} + S_{zz}) = p_{zz} ,$$

$$p_{xy} = p_{yx} = -\frac{p_0}{\Omega_+} S_{xy} = -p_{xz} = -p_{zx} ,$$

$$p_{yz} = -p_{zy} = \frac{p_0}{4\Omega_+}(S_{yy} - S_{zz}) .$$

III. – Kinetic Theory of Plasma. The Transport Equation.

Introduction.

Having shown how the Boltzmann equation leads to the appearance of transport coefficients and to phenomena associated with « real » fluids, we turn to the prior question, that of determining the correct form of the transport equation. Our procedure is first to show how the Boltzmann equation can be expanded for small-angle scattering, which dominates the collision process. Having done this we are free to discuss the effect of long-range correlations on the small-angle scattering and to develop forms for the transport equation which are valid in this region. We may then compare the expanded B.E. with the long-range equation of transport, and from this justify our use of the

B.E. in those situations for which the meaning of the transport coefficients is unambiguous.

1. – An expansion of the B.E.

Consider the Boltzmann collision integral for an ionized gas of like particles (for the moment). Using

$$\sigma = \sigma_R = \left(\frac{e}{2m}\right)^2 \left(g \sin\frac{\theta}{2}\right)^{-4},$$

where $\boldsymbol{g} = \boldsymbol{v} - \boldsymbol{v}'$, and further, using the result (II.3.4) to write

$$\bar{\boldsymbol{v}} = \boldsymbol{v} - \tfrac{1}{2}\Delta \boldsymbol{g},$$

$$\bar{\boldsymbol{v}}' = \boldsymbol{v}' + \tfrac{1}{2}\Delta \boldsymbol{g},$$

where $\Delta\boldsymbol{g}$ = change in $\boldsymbol{g}$ on collision, *i.e.* $\bar{\boldsymbol{g}} - \boldsymbol{g} = \Delta\boldsymbol{g}$, enables this to be written

$$\text{(III.1.1)} \qquad I = \left(\frac{e^2}{m}\right)^2 \int \mathrm{d}^3 v' \int \mathrm{d}\varphi\, \mathrm{d}\vartheta \sin\theta \, \frac{g}{g^4 \sin^4(\theta/2)} \cdot \left[f\left(\boldsymbol{v}' + \frac{1}{2}\Delta\boldsymbol{g}\right) f\left(\boldsymbol{v} - \frac{1}{2}\Delta\boldsymbol{g}\right) - f(\boldsymbol{v}') f(\boldsymbol{v}) \right].$$

If $\boldsymbol{m}$, $\boldsymbol{n}$ are unit vectors orthogonal to $\boldsymbol{g} = g\hat{\boldsymbol{g}}$, we may write

$$\text{(III.1.2)} \qquad \Delta\boldsymbol{g} = 2g \sin\left(\frac{\theta}{2}\right) \left\{ -\hat{\boldsymbol{g}} \sin\frac{\theta}{2} + \boldsymbol{m} \cos\frac{\theta}{2} \cos\varphi + \boldsymbol{n} \cos\left(\frac{\theta}{2}\right) \sin\varphi \right\}.$$

Now we may expand the quantity within the square brackets thus:

$$\text{(III.1.3)} \qquad f\left(\boldsymbol{v}' + \frac{1}{2}\Delta\boldsymbol{g}\right) f\left(\boldsymbol{v} - \frac{1}{2}\Delta\boldsymbol{g}\right) - f(\boldsymbol{v}') f(\boldsymbol{v}) =$$

$$= \left[f(\boldsymbol{v}) \frac{\partial f(\boldsymbol{v}')}{\partial \boldsymbol{v}'} - f(\boldsymbol{v}') \frac{\partial f(\boldsymbol{v})}{\partial \boldsymbol{v}} \right] \cdot \frac{1}{2} \Delta\boldsymbol{g} + \frac{1}{2} \left[f \frac{\partial^2 f(\boldsymbol{v})}{\partial \boldsymbol{v}' \partial \boldsymbol{v}'} + f(\boldsymbol{v}') \frac{\partial^2 f(\boldsymbol{v})}{\partial \boldsymbol{v} \partial \boldsymbol{v}} - \right.$$

$$\left. - 2 \frac{\partial f(\boldsymbol{v}')}{\partial \boldsymbol{v}'} \frac{\partial f(\boldsymbol{v})}{\partial \boldsymbol{v}} \right] : \frac{1}{4} \Delta\boldsymbol{g} \Delta\boldsymbol{g} + 0(\Delta\boldsymbol{g})^3 .$$

If we are interested only in scattering at small $\Delta\boldsymbol{g}$, *i.e.* small-angle scattering, this Taylor expansion may be substituted in the collision integral (III.1.1) and if the series is cut off at the second term, the expansion (III.1.2) used for $\Delta\boldsymbol{g}$

and the integral over φ performed, there results

$$\text{(III.1.4)}\quad I = 8\pi\left(\frac{e^2}{m}\right)^2\int d^3v'\int\limits^{\pi/2} d\left(\frac{\theta}{2}\right)\cos\left(\frac{\theta}{2}\right)\left\{-\frac{\boldsymbol{g}}{g^3\sin(\theta/2)}\left[f(\boldsymbol{v})\frac{\partial f(\boldsymbol{v}')}{\partial\boldsymbol{v}} - f(\boldsymbol{v}')\frac{\partial f(\boldsymbol{v})}{\partial\boldsymbol{v}}\right] + \right.$$

$$+\frac{1}{2}\left[f(\boldsymbol{v})\frac{\partial^2 f(\boldsymbol{v}')}{\partial\boldsymbol{v}\,\partial\boldsymbol{v}'} + f(\boldsymbol{v}')\frac{\partial^2 f(\boldsymbol{v})}{\partial\boldsymbol{v}\,\partial\boldsymbol{v}} - 2\frac{\partial f(\boldsymbol{v}')}{\partial\boldsymbol{v}}\frac{\partial f(\boldsymbol{v})}{\partial\boldsymbol{v}}\right]\cdot$$

$$\left.\cdot\frac{1}{g}\left[\frac{\boldsymbol{gg}}{g^2}\sin\frac{\theta}{2} + \frac{1}{2}(\boldsymbol{mm}+\boldsymbol{nn})\left(\frac{1}{\sin(\theta/2)} - \sin\frac{\theta}{2}\right)\right]\right\}.$$

Now

$$\int\limits_0^{\pi/2} d\left(\frac{\theta}{2}\right)\cos\left(\frac{\theta}{2}\right)\sin\left(\frac{\theta}{2}\right) = 1\,,\qquad \text{while}\qquad \int\limits_0^{\pi/2} d\left(\frac{\theta}{2}\right)\cos\left(\frac{\theta}{2}\right)\Big/\sin\left(\frac{\theta}{2}\right) = \log\sin\left(\frac{\theta}{2}\right)\Big|_0^{\pi/2}$$

diverges; however, the divergence arises from the lower limit $\theta/2 = 0$, and this represents just those long-range effects where co-operative phenomena may be important. The scattering angle at long ranges may be related to the impact parameter b: indeed, using the impulse approximation

$$\frac{v_\perp}{v_0} = \sin\theta = \frac{1}{v_0}\int F_\perp\,dt = \frac{1}{v_0^2}\int F_\perp\,dx =$$

$$= \frac{e^2}{m_r}v_0^2\int\limits_{-\infty}^{\infty}\frac{b}{(x^2+b^2)^{\frac{3}{2}}}\,dx = \frac{1}{b}\frac{e^2}{m_r v_0^2}\int\limits_{-\infty}^{\infty}\frac{dt}{(1+t^2)^{\frac{3}{2}}} = \frac{2e^2}{bm_r v_0^2}\,,$$

$$\therefore\ \sin\left(\frac{\theta}{2}\right) \simeq \frac{\theta}{2} \simeq \frac{1}{2}\sin\theta \simeq \frac{e^2}{bm_r v_0^2}\,.$$

To approximate the maximum impact parameter for which collective effects are unimportant, we may observe, in the most primitive fashion that the plasma acts as a dielectric medium, for which, roughly, $\varepsilon = 1 - \omega_0^2/\omega^2$, where $\omega_0^2 = 4\pi ne^2/m$ is the plasma frequency. If also, we Fourier-analyze the potential due to a particle passing the test particle at a distance b, *i.e.*

$$\phi(\omega) = \int\limits_{-\infty}^{\infty}\frac{\exp[i\omega t]}{[b^2+v^2t^2]^{\frac{1}{2}}}\,dt = \frac{1}{v}K_0\left(\frac{\omega b}{v}\right),$$

this is exponentially small for values of $\omega \gg v/b$. However, frequencies $< \omega_0$ are not transmitted, but exponentially damped in the plasma, hence if $v/b < \omega_0$, the interaction is screened; this gives $b_{\max} \simeq v/\omega_0$ and

$$\text{(III.1.5)}\quad -\log\left(\sin\frac{\theta}{2}\right)_{\min} = \log\Lambda \simeq \log\left[\frac{e^2}{b_{\max}mv^2}\right] = \log\left(\frac{e^2\omega_0}{mv^3}\right) \simeq \log\left[\frac{\pi ne^6}{(kT)^2}\right]^{\frac{1}{2}} \simeq$$

$$\simeq \log\left\{2\sqrt{\pi}\left[\frac{n^{\frac{1}{3}}e^2}{kT}\right]^{\frac{3}{2}}\right\} \simeq \log\left\{2\sqrt{\pi}\left(\frac{V}{T}\right)^{\frac{3}{2}}\right\},$$

where V and T are the average interparticle potential and the average kinetic energy. The parameter Λ or better (V/T) is usually small and plays an important role in our theory.

(III.1.4) may be reduced to

$$I = 4\pi\left(\frac{e^2}{m}\right)^2 \log\Lambda\int \mathrm{d}^3v' \left\{-\frac{\boldsymbol{g}}{g^3}\cdot\left[f(\boldsymbol{v})\frac{\partial f(\boldsymbol{v}')}{\partial \boldsymbol{v}'} - f(\boldsymbol{v}')\frac{\partial f(\boldsymbol{v})}{\partial \boldsymbol{v}}\right] + \right.$$

$$\left. + \frac{1}{2}\left[f(\boldsymbol{v})\frac{\partial^2 f(\boldsymbol{v}')}{\partial \boldsymbol{v}'\partial \boldsymbol{v}'} + f(\boldsymbol{v}')\frac{\partial^2 f(\boldsymbol{v})}{\partial \boldsymbol{v}\partial \boldsymbol{v}} - 2\frac{\partial f(\boldsymbol{v}')}{\partial \boldsymbol{v}'}\frac{\partial f(\boldsymbol{v})}{\partial \boldsymbol{v}}\right] : \frac{1}{2}\frac{\boldsymbol{mm}+\boldsymbol{nn}}{g}\right\},$$

$\boldsymbol{mm}+\boldsymbol{nn}$ is the unit tensor normal to $\hat{\boldsymbol{g}}$, *i.e.* $\boldsymbol{mm}+\boldsymbol{nn} = (\mathbf{1}-\hat{\boldsymbol{g}}\hat{\boldsymbol{g}})$.

If

$$\text{(III.1.6)}\qquad \boldsymbol{w} = \frac{1}{g}(1-\hat{\boldsymbol{g}}\hat{\boldsymbol{g}}) = \frac{\boldsymbol{mm}+\boldsymbol{nn}}{g};\qquad \boldsymbol{w} = \frac{\partial}{\partial \boldsymbol{g}}\frac{\partial}{\partial \boldsymbol{g}}|g|,$$

and

$$\frac{\partial}{\partial \boldsymbol{g}}\cdot \boldsymbol{w} = -2\frac{\boldsymbol{g}}{g^3}.$$

I may be written

$$I = 8\pi\left(\frac{e^2}{m}\right)^2 \log\Lambda\int \mathrm{d}^3v' \,\frac{1}{2}\frac{\partial\cdot \boldsymbol{w}}{\partial \boldsymbol{g}}\cdot\left[f\frac{\partial f(\boldsymbol{v}')}{\partial \boldsymbol{v}'} - f(\boldsymbol{v}')\frac{\partial f(\boldsymbol{v})}{\partial \boldsymbol{v}}\right] +$$

$$+ \frac{1}{2}\boldsymbol{w} : \frac{\partial}{\partial \boldsymbol{v}}\left[f(\boldsymbol{v}')\frac{\partial f(\boldsymbol{v})}{\partial \boldsymbol{v}} - f(\boldsymbol{v})\frac{\partial f(\boldsymbol{v}')}{\partial \boldsymbol{v}'}\right] + \frac{1}{4}\boldsymbol{w} : \left[f(\boldsymbol{v})\frac{\partial^2 f(\boldsymbol{v}')}{\partial \boldsymbol{v}'\partial \boldsymbol{v}'} - \frac{\partial f(\boldsymbol{v}')}{\partial \boldsymbol{v}'}\frac{\partial f(\boldsymbol{v})}{\partial \boldsymbol{v}}\right],$$

the last term here may be integrated by parts, and if we use $\partial \boldsymbol{w}/\partial \boldsymbol{g} = -\partial \boldsymbol{w}\,\partial \boldsymbol{v}'$, there results

$$\text{(III.1.7)}\qquad I = 2\pi\left(\frac{e^2}{m}\right)^2 \log\Lambda\,\frac{\partial}{\partial v_j}\int \mathrm{d}^3v' \left[f(v')\frac{\partial f(v)}{\partial v_i} - f(v)\frac{\partial f(v')}{\partial v_i'}\right] w_{ij},$$

(cf. LANDAU). This may also be written as

$$\text{(III.1.8)} \quad I = -2\pi\left(\frac{e^2}{m}\right)^2 \log\Lambda \frac{\partial}{\partial v_i}\left[\int \mathrm{d}^3v' \left\{\frac{\partial f(v')}{\partial v'_j} w_{ij} - f(v')\frac{\partial w_{ij}}{\partial v'_j}\right\} f(v)\right] + \\ + 2\pi\left(\frac{e^2}{m}\right)^2 \log\Lambda \frac{\partial^2}{\partial v_i \partial v_j}\left[\int \mathrm{d}^3v' \left\{f(v') w_{ij}\right\} f(v)\right],$$

which has the form

$$\text{(III.1.9)} \qquad I(f) = \frac{\partial}{\partial \boldsymbol{v}}\cdot(\boldsymbol{D}f) + \frac{\partial^2}{\partial \boldsymbol{v}\,\partial \boldsymbol{v}}:(\boldsymbol{D}f)\,,$$

of the Fokker-Planck equation.

2. – The Fokker-Planck equation.

The form of I, (III.1.9) may be approached by a somewhat different route and arises from a study of the rate of change of a Markovian probability distribution. Suppose that $P(x, t)$ represents the distribution function for a quantity x at time t: and suppose moreover, that during the interval $t \to t + \delta t$, x changes to x' with a probability $w(x, x' - x)$. Then, $P(x, t + \delta t)$ may be expressed in terms of $P(x', t)$ as

$$P(x, t + \delta t) = \int \mathrm{d}x'\, w(x', x - x')\, P(x', t) = \int \mathrm{d}\xi\, w(x - \xi, \xi)\, P(x - \xi, t)\,.$$

If w is a rapidly decreasing function of ξ, *i.e.* if P develops by small steps, then we may use a Taylor expansion on the R.H.S., so that

$$P(x, t + \delta t) = \int \mathrm{d}\xi w(x, \xi)\, P(x, t) - \frac{\partial}{\partial x}\int \mathrm{d}\xi\xi\, w(x, \xi)\, P(x, t) + \\ + \frac{1}{2}\frac{\partial}{\partial x}\frac{\partial}{\partial x}\int \mathrm{d}\xi w(x, \xi)\xi\xi\, P(x, t)\,.$$

Now, the first integral here $= 1$, since

$$\int w(x, \xi)\, \mathrm{d}^3\xi = 1\,,$$

the second $= \langle\xi\rangle$, the third $\langle\xi\xi\rangle$, thus,

$$\frac{\partial P}{\partial t} = \frac{P(x, t + \delta t) - P(x, t)}{\delta t} = -\frac{\partial}{\partial x}[D_1 P(x, t)] + \frac{1}{2}\frac{\partial^2}{\partial x\,\partial x}\cdot D_2 P(x, t) + \ldots\,,$$

where

$$D_1 = \frac{1}{\tau}\langle\xi\rangle\,, \qquad D_2 = \frac{1}{\tau}\langle\xi\xi\rangle\,, \qquad \tau = \delta t\,.$$

For $I(f, f)$ we write

$$I(f, f) = -\frac{\partial}{\partial \boldsymbol{v}}\cdot D_1 f(v) + \frac{1}{2}\frac{\partial^2}{\partial \boldsymbol{v}\,\partial \boldsymbol{v}}:D_2 f(v)\,, \tag{III.2.1}$$

where

$$\boldsymbol{D}_1 = \frac{1}{\tau}\langle\Delta\boldsymbol{v}\rangle\,, \qquad \boldsymbol{D}_2 = \frac{1}{\tau}\langle\Delta\boldsymbol{v}\,\Delta\boldsymbol{v}\rangle\,.$$

From an unmagnetized plasma the quantities $\boldsymbol{D}_1$, $\boldsymbol{D}_2$ can be expressed in terms of the fluctuating microfield within the plasma, *i.e.* the change in velocity in a time τ of a particle initially at $\boldsymbol{x}$ having velocity $\boldsymbol{v}$ is

$$\Delta\boldsymbol{v} = \frac{e}{m}\int_{t-\tau}^{t}\boldsymbol{E}(x', t')\,\mathrm{d}t'\,,$$

where

$$\boldsymbol{x}' = \boldsymbol{x}_0 + \int \boldsymbol{v}(x', t')\,\mathrm{d}t' \qquad \text{and} \qquad \boldsymbol{v}(x, t) = \boldsymbol{v}_0 + \frac{e}{m}\int \boldsymbol{E}(x', t')\,\mathrm{d}t'\,.$$

If $(e/m)\boldsymbol{E}$ is small, we may expand $\Delta\boldsymbol{v}$ thus

$$\Delta v_i = \frac{e}{m}\int_{t-\tau}^{t}\mathrm{d}t'\left\{E_i(x_0 + v_0 t', t') + \frac{\partial E_i}{\partial x_j}\int_{t-\tau}^{t'}\mathrm{d}t''\int_{t-\tau}^{t''}\frac{e}{m}E_j(x_0 + v_0 t''', t''')\,\mathrm{d}t'''\right\}, \tag{III.2.2}$$

while to the same order in eE/m

$$\Delta\boldsymbol{v}\,\Delta\boldsymbol{v} = \frac{e^2}{m^2}\int_{t-\tau}^{t}\mathrm{d}t'\,\boldsymbol{E}(x_0 + v_0 t', t')\int_{t-\tau}^{t}\mathrm{d}t''\,\boldsymbol{E}(x_0 + v_0 t'', t'') = \tag{III.2.3}$$

$$= \frac{e^2}{m^2}\int_{t-\tau}^{t}\mathrm{d}t'\int_{0}^{\tau}\mathrm{d}s\,\boldsymbol{E}(x_0 + v_0 t', t')\,\boldsymbol{E}[x_0 + v_0(t' - s), t - s]\,.$$

This equation may be derived in an alternative way which requires, however, a return to the Liouville equation.

$$\frac{\partial F}{\partial t} + \boldsymbol{v}_i\cdot\frac{\partial F}{\partial \boldsymbol{x}_i} + \frac{e}{m}\boldsymbol{E}_i\cdot\frac{\partial}{\partial \boldsymbol{v}_i}F = 0\,. \tag{III.2.4}$$

Now, the Liouville function F can be written $F_1(x_1)\Psi(x_2 \dots x_N; x_1)$ and by integrating over $x_2 \dots x_N$, an equation for $F_1(x_1)$ may be deduced.

Since

$$\boldsymbol{E}_i = -\frac{\partial}{\partial \boldsymbol{x}_i}\phi_i = -\frac{\partial}{\partial \boldsymbol{x}_i}\sum_j \frac{e}{|\boldsymbol{x}_i - \boldsymbol{x}_j|}$$

depends upon the $\boldsymbol{x}_j$, however, the integration cannot be carried out explicitly, but involves a term of the form

$$-\frac{\partial}{\partial \boldsymbol{v}}\cdot\sum_j\int\frac{e^2}{m}\frac{\partial}{\partial \boldsymbol{x}_1}\phi(\boldsymbol{x}_1, \boldsymbol{x}_j)\Psi(\boldsymbol{x}_2 \dots \boldsymbol{x}_N; \boldsymbol{x}_1)F_1(\boldsymbol{x}_1)\,\mathrm{d}x_2 \dots \mathrm{d}x_N\,, \qquad i.e. \quad \frac{e^2}{m}\frac{\partial}{\partial v}\cdot \boldsymbol{E}F\,.$$

For any *given complexion*

$$F = \prod \delta[\boldsymbol{x}_i - \boldsymbol{X}_i(t)]\,\delta[\boldsymbol{v}_i - \boldsymbol{V}_i(t)]$$

is a rapidly varying function, and we prefer to work with a smoothly varying quantity, f_1, which represents the probability of a particle being at $\boldsymbol{x}_1, \boldsymbol{v}_1$ given some initial probability distribution $p_0(x_1 \dots x_N, v_1 \dots v_N)$ for the entire distribution.

Then, F has the form

$$p_0(\boldsymbol{x}_1^0 \dots \boldsymbol{x}_N^0, \boldsymbol{v}_1^0 \dots \boldsymbol{v}_N^0)\prod_i \delta[\boldsymbol{x} - \boldsymbol{X}_i(t)]\,\delta[\boldsymbol{v} - \boldsymbol{V}_i(t)]\,,$$

where

$$\boldsymbol{X}(t) = \boldsymbol{x}_1^0 + \int \boldsymbol{V}_i(t')\,\mathrm{d}t'\,,$$

$$\boldsymbol{V}(t) = \boldsymbol{v}(0) + \frac{e}{m}\int \boldsymbol{E}(\boldsymbol{X}(t'), t')\,\mathrm{d}t'\,.$$

From this we may form F_1, again by integrating over $\mathrm{d}x_2 \dots \mathrm{d}x_N$ and noting that if p_0 is smooth, we expect F_1 to be a slowly varying function. In fact we expect F_1 to satisfy an equation of the Boltzmann type, and to change significantly in times of order τ_c, the collision period. This means that instead of considering the equation of motion for F_1 we obtain an adequate description by considering the motion of the coarse-grained distribution

$$f = \frac{1}{\tau}\int_{t-\tau}^{t} F_1(t')\,\mathrm{d}t'\,,$$

where $\tau \ll \tau_c$. This satisfies

$$\frac{\partial f}{\partial t} + \boldsymbol{v}\cdot\frac{\partial f}{\partial \boldsymbol{x}} + \frac{e}{m}\frac{\partial}{\partial \boldsymbol{v}}\cdot\frac{1}{\tau}\int_{t-\tau}^{t} \boldsymbol{E}(t', x(t'))\,F_1(t')\,\mathrm{d}t' = 0\,,$$

i.e.

$$\text{(III.2.5)}\qquad \frac{\partial f}{\partial t} + \boldsymbol{v}\cdot\frac{\partial f}{\partial \boldsymbol{x}} + \frac{e}{m}\frac{\partial}{\partial \boldsymbol{v}}\cdot\frac{1}{\tau}\int_{t-\tau}^{t'} \mathrm{d}t'\,\boldsymbol{E}(t', \boldsymbol{x})\,\delta[\boldsymbol{x}-\boldsymbol{X}(t')]\,\delta[\boldsymbol{v}-\boldsymbol{V}(t')]F_1(t'-\tau)\,,$$

where the last term is explicitly

$$\text{(III.2.6)}\qquad -\frac{e^2}{m}\frac{\partial}{\partial \boldsymbol{v}}\cdot\frac{1}{\tau}\int_{t-\tau}^{t} \mathrm{d}t'\,\mathrm{d}^3x_2\ldots\mathrm{d}^3x_N\,\mathrm{d}^3v_2\ldots\mathrm{d}^3v\sum_j\frac{\partial\phi}{\partial \boldsymbol{x}_i}(\boldsymbol{x}_i, \boldsymbol{x}_j)\cdot$$

$$\cdot\,\psi(\boldsymbol{x}_2\ldots\boldsymbol{x}_N, \boldsymbol{v}_2\ldots\boldsymbol{v}_N; x_1)\,F_1(\boldsymbol{x}_1, \boldsymbol{v}_1, t-\tau)\,\delta[\boldsymbol{x}_1-\boldsymbol{X}_1(t')]\,\delta[\boldsymbol{v}_1-\boldsymbol{V}_1(t')]\,.$$

Now the electric field may contain some mean part $\boldsymbol{E}_0$, but will certainly contain a rapidly fluctuating part $\boldsymbol{E}_f$. $\boldsymbol{E}_0$ is easily handled, contributing a term $(e/m)\boldsymbol{E}_0\cdot(\partial f/\partial \boldsymbol{v})$, and to treat $\boldsymbol{E}_f$ we may observe that for most plasmas, where the ratio of kinetic to potential energy is small, the field-dependent quantities $\boldsymbol{X}$ and $\boldsymbol{V}$ may be calculated in perturbation theory: thus

$$\boldsymbol{V}(t) = \boldsymbol{v}_0 + \frac{e}{m}\int^{t} \boldsymbol{E}(\boldsymbol{x}_0+\boldsymbol{v}_0 t', t')\,\mathrm{d}t'\,,$$

$$\boldsymbol{X}(t) = \boldsymbol{x}_0 + \boldsymbol{v}_0 t + \frac{e}{m}\int_0^{t}\mathrm{d}t'\int_0^{t'}\mathrm{d}t''\,E(\boldsymbol{x}_0+\boldsymbol{v}_0 t'', t'')\,,$$

and the δ functions in (III.2.5) may similarly be expanded, thus

$$\delta(\boldsymbol{x}-\boldsymbol{X}(t'))\,\delta(\boldsymbol{v}-\boldsymbol{V}(t')) = \delta(\boldsymbol{x}-\boldsymbol{x}_0-\boldsymbol{v}_0 t')\,\delta(\boldsymbol{v}-\boldsymbol{v}_0) -$$

$$-\,\delta(\boldsymbol{x}-\boldsymbol{x}_0-\boldsymbol{v}_0 t')\frac{\partial}{\partial \boldsymbol{v}}\delta(\boldsymbol{v}-\boldsymbol{v}_0 t', t')\cdot\frac{e}{m}\int^{t'}\boldsymbol{E}(x_0+v_0 t'', t'')\,\mathrm{d}t'' -$$

$$-\,\delta(\boldsymbol{v}-\boldsymbol{v}_0)\frac{\partial}{\partial \boldsymbol{x}}\delta(\boldsymbol{x}-\boldsymbol{x}_0-\boldsymbol{v}_0 t')\cdot\frac{e}{m}\int^{t'}\mathrm{d}t''\int^{t''}\mathrm{d}t'''\,\boldsymbol{E}(x_0+v_0 t''', t''')\,.$$

These may now be handled, as usual, by a partial integration, and use made of the fact that F_1 is slowly varying, so that $F(t-\tau, v) = f(t, v)$ whereupon (3.3) becomes

$$\text{(III.2.7)} \quad \frac{\partial f}{\partial t} + \boldsymbol{v}\cdot\frac{\partial f}{\partial \boldsymbol{x}} + \frac{e}{m}\boldsymbol{E}_0\cdot\frac{\partial f}{\partial \boldsymbol{v}} - \frac{e^2}{m^2}\frac{\partial}{\partial v_i}\left\{\frac{1}{\tau}\int_{t-\tau}^{t}\mathrm{d}t'\,\frac{\partial}{\partial x_j}E_i(\boldsymbol{x}_0+\boldsymbol{v}_0 t', t')\cdot\right.$$

$$\left.\cdot\int_{t-\tau}^{t'}\mathrm{d}t''\int_{t-\tau}^{t''}E_j(\boldsymbol{x}_0+\boldsymbol{v}_0 t''', t''')f\right\} - \frac{e^2}{m^2}\frac{\partial}{\partial v_i}\frac{\partial}{\partial v_j}\left\{\frac{1}{\tau}\int_{t-\tau}^{t}\mathrm{d}t'\,E_i(\boldsymbol{x}_0+\boldsymbol{v}_0 t', t')\int_{t-\tau}^{t}\mathrm{d}t''\,E_j(\boldsymbol{x}_0+\boldsymbol{v}_0 t'', t'')f(v)\right\},$$

an equation of the Fokker-Planck form with the coefficient given in (III.2.2) and (III.2.3). These expanded forms may be written exactly as those in (III.2.3), *i.e.*

$$\left\langle \boldsymbol{E}\int_{t-\tau}^{t}\boldsymbol{E}\,\mathrm{d}t'\right\rangle = \int \mathrm{d}^3x_2\ldots\mathrm{d}^3x_N,\,\mathrm{d}^3v_2\ldots\mathrm{d}^3v_N\sum_j\sum_k\frac{\partial\phi}{\partial\boldsymbol{x}_1}\,\partial\phi(x_1, x_j)\int_{t-\tau}^{t}\mathrm{d}t'\,\frac{\partial\phi}{\partial\boldsymbol{x}_1}(x_1, x_k)\cdot$$

$$\cdot\,\psi(X_2\ldots X_N,\, V_2\ldots V_N,\, x_1)\,,$$

i.e. as mean values $\langle E_i E_j\rangle$.

Now, the correlation functions

$$\text{(III.2.8)} \qquad \left\langle \phi(t)\int_{t-\tau}^{t}\phi(t')\,\mathrm{d}t'\right\rangle = \left\langle\int_0^{\tau}\phi(t)\,\phi(t-s)\,\mathrm{d}s\right\rangle$$

may be simplified somewhat by assuming a property of $\langle \boldsymbol{E}(t)\boldsymbol{E}(t-s)\rangle$ which we will be able to demonstrate, namely, that the fields are strongly correlated for small times, but that the product $\boldsymbol{E}\cdot\boldsymbol{E}$ becomes small and fluctuates about zero for long times; indeed, in times $\sim 1/\omega_0$ the correlation is already small. This suggests that if the upper limit of integration is taken as $\tau \gg \omega_0^{-1}$, which implies $\tau_c \gg \omega_0^{-1}$, a condition which is usually satisfied: then the upper limit in (III.2.8) may be extended to infinity and the required quantities become

$$\left\langle\int_0^{\infty}\phi(t)\,\phi(t-s)\,\mathrm{d}s\right\rangle,$$

and, if the field is derivable from a potential,

$$\langle E_i E_j\rangle = \left\langle\frac{\partial\phi}{\partial x_i}\frac{\partial\phi}{\partial x_j}\right\rangle.$$

The correlation function may be expressed in terms of the spectrum of ϕ, for, if

$$\phi(\boldsymbol{x}, t) = (2\pi)^{-4}\int \mathrm{d}^3k\,\mathrm{d}\omega \exp[i\boldsymbol{k}\cdot\boldsymbol{x}]\exp[i\omega t]\phi(\boldsymbol{k}, \omega)\,,$$

then

$$\left\langle \int_0^\infty \phi(\boldsymbol{x}, t)\phi(\boldsymbol{x}-\boldsymbol{v}s, t-s)\,\mathrm{d}s \right\rangle = R\,(2\pi)^{-8}\left\langle \int_0^\infty \mathrm{d}^3k\,\mathrm{d}\omega \int \mathrm{d}^3k'\,\mathrm{d}\omega' \cdot \right.$$

$$\left. \cdot \int \mathrm{d}s \exp[i\boldsymbol{k}\cdot\boldsymbol{x}]\exp[i\omega t]\exp[i\boldsymbol{k}'\cdot(\boldsymbol{x}-\boldsymbol{v}s)]\exp[i\omega'(t-s)]\,\phi(\boldsymbol{k}, \omega)\phi(\boldsymbol{k}', \omega')\right\rangle =$$

$$= R\left\langle (2\pi)^{-4}\int \mathrm{d}s \int \mathrm{d}^3k\,\mathrm{d}\omega \exp[-i(\boldsymbol{k}\cdot\boldsymbol{v}+\omega)s](2\pi)^{-4}\int \mathrm{d}^3k'\,\mathrm{d}\omega' \exp[i(\boldsymbol{k}+\boldsymbol{k}')\cdot\boldsymbol{x}]\cdot \right.$$

$$\left. \cdot \exp[i(\omega+\omega')t]\,\phi(\boldsymbol{k}, \omega)\phi(\boldsymbol{k}', \omega')\right\rangle .$$

The inner integral however, is the energy spectrum, $\langle\phi(\boldsymbol{k}, \omega)\phi(\boldsymbol{k}, \omega)\rangle$ hence, the correlation function

$$\langle\phi(\boldsymbol{x}, t)\,\phi(\boldsymbol{x}-\boldsymbol{v}s, t-s)\rangle = R(2\pi)^{-4}\int \mathrm{d}^3k\,\mathrm{d}\omega\,\phi(\boldsymbol{k}, \omega)\phi(\boldsymbol{k}, \omega)\exp[-i(\boldsymbol{k}\cdot\boldsymbol{v}-\omega)s] =$$

$$= \frac{1}{(2\pi)^4}\left\langle \int \mathrm{d}^3k\,\mathrm{d}\omega\,\phi(\boldsymbol{k}, \omega)\phi(\boldsymbol{k}, \omega)\cos(\boldsymbol{k}\cdot\boldsymbol{v}+\omega)s \right\rangle$$

is the cosine transform of the energy spectrum (the Wiener-Kinchin theorem). It follows that

$$\text{(III.2.9)}\quad \left\langle \int_0^\infty \mathrm{d}s\,\phi(\boldsymbol{x},t)\phi(\boldsymbol{x}-\boldsymbol{v}s, t-s)\right\rangle = (2\pi)^{-4}\left\langle \int \mathrm{d}^3k\,\mathrm{d}\omega\,\phi(k,\omega)\phi(\boldsymbol{k},\omega)\delta(\omega+\boldsymbol{k}\cdot\boldsymbol{v})\right\rangle .$$

3. – Calculation of the spectrum.

To calculate the spectrum we can again assume that the electric field within the plasma is weak, and the interactions are small. At the same time, the effect of the field on the distribution function must be retained in any calculation of the field, which otherwise diverges. Our first object then must be to calculate the response of the plasma to a field in that approximation in which the particle interaction is neglected. This however, requires a solution to the Vlasov equation,

$$\text{(III.3.1)}\qquad \frac{\partial f}{\partial t} + \boldsymbol{v}\cdot\frac{\partial f}{\partial \boldsymbol{x}} + \frac{e}{m}\boldsymbol{E}\cdot\frac{\partial f}{\partial \boldsymbol{v}} = 0\,,$$

and since the fields are assumed small, we may use a perturbation solution of this about some distribution f_0, assumed known, — whereupon, on Fourier-transforming,

$$\text{(III.3.2)} \qquad f = f_0 + f', \qquad f'(\omega, \boldsymbol{k}) = -\frac{e}{m}\frac{\boldsymbol{E}(\omega, \boldsymbol{k})\cdot(\partial/\partial\boldsymbol{v})f_0}{i(\omega + \boldsymbol{k}\cdot\boldsymbol{v})},$$

or, if $\boldsymbol{E}$ is derivable from a potential $\boldsymbol{E} = -ik\phi$,

$$\text{(III.3.3)} \qquad f' = \frac{e}{m}\frac{\boldsymbol{k}\cdot(\partial f_0/\partial\boldsymbol{v})}{(\omega + \boldsymbol{k}\cdot\boldsymbol{v})}\phi,$$

The charge induced by a potential ϕ, then becomes, if f_0^-, f_0^+ are the unperturbed distributions of electrons and ions,

$$q_{\text{ind}}(\omega, \boldsymbol{k}) = \int \mathrm{d}^3v \frac{1}{(\omega + \boldsymbol{k}\cdot\boldsymbol{v})}\boldsymbol{k}\cdot\frac{\partial}{\partial\boldsymbol{v}}\left[\frac{e_-^2}{m_-}f_0^- + \frac{e_+^2}{m_+}f_0^+\right]\cdot\phi(\omega, \boldsymbol{k}) = -k^2K(\omega, \boldsymbol{k})\phi(\omega, \boldsymbol{k}).$$

If now, a test charge e_1 is introduced into a plasma, the charge produced is

$$\text{(III.3.4)} \quad q^*(\omega, k) = e_1\int \exp[-i(\omega t + \boldsymbol{k}\cdot\boldsymbol{x})]\,\delta(\boldsymbol{x} - \boldsymbol{v}\cdot t)\,\mathrm{d}^3x\mathrm{d}t = e_1\,\delta(\omega + \boldsymbol{k}\cdot\boldsymbol{v}),$$

and the induced potential may be written, from

$$\text{(III.3.5)} \qquad k^2\phi = 4\pi q = 4\pi(q_{\text{ind}} + q^*) = -4\pi k^2K\phi + 4\pi e_1\,\delta(\omega + \boldsymbol{k}\cdot\boldsymbol{v}),$$

whence

$$\phi = 4\pi\frac{e_1\,\delta(\omega + \boldsymbol{k}\cdot\boldsymbol{v})}{k^2(1 + 4\pi K)} = 4\pi\frac{e_1\,\delta(\omega + \boldsymbol{k}\cdot\boldsymbol{v})}{k^2\varepsilon(\omega, \boldsymbol{k})},$$

introducing the dielectric coefficient,

$$\text{(III.3.6)} \qquad \varepsilon(\boldsymbol{k}, \omega) = 1 + 4\pi K(\boldsymbol{k}, \omega) = 1 + \frac{\omega_0^2}{k^2}\int\frac{\mathrm{d}v_\parallel}{(v_\parallel - v_p)}\frac{\partial}{\partial_\parallel}\left\{g_- + \frac{\omega_+^2}{\omega_0^2}g_+\right\},$$

where $-v_p = \omega/k$, $\omega_0^2 = 4\pi n_- e^2/m_-$, $\omega_+^2 = 4\pi n_+ e_+^2/m_+$

$$\text{(III.3.7)} \qquad g_\pm = \frac{1}{n}\int \mathrm{d}^2v_\perp f_0^\pm(\boldsymbol{v}), \qquad \boldsymbol{v}_\perp = \boldsymbol{v} - v_\parallel\hat{\boldsymbol{k}}, \qquad v_\parallel = \hat{\boldsymbol{k}}\cdot\boldsymbol{v}.$$

To handle the singular integrals in (III.3.6) requires some care; however several different arguments, *e.g.* solving an initial value problem, considering a perturbation which is adiabatically switched on, or considering a model of residual collision process as in (II.4.1), all lead to the conclusion that, with $P\int =$ Cauchy principle value,

$$\int \mathrm{d}t \frac{g(t)}{t-x} = P\!\!\int \mathrm{d}t \frac{g(t)}{t-x} + i\pi g(x) = \int_L \mathrm{d}t \frac{g(t)}{t-x}, \tag{III.3.8}$$

hence ε is complex. In normal systems an imaginary part to ε represents the loss due to collisions, here the loss process is Landau damping.

The field at a charge e_i introduced by its own presence is

$$E(\boldsymbol{v}t, t) = \mathrm{Re}\int \frac{\mathrm{d}^3k\,\mathrm{d}\omega}{(2\pi)^4} 4\pi e_1 i\boldsymbol{k} \exp[i(\boldsymbol{k}\cdot\boldsymbol{x} + \omega t)] \frac{\delta(\omega + \boldsymbol{k}\cdot\boldsymbol{v})}{k^2\varepsilon(\omega, \boldsymbol{k})} \tag{III.3.9}$$

$$= \frac{1}{4\pi^3} e_1 \int \boldsymbol{k}\, \mathrm{Im} \frac{\{k^2\varepsilon(\omega, \boldsymbol{k})\}}{|k^2\varepsilon(\omega, \boldsymbol{k})^2|} \delta(\omega + \boldsymbol{k}\cdot\boldsymbol{v})\,\mathrm{d}^3k\,\mathrm{d}\omega\,. \tag{III.3.10}$$

The fluctuating field in the plasma may be calculated by noting that each charge in the plasma itself produces a polarization, like that of a test charge; hence, if particles are distributed at points $\boldsymbol{X}_i(t)$:

$$\begin{cases} q(\boldsymbol{x}, t) = \sum e_i\, \delta(\boldsymbol{x} - \boldsymbol{X}_i(t))\,, \\ q(\omega, k) = \sum_i e_i \int \mathrm{d}t \exp[-i\boldsymbol{k}\cdot X_i(t) - i\omega t]\,, \\ \phi(\omega, k) = 4\pi \sum_i e_i \int \mathrm{d}t \dfrac{\exp[-i[\boldsymbol{k}\cdot X_i(t) + \omega t]]}{k^2\varepsilon(\omega, \boldsymbol{k})}\,. \end{cases} \tag{III.3.11}$$

The motion of each particle is approximately constant, for times of order τ_c, provided the mean field is small, *i.e.* $n^{\frac{1}{3}}e^2 \ll kT$, and provided no close collision occurs; hence if

$$t' = t + s\,, \qquad \boldsymbol{x}(t') = \boldsymbol{x}(t) + \boldsymbol{v}s\,,$$

and

$$\phi(\omega, \boldsymbol{k})\phi(\omega', \boldsymbol{k}') = \frac{(4\pi)^2}{k^2\varepsilon(\omega, \boldsymbol{k})k'^2\varepsilon(\omega', \boldsymbol{k}')} \sum_{i,j} \int \mathrm{d}t \int \mathrm{d}s \exp[i(\boldsymbol{k}\cdot\boldsymbol{x}_i(t) + \omega t)]\cdot$$
$$\cdot\exp[i(\boldsymbol{k}'\cdot\boldsymbol{x}_j(t) + \boldsymbol{V}s)] \exp[i\omega'(t+s)]\,.$$

Further, because the correlation between any pair of particles is small the random phase approximation may be used to reduce the double sum to a single one:

$$\langle \sum_i \exp [i(\boldsymbol{k}' + \boldsymbol{k}) \cdot \boldsymbol{x}_i(t)] \rangle ,$$

and the mean value of this (only quantity involving $\boldsymbol{x}_i$)

$$= \frac{1}{v} \int d^3x_i \exp [i(\boldsymbol{k}' + \boldsymbol{k}) \cdot x_i] = (2\pi)^3 \, \delta(\boldsymbol{k} + \boldsymbol{k}') ,$$

∴ further

$$\left\langle \int \exp [-i(\omega + \omega')t] \, dt \right\rangle = 2\pi\delta(\omega + \omega') ,$$

$$\int ds \exp [i(\omega' + \boldsymbol{k} \cdot \boldsymbol{v})s] = 2\pi \, \delta(\omega + \boldsymbol{k} \cdot \boldsymbol{v}) ,$$

while

$$\sum_i \psi(\boldsymbol{v}_i) = \int d^3v \, f(\boldsymbol{v}) \, \psi(\boldsymbol{v}) .$$

Therefore

$$\langle \phi(\boldsymbol{k}, \omega) \phi(\boldsymbol{k}', \omega') \rangle = (2\pi)^5 \, (4\pi) \, \frac{\delta(\boldsymbol{k} + \boldsymbol{k}') \, \delta(\omega + \omega')}{|k^2 \varepsilon(\omega, \boldsymbol{k})|^2} \, e^2 \int d^3v \, \delta(\omega + \boldsymbol{k} \cdot \boldsymbol{v}) \, f(v) .$$

The power spectrum is then

(III.3.12)
$$\frac{1}{(2\pi)^4} \int d^3k' \, d\omega' \langle \phi(k, \omega) \phi(\boldsymbol{k}', \omega') \rangle =$$
$$= \frac{32\pi^3}{|k^2 \varepsilon(\boldsymbol{k}, \omega)|^2} \cdot \int d^3v \, \delta(\omega + \boldsymbol{k} \cdot \boldsymbol{v}) \sum_\pm e_\pm^2 \, f_\pm \, (\boldsymbol{v}) ,$$

and the diffusion coefficient, using (III.2.7) and (III.2.9), becomes

(III.3.13)
$$D_{ij} = 32\pi^3 \frac{e^2}{m^2} \int d^3v' \int \frac{d^3k \, d\omega \, k_i \, k_j}{(2\pi)^4} \, \frac{\delta(\omega + \boldsymbol{k} \cdot \boldsymbol{v})}{|k^2 \varepsilon(\boldsymbol{k}, \omega)|^2} \cdot \delta(\omega + \boldsymbol{k} \cdot \boldsymbol{v}') \sum_\pm e_\pm^2 \, f_\pm(\boldsymbol{v}') .$$

The friction involves a term of the form

$$-\frac{e^2}{m^2} \left\langle \frac{\partial}{\partial x_j} \, E_i \int dt' \int^{t'} dt'' \, E_j(t'') \right\rangle ,$$

in using the property that the quantity under the first integral sign, $\varepsilon =$ function of $t-s$, we may write this as

$$-\frac{e^2}{m^2}\left\langle\int \mathrm{d}s\, E_j(x,t)s\,\frac{\partial}{\partial x_j}\,E_i(\boldsymbol{x}-\boldsymbol{v}\,s,\,t-s)\right\rangle =$$

$$=\frac{e^2}{m^2}\frac{\partial}{\partial v_j}\cdot\left\langle\int \mathrm{d}s\, E_j(\boldsymbol{x},t)\,E_i(\boldsymbol{x}-\boldsymbol{v}),\,t-s)\right\rangle = \frac{\partial}{\partial v_j}\,D_{ij}\,.$$

4. – The dominant approximation.

The integrals required to evaluate the D_{ij} take the form

from (III.3.10) $$\int \mathrm{d}^3g\int \mathrm{d}^3k\,\mathrm{d}\omega\,\boldsymbol{k}\,\frac{\delta(\omega+\boldsymbol{k}\cdot\boldsymbol{v})}{|k^2\varepsilon(\boldsymbol{k},\omega)|^2}\,\delta(\boldsymbol{k}\cdot\boldsymbol{g})\boldsymbol{k}\cdot\frac{\partial}{\partial\boldsymbol{v}}\,f(\boldsymbol{v}+\boldsymbol{g})\,,$$

from (III.3.13) $$\int \mathrm{d}^3g\int \mathrm{d}^3k\,\mathrm{d}\omega\,\boldsymbol{k}\boldsymbol{k}\,\frac{\delta(\omega+\boldsymbol{k}\cdot\boldsymbol{v})}{|k^2\varepsilon(\boldsymbol{k},\omega)|^2}\,\delta(\boldsymbol{k}\cdot\boldsymbol{g})\,f(\boldsymbol{v}+\boldsymbol{g})\,,$$

where $\boldsymbol{g}=\boldsymbol{v}'-\boldsymbol{v}$.
The integral over $\mathrm{d}\omega$ is trivial, and on splitting $\mathrm{d}^3k=k^2\mathrm{d}k\mathrm{d}\Omega$ these become

(III.4.1) $$\int \mathrm{d}^3g\int \mathrm{d}\Omega\,\delta(\hat{\boldsymbol{k}}\cdot g)\hat{\boldsymbol{k}}\hat{\boldsymbol{k}}\int \mathrm{d}k\,\frac{k^3}{|k^2\varepsilon(\boldsymbol{k},-\boldsymbol{k}\cdot\boldsymbol{v})|^2}\,.$$

Now

$$k^2\varepsilon(\boldsymbol{k},\omega)=k^2+\omega_0^2\int_L\frac{\mathrm{d}v_\parallel}{(v_\parallel-v_p)}\cdot\frac{\partial}{\partial v_\parallel}\left\{g_-+\frac{\omega_+^2}{\omega_0^2}\,g_+\right\}=k^2+k_D^2\left[X\left(\frac{\omega}{kv_\theta}\right)+i\,Y\left(\frac{\omega}{kv_\theta}\right)\right]$$

$$\int\frac{\mathrm{d}k k^3}{(k^2+k_0^2X)^2+k_0^4Y^2}=\log\left(\frac{k_{\max}}{k_0}\right)+\frac{1}{4}\log\left[\frac{[1+(k_0^2/k_{\max}^2)X]^2+Y^2}{X^2+Y^2}\right]-\frac{X}{4}\left\{\mathrm{tg}^{-1}\left|\frac{X}{Y}\right|-\frac{\pi}{2}\right\}.$$

In this integral $k_{\max}$ is an upper limit which is forced upon us by the existence of close encounters, for which the field strength becomes large, and the linearization in $\boldsymbol{E}$ underlying this treatment breaks down. Since

$$\left(\frac{n^{\frac{1}{3}}}{k_D}\right)^2=n^{\frac{2}{3}}\,\frac{mv_\theta^2}{4\pi ne^2}=\frac{1}{2\pi}\,\frac{\frac{1}{2}mv_\theta^2}{n^{\frac{1}{3}}e^2}=\frac{1}{2\pi}\,\frac{T}{V}\,,$$

hence if $k_{\max}\geqslant n^{\frac{1}{3}}$ the large value of the ratio T/V insures that $(k_{\max}/k)$ is large.

In the *dominant approximation* only this term is retained; and the integrals become

$$\text{(III.4.4)} \qquad \frac{1}{4}\int d^3g \int d\Omega\, \hat{\boldsymbol{k}}\,\delta(\hat{\boldsymbol{k}}\cdot\boldsymbol{g})\hat{\boldsymbol{k}}\cdot\frac{\partial f}{\partial \boldsymbol{v}}(\boldsymbol{v}+\boldsymbol{g})\,\frac{1}{2}\log\Lambda\,,$$

$$\text{(III.4.5)} \qquad \frac{1}{4}\int d^3g \int d\Omega\, \hat{\boldsymbol{k}}\hat{\boldsymbol{k}}\,\delta(\hat{\boldsymbol{k}}\cdot\boldsymbol{g})\, f(\boldsymbol{v}+\boldsymbol{g})\,\frac{1}{2}\log\Lambda\,.$$

Note that (III.4.4) $= \partial/\partial\boldsymbol{v}\cdot$(III.4.5), $\therefore (e_1/m_1)\boldsymbol{E} = -\,\partial/\partial\boldsymbol{v}\cdot\boldsymbol{D}$, while in (III.4.4) the angular integration merely selects those parts of $\boldsymbol{kk}$ orthogonal to $\boldsymbol{g}$, *i.e.*

$$\int d\Omega\,\hat{\boldsymbol{k}}\hat{\boldsymbol{k}}\,\delta(\hat{\boldsymbol{k}}\cdot\boldsymbol{g}) = \frac{\pi}{g}(1-\hat{\boldsymbol{g}}\hat{\boldsymbol{g}}) = \boldsymbol{w}\,,$$

cf. (III.1.6), hence

$$D_{ij} = 2\pi\left(\frac{e^2}{m}\right)^2 \log\Lambda\int d^3v'\, w_{ij} f(\boldsymbol{v}')\,.$$

Now, using the relations displayed between the integrals, the F.P.E. may be written

$$\text{(III.4.6)} \qquad -\frac{\partial D_{ij}}{\partial v_i}\frac{\partial f}{\partial v_j} - \frac{\partial}{\partial v_i}\left(\frac{\partial D_{ij}}{\partial v_j}\right) f + \frac{\partial^2}{\partial v_i \partial v_j}(D_{ij} f) =$$

$$= -2\frac{\partial}{\partial v_i} D_{ij}\frac{\partial f}{\partial v_j} - \frac{\partial^2 D_{ij}}{\partial v_i \partial v_j} f + \frac{\partial^2 D_{ij}}{\partial v_i \partial v_j} f + \frac{\partial D_{ij}}{\partial v_i}\frac{\partial f}{\partial v_j} + D_{ij}\frac{\partial^2 f}{\partial v_i \partial v_j} =$$

$$= D_{ij}\frac{\partial f}{\partial v_i \partial v_j} - \frac{\partial D_{ij}}{\partial v_j}\frac{\partial f}{\partial v_i} =$$

$$= 2\pi\left(\frac{e^2}{m}\right)^2 \log\Lambda\,\frac{\partial}{\partial v_j}\int d^3v\, w_{ij}\left[\frac{\partial f(\boldsymbol{v})'}{\partial v_i'} f(\boldsymbol{v}) - f(\boldsymbol{v}')\frac{\partial f(v)}{\partial v_i}\right].$$

This however is Landau's form (III.1.8), which thus represents the *dominant* approximation to a kinetic equation which includes both the *static* correlation effects which produce screening with the *dynamic* effects that represent the production of plasma oscillations. It is of particular interest to observe that our approach to this has required that f_0 be substantially uniform over times $\gg w_0^{-1}$; thus, if we wish to discuss the attenuation of radio-frequency oscillations propagating through the plasma, the usual Boltzmann treatment is inadequate; instead correlation effects must be determined for the distribution function perturbed by the incident $r-f$ field.

IV. – The Kinetic Equation in a Magnetic Field.

1. – Fokker-Planck equation in a magnetic field.

As in the absence of a magnetic field, it is possible to use the Liouville equation to describe the dynamics of a complete system, and again one may integrate this to obtain the Boltzmann equation

$$\frac{\partial f}{\partial t}+\boldsymbol{v}\cdot\frac{\partial f}{\partial \boldsymbol{x}}+\frac{e}{m}\left[\boldsymbol{E}+\frac{\boldsymbol{v}}{c}\times\boldsymbol{B}\right]\cdot\frac{\partial f}{\partial \boldsymbol{v}}+\frac{ne}{m}\frac{\partial}{\partial \boldsymbol{v}}\cdot\int \boldsymbol{F}_{12}f(\boldsymbol{v}_1,\boldsymbol{x}_1;\boldsymbol{v}_2,\boldsymbol{x}_2)\,\mathrm{d}^3x_2\,\mathrm{d}^3v_2\,, \tag{IV.1.1}$$

where now the interaction forces $\boldsymbol{F}_{12}$ include, in general, terms of the form

$$\frac{\partial}{\partial \boldsymbol{r}_1}\frac{e_2}{|\boldsymbol{r}_1-\boldsymbol{r}_2|} \qquad \text{and} \qquad \frac{\boldsymbol{v}_1}{c}\times\frac{\partial}{\partial \boldsymbol{r}_1}\times\frac{e_2(\boldsymbol{v}_2/c)}{|\boldsymbol{r}_1-\boldsymbol{r}_2|}\,,$$

the first representing electric and the second magnetic interactions. Note that the ratio of the 2nd to the 1st is v^2/c^2, hence the 2nd is negligible except for relativistic systems, for which the Coulomb representation of the field is inadequate; thus retardation effects become significant wherever the (local) magnetic interaction matters. For temperatures $T > 500$ keV the entire calculation should be made relativistic — a complication I propose to avoid.

If the magnetic interaction is omitted (IV.1.1) may be coarse-grained as before, and the last term written

$$\frac{\partial}{\partial \boldsymbol{v}}\frac{1}{\tau}\int_{t-\tau}^{t}\mathrm{d}t'\,\frac{e}{m}\,\boldsymbol{E}(\boldsymbol{x}',t')\,\delta[\boldsymbol{x}'-\boldsymbol{X}(t')]\,\delta[\boldsymbol{v}'-\boldsymbol{X}(t')]\,\mathrm{d}^3x'\,\mathrm{d}^3v'\,f(\boldsymbol{v})\,, \tag{IV.1.2}$$

and again the δ functions may be expanded to first order in $\boldsymbol{E}$, (IV.1.2) becoming

$$\frac{e}{m}\frac{\partial}{\partial \boldsymbol{v}}\cdot\frac{1}{\tau}\int\mathrm{d}t'\int\mathrm{d}^3x'\,\mathrm{d}^3v'\cdot\delta[\boldsymbol{x}'-\boldsymbol{X}_0(t')]\,\delta[\boldsymbol{v}'-\boldsymbol{V}_0(t')]\cdot$$
$$\cdot\left\{\boldsymbol{E}(x',t')+\frac{\partial}{\partial \boldsymbol{x}}\cdot\Delta\boldsymbol{x}\,\boldsymbol{E}(x',t')+\frac{\partial}{\partial \boldsymbol{v}}\cdot\Delta\boldsymbol{v}\,\boldsymbol{E}(x',t')\right\}. \tag{IV.1.3}$$

The complications introduced by the magnetic field are two-fold. In the first place $\boldsymbol{X}_0(t)$, $\boldsymbol{V}_0(t)$ are no longer linear, but represent helical motions; *i.e.*, if

$V(0) = (V_\parallel, V_\perp \cos\varphi, V_\perp \sin\varphi)$

$$V(t) = [V_\parallel, V_\perp \cos(\Omega t + \varphi), V_\perp \sin(\Omega t + \varphi)]\,, \tag{IV.1.4}$$

and if $X(0) = (z_0, x_0, y_0)$

$$X_0(t) = \tag{IV.1.5}$$
$$= \left[z_0 + V_\parallel t,\ x_0 + \frac{V_\perp}{\Omega}\left(\sin(\Omega t + \varphi) - \sin\varphi\right),\ y_0 - \frac{V_\perp}{\Omega}\left(\cos(\Omega t + \varphi) - \cos\varphi\right)\right].$$

Furthermore, the quantities $\Delta\boldsymbol{v}, \Delta\boldsymbol{x}$ must be determined by

$$\dot{\boldsymbol{v}} + \Omega\hat{\boldsymbol{b}} \times \boldsymbol{v} = \frac{e}{m}\boldsymbol{E}(\boldsymbol{x}, t)\,,$$

i.e.

$$\left\{\begin{aligned} \Delta v_\parallel &= \frac{e}{m}\int_0^t E_\parallel\,\mathrm{d}t'\,,\\ \Delta\boldsymbol{v}_\perp &= \operatorname{Re}\exp[i\Omega t]\int_0^t \frac{e}{m}(\boldsymbol{E}_\perp + i\boldsymbol{b}\times\boldsymbol{E})\exp[-i\Omega t']\,\mathrm{d}t' =\\ &= \frac{e}{m}\int_0^t \boldsymbol{E}_\perp\cos\Omega(t-t') + \boldsymbol{b}\times\boldsymbol{E}\sin\Omega(t-t')\,\mathrm{d}t'\,, \end{aligned}\right. \tag{IV.1.6}$$

and

$$\left\{\begin{aligned} \Delta x_\parallel &= \frac{e}{m}\int_0^t \mathrm{d}t'\int_0^{t'}\mathrm{d}t''\,E_\parallel\,,\\ \Delta\boldsymbol{x}_\perp &= \frac{c}{B}\int_0^t \mathrm{d}t'\,\boldsymbol{E}_\perp(t')\sin\Omega(t-t') + \boldsymbol{b}\times\boldsymbol{E}(t')[\cos\Omega(t-t') - 1]\,. \end{aligned}\right. \tag{IV.1.7}$$

Now, if the correlation time $\tau \ll \Omega^{-1}$, the term introduced by the field $\sim \exp[i\Omega t] \approx 1$, and these expressions reduce to those in the absence of a magnetic field. Since the correlation time $\sim \tau \sim \omega_0^{-1}$ this will be true whenever $\omega_p \gg \Omega$ *i.e.* $\lambda_D \ll r_L$, (as might be expected!) — *i.e.* whenever

$$\frac{4\pi n e^2}{m}\bigg/\frac{e^2 B^2}{m^2 c^2} = \frac{4\pi n m c^2}{B^2} \gg 1\,.$$

For a field of 5 kG this holds for electron densities $> 5\cdot 10^{10}$ and ion density

$n > 2 \cdot 10^8$, thus for most problems in which kinetic theory results are useful, the Landau form of the F.P.E. forms an adequate description. For very diffuse plasma, however, modifications are required.

(IV.1.3) may be written

$$\text{(IV.1.8)}\qquad \frac{\partial}{\partial \boldsymbol{v}}\cdot\langle e\,\boldsymbol{E} f\rangle + \left(\frac{e}{m}\right)^2 \frac{\partial}{\partial \boldsymbol{v}}\cdot\left\langle \frac{\partial}{\partial x_\parallel}\,\boldsymbol{E}\int \mathrm{d}t' \int^{t'} \mathrm{d}t''\, E_\parallel(t'') f \right\rangle +$$

$$+ \frac{e}{m}\frac{\partial}{\partial \boldsymbol{v}}\cdot\left\langle \frac{\partial}{\partial \boldsymbol{x}_\perp}\cdot \boldsymbol{E}\,\frac{c}{B}\int \mathrm{d}t'\, \boldsymbol{E}(t') \sin\Omega(t-t') + \boldsymbol{b}\times\boldsymbol{E}(t')[\cos\Omega(t-t') - 1] f \right\rangle +$$

$$+ \left(\frac{e}{m}\right)^2 \frac{\partial}{\partial \boldsymbol{v}}\,\frac{\partial}{\partial v_\parallel}\left\langle \boldsymbol{E}\int E_\parallel\,\mathrm{d}t\right\rangle f + \left(\frac{e}{m}\right)^2 \frac{\partial}{\partial \boldsymbol{v}}\,\frac{\partial}{\partial \boldsymbol{v}_\perp}\cdot$$

$$\cdot\left\langle \boldsymbol{E}\int \boldsymbol{E}_\perp \cos\Omega(t-t') + \boldsymbol{b}\times\boldsymbol{E}\sin\Omega(t'-t) f\right\rangle .$$

Now, the term

$$\left\langle \frac{\partial}{\partial x_\parallel}\,\boldsymbol{E}\int \mathrm{d}t' \int^{t'} \mathrm{d}t''\, E_\parallel(t'')\right\rangle f$$

may be handled by using a Fourier representation of **E**, *i.e.*

$$\frac{\partial}{\partial x_\parallel}\exp[i\boldsymbol{k}\cdot\boldsymbol{x}]\,\boldsymbol{E}(\boldsymbol{k},\omega)\int \mathrm{d}t' \int \mathrm{d}t''\, E_\parallel \exp[i\boldsymbol{k}'\cdot\boldsymbol{x}]\exp[i\boldsymbol{k}_\parallel v_\parallel(t-t')]\exp[i\boldsymbol{k}_\perp\cdot\Delta\boldsymbol{x}_\perp] .$$

On integrating by parts

$$\frac{\partial}{\partial \boldsymbol{v}}\cdot\frac{\partial}{\partial x_\parallel}\,\boldsymbol{E}\int \mathrm{d}t' \int \mathrm{d}t''\, E_\parallel = \frac{\partial}{\partial \boldsymbol{v}}\cdot\frac{\partial}{\partial x_\parallel}\,\boldsymbol{E}\int \mathrm{d}t'(t-t') E_\parallel = \frac{\partial}{\partial \boldsymbol{v}}\,\frac{\partial}{\partial v_\parallel}\,\boldsymbol{E}\int E_\parallel\,\mathrm{d}t ,$$

and (IV.1.8) becomes

$$\text{(IV.1.9)}\qquad \frac{e}{m}\frac{\partial}{\partial \boldsymbol{v}}\cdot\langle\boldsymbol{E}\rangle f + \frac{e}{m}\frac{c}{B}\frac{\partial}{\partial \boldsymbol{v}}\cdot$$

$$\cdot\left\langle \frac{\partial}{\partial \boldsymbol{x}_\perp}\cdot\boldsymbol{E}\int \mathrm{d}t'\,\boldsymbol{E}_\perp \sin\Omega(t-t') + \boldsymbol{b}\times\boldsymbol{E}(\cos\Omega(t-t') - 1)\right\rangle f +$$

$$+ 2\left(\frac{e}{m}\right)^2 \frac{\partial}{\partial \boldsymbol{v}}\,\frac{\partial}{\partial v_\parallel}\cdot\left\langle \boldsymbol{E}\int \mathrm{d}t'\, E_\parallel\right\rangle f + \left(\frac{e}{m}\right)^2 \frac{\partial}{\partial \boldsymbol{v}}\cdot\frac{\partial}{\partial \boldsymbol{v}_\perp}\cdot$$

$$\cdot\left\langle \boldsymbol{E}\int \mathrm{d}t'\,\boldsymbol{E}_\perp \cos\Omega(t-t') + \boldsymbol{b}\times\boldsymbol{E}\sin\Omega(t-t')\right\rangle f .$$

As before E may be expressed in terms of a potential φ, and φ may be Fourier-analysed, whereupon the required correlation functions become

$$\left\langle \boldsymbol{E}\cdot\int \boldsymbol{E}(t-s,\, x(t-s))\,\mathrm{d}s \right\rangle = \int \frac{\mathrm{d}^3\boldsymbol{k}\,\mathrm{d}\omega}{(2\pi)^4}\,\boldsymbol{kk}\,\langle \phi(\boldsymbol{k},\,\omega)\phi^*(\boldsymbol{k},\,\omega)\rangle\cdot$$

$$\cdot\int \mathrm{d}s\,\exp[-i\boldsymbol{k}\cdot\Delta\boldsymbol{x}(s)]\exp[-i\omega s] = \int \frac{\mathrm{d}^3k\,\mathrm{d}\omega}{(2\pi)^4}\,\boldsymbol{kk}\,P(\boldsymbol{k},\,\omega)\,R(\boldsymbol{k},\,\omega)\,,$$

and

$$\left\langle \boldsymbol{E}\exp[i\Omega t]\int \boldsymbol{E}(t-s)\exp[-i\Omega(t-s)] \right\rangle = \int \frac{\mathrm{d}^3k\,\mathrm{d}\omega}{(2\pi)^4}\,\boldsymbol{kk}\,P(\boldsymbol{k},\,\omega)\,R(\boldsymbol{k},\,\omega+\Omega)\,.$$

The term

$$\frac{\partial}{\partial \boldsymbol{v}}\frac{\partial}{\partial \boldsymbol{x}_\perp}:\left\langle \boldsymbol{E}\int[\boldsymbol{E}_\perp\times\boldsymbol{b}(1-\cos\Omega s)-\boldsymbol{E}_\perp\sin\Omega s]\right\rangle \mathrm{d}s\,,$$

may be written

$$-\frac{\partial}{\partial \boldsymbol{v}}\left\langle \frac{\partial}{\partial \boldsymbol{x}_\perp}:\int \frac{\mathrm{d}^3k\,\mathrm{d}\omega}{(2\pi)^4}\,\frac{\mathrm{d}^3k'\,\mathrm{d}\omega'}{(2\pi)^4}\,\phi(\boldsymbol{k})\phi(\boldsymbol{k}')\exp[i(\boldsymbol{k}+\boldsymbol{k}')\cdot\boldsymbol{x}]\exp[i(\omega+\omega')t]\cdot\right.$$

$$\left.\cdot\int \mathrm{d}s\,\boldsymbol{kk}'\times\boldsymbol{b}\exp[i(\boldsymbol{k}\cdot\Delta\boldsymbol{x}+\omega s)](1-\cos\Omega s)-\boldsymbol{kk}'_\perp\exp[i(\boldsymbol{k}\cdot\Delta\boldsymbol{x}+\omega s)]\sin\Omega s\right\rangle.$$

Since

$$\langle\exp[i(\boldsymbol{k}+\boldsymbol{k}')\cdot\boldsymbol{x}]\rangle = (2\pi)^3\,\delta(\boldsymbol{k}+\boldsymbol{k}')\,,$$

this vanishes and (IV.1.9) becomes

$$\text{(IV.1.10)}\quad \frac{e}{m}\frac{\partial}{\partial \boldsymbol{v}}\cdot\boldsymbol{E}f + 2\left(\frac{e}{m}\right)^2\frac{\partial}{\partial \boldsymbol{v}}\frac{\partial}{\partial v_\parallel}\int \frac{\mathrm{d}^3k\,\mathrm{d}\omega}{(2\pi)^4}\,P(\boldsymbol{k},\,\omega)\,R(\boldsymbol{k},\,\omega)\boldsymbol{k}k_\parallel f +$$

$$+\left(\frac{e}{m}\right)^2\frac{\partial}{\partial \boldsymbol{v}}\frac{\partial}{\partial \boldsymbol{v}_\perp}\int \frac{\mathrm{d}^3k\,\mathrm{d}\omega}{(2\pi)^4}\,P(\boldsymbol{k},\,\omega)\left[\frac{1}{2}(\boldsymbol{k}-i\boldsymbol{b}\times\boldsymbol{k})\boldsymbol{k}R(\boldsymbol{k},\,\omega+\Omega)+\right.$$

$$\left.+\frac{1}{2}(\boldsymbol{k}+i\boldsymbol{b}\times\boldsymbol{k})\boldsymbol{k}R(\boldsymbol{k},\,\omega-\Omega)\right]f\,.$$

The problem now is reduced to that of evaluating R and P and, of course, carrying out the required integrations.

Now

$$\text{(IV.1.11)}\quad R(\omega, \boldsymbol{k}) = \int \mathrm{d}s \exp[i(\boldsymbol{k}\cdot\Delta\boldsymbol{x} + \omega s) =$$

$$= \int \mathrm{d}s \exp\left[i\left\{(k_\parallel v_\parallel + \omega)s + \frac{k_x v_\perp}{\Omega}[\sin(\Omega s + \varphi) - \sin\varphi)] - \right.\right.$$

$$\left.\left. - \frac{k_y v_\perp}{\Omega}[\omega s(\Omega s + \varphi) - \cos\varphi]\right\}\right].$$

Write $k_x = k_\perp \cos\psi$, $k_y = k_\perp \sin\psi$, then

$$\text{(IV.1.12)}\quad R = \int \mathrm{d}s \exp i(k_\parallel v_\parallel + \omega)s + i\frac{k_\perp v_\perp}{\Omega}[\sin(\Omega s + \varphi - \psi) - \sin(\varphi - \psi)] =$$

$$= \sum_m \exp\left[-i\frac{k_\perp v_\perp}{\Omega}\sin(\varphi - \psi)\right]\cdot\exp\left[\sin(\varphi - \psi)\cdot\int \mathrm{d}s J_n\left(\frac{k_\perp v_\perp}{\Omega}\right)\right]\cdot$$

$$\cdot \exp[i(k_\parallel v_\parallel + \omega + n\Omega)s] = \sum_{n,m} J_n\left(\frac{k_\perp v_\perp}{\Omega}\right)\cdot J_m\left(\frac{k_\perp v_\perp}{\Omega}\right)\cdot\exp[i(n-m)(\varphi - \psi)]\cdot$$

$$\cdot\left[2\pi\delta(\omega + n\Omega + k_\parallel v_\parallel) - \frac{i}{k_\parallel v_\parallel + \omega + n\Omega}\right].$$

We will evaluate the power spectrum exactly as in the nonmagnetic case, except that now our test particles must be allowed to move along helical orbits instead of straight lines; and the dielectric coefficient becomes complicated by the presence of the magnetic field.

2. – The dielectric coefficient of a magnetized plasma.

We will require the field induced in a plasma by the presence of charge, q, and as a preliminary, we may ask what charge is induced in a magnetized plasma by a potential φ. This can be discovered from the relevant Vlasov equation

$$\text{(IV.2.1)}\qquad \frac{\partial f}{\partial t} + \boldsymbol{v}\cdot\frac{\partial f}{\partial \boldsymbol{x}} + \frac{e}{m}\left(\frac{\boldsymbol{v}}{c}\times\boldsymbol{B}\right)\cdot\frac{\partial f}{\partial \boldsymbol{v}} - \frac{e}{m}\nabla\phi\cdot\frac{\partial f_0}{\partial \boldsymbol{v}} = 0\,,$$

This has as its solution

$$f' = \frac{e}{m}\int \mathrm{d}t'\,\phi i\boldsymbol{k}\cdot\frac{\partial f_0}{\partial \boldsymbol{v}}\,,$$

i.e.

$$f' = i\frac{e}{m}\phi(\boldsymbol{k},\omega)\exp[i(\boldsymbol{k}\cdot\boldsymbol{x}+\omega t)]\int_0^\infty \mathrm{d}s\exp[i\boldsymbol{k}\cdot\boldsymbol{x}(s)+i\omega s]\cdot$$

$$\cdot\left\{k_\parallel\frac{\partial f}{\partial v_\parallel}+k_x\frac{\partial f}{\partial v_x}+k_y\frac{\partial f}{\partial v_y}\right\}.$$

To carry out the integrals in (IV.2.1), we will generally need to know the phase-dependence of f: for

$$\boldsymbol{k}\cdot(\partial/\partial\boldsymbol{v})f = k_\parallel\frac{\partial f}{\partial v_\parallel}+\frac{\boldsymbol{k}\cdot\boldsymbol{v}_\perp}{v_\perp}\frac{\partial f}{\partial v_\perp}+\boldsymbol{b}\cdot\frac{\boldsymbol{k}\times\boldsymbol{v}_\perp}{v_\perp^2}\frac{\partial f}{\partial\varphi},$$

however, it is usually true that the distribution function f may be described as only weakly dependent upon the phase, and in the dielectric coefficient we will ignore this phase-dependence; whereupon, using the notation of (IV.1.1)

$$\text{(IV.2.2)}\qquad f' = i\frac{e}{m}\phi(\boldsymbol{k},\omega)\exp[i(\boldsymbol{k}\cdot\boldsymbol{x}+\omega t)]\int_0^\infty \exp[i(\boldsymbol{k}\cdot\Delta\boldsymbol{x}+\omega s)]\cdot$$

$$\cdot\left[k_\parallel\frac{\partial f}{\partial v_\parallel}+k_\perp v_\perp\cos(\Omega t+\varphi-\psi)\frac{\partial f}{\partial v_\perp}\right]=$$

$$= i\frac{e}{m}\phi(\boldsymbol{k},\omega)\exp[i\boldsymbol{k}\cdot\boldsymbol{x}+\omega t]\cdot R(\boldsymbol{k},\omega)k_\parallel\frac{\partial f}{\partial v_\parallel}+$$

$$+\frac{1}{2}[\exp[i(\varphi-\psi)]R(\omega+\Omega,\boldsymbol{k})+\exp[-i(\varphi-\psi)]R(\omega-\Omega,\boldsymbol{k})\,k_\perp\frac{\partial f}{\partial v_\perp}.$$

The induced charge may be written

$$q = i\frac{e^2}{m}\phi\exp[i(\boldsymbol{k}\cdot\boldsymbol{x}+\omega t)]\int \mathrm{d}^3v\,R(\boldsymbol{k},\omega)k_\parallel\frac{\partial f}{\partial v_\parallel}+\frac{1}{2}\Big[\exp[i(\varphi-\psi)]R(\omega+\Omega,\boldsymbol{k})+$$

$$+\exp[-i(\varphi-\psi)]R(\omega-\Omega,\boldsymbol{k})\Big]k_\perp\frac{\partial f}{\partial v_\perp}.$$

Carrying out the integral over the phases yields

$$\int \mathrm{d}\varphi\,R(\boldsymbol{k},\omega)k_\parallel\frac{\partial f}{\partial v_\parallel} = \sum_n J_n^2 2\pi\delta(\omega+n\Omega+k_\parallel v_\parallel)-\frac{i}{k_\parallel v_\parallel+\omega+n\Omega}=$$

$$=\sum J_n^2\left(\frac{k_\perp v_\perp}{\Omega}\right)2\pi\delta_+(\omega+n\Omega+k_\parallel v_\parallel),$$

$$\int \mathrm{d}\varphi R(\boldsymbol{k}, \omega + \Omega) \exp[i(\varphi - \psi)] = 2\pi \sum J_n \left(\frac{k_\perp v_\perp}{\Omega}\right) J_{n+1}\left(\frac{k_\perp v_\perp}{\Omega}\right) \cdot$$

$$\cdot \delta_+(\omega + (n+1)\Omega + k_\parallel v_\parallel) = 2\pi \sum_n J_n J_{n-1} \delta_+(\omega + n\Omega + k_\parallel v_\parallel)\,,$$

$$\int \mathrm{d}\varphi R(\boldsymbol{k}, \omega - \Omega) \exp[-i(\varphi - \psi)] = 2\pi \sum_n J_n J_{n+1} \delta_+(\omega + n\Omega + k_\parallel v_\parallel)\,,$$

hence on using

$$J_{n+1}(x) + J_{n-1}(x) = \frac{2nJ_n}{x}\,,$$

$$q = \frac{i2\pi e^2}{m} \phi \exp[i(\boldsymbol{k}\cdot\boldsymbol{x} + \omega t)]\cdot \int \mathrm{d}v_\parallel\, \mathrm{d}v_\perp v_\perp \cdot \sum_n J_n^2 \left(\frac{k_\perp v_\perp}{\Omega}\right) \cdot \delta_+(\omega + n\Omega + k_\parallel v_\parallel)\cdot$$

$$\cdot \left[k_\parallel \frac{\partial f}{\partial v_\parallel} + \frac{n\Omega}{v_\perp} \frac{\partial f}{\partial v_\perp}\right].$$

The integral here is singular and must be treated by the Landau procedure, so that the dielectric coefficient becomes complex; although now, each term has distinct zeros, and a large number of resonances appear.

The potential due to a test charge $q^*(\boldsymbol{k}, \omega)$ is then given by

$$\Phi = \frac{4\pi q^*(\boldsymbol{k}, \omega)}{k^2 - 2\pi i \omega_0^2 \int \mathrm{d}v_\parallel\, \mathrm{d}v_\perp v_\perp \sum J_n^2 \left(\frac{k_\perp v_\perp}{\Omega}\right) \delta_+(\omega + n\Omega + k_\parallel v_\parallel) \left[k_\parallel \frac{\partial f}{\partial v_\parallel} + \frac{n\Omega}{v_\perp} \frac{\partial f}{\partial v_\perp}\right]}.$$

3. – Field of a test particle.

We may now calculate the field due to a particle of charge e moving through the plasma with a velocity

$$\boldsymbol{v} = (v_\parallel, v_\perp \cos(\Omega t + \varphi), v_\perp \sin(\Omega t + \varphi)]$$

and position

$$\boldsymbol{x} = \left\{z_0 + v_\parallel t, x_0 + \frac{v_\perp}{\Omega}[\sin(\Omega t + \varphi) - \sin\varphi], y_0 - \frac{v_\perp}{\Omega}[\cos(\Omega t + \varphi) - \cos\varphi]\right\},$$

so that

$$\text{(IV.3.1)}\qquad q^* = e\,\delta[x - X_0(t)] =$$

$$= e\exp[-i\boldsymbol{k}\cdot\boldsymbol{x}_0]\int_{-\infty}^{\infty}\exp\left[-i\left\{k_\parallel v_\parallel t + \omega t + \frac{k_\perp v_\perp}{\Omega}[\sin(\Omega t + \varphi - \psi) - \sin(\varphi - \psi)]\right\}\right] =$$

$$= e\exp[-i\boldsymbol{k}\cdot\boldsymbol{x}_0]\sum J_n\frac{k_\perp v_\perp}{\Omega}\exp[-in(\varphi-\psi)]\exp\left[i\frac{k_\perp v_\perp}{\Omega}\sin(\varphi-\psi)\right]\cdot$$

$$\cdot 2\pi\,\delta(\omega + n\Omega + k_\parallel v_\parallel),$$

$\therefore$ the potential induced by a test particle is

$$\Phi = \frac{8\pi^2 e\exp[-i\boldsymbol{k}\cdot\boldsymbol{x}_0]\sum\limits_{n,m} J_n J_m\left(\dfrac{k_\perp v_\perp}{\Omega}\right)\exp[i(m-n)(\varphi-\psi)]\,\delta(\omega + n\Omega + k_\parallel v_\parallel)}{k^2 - 2\pi i\omega_0^2\displaystyle\int \mathrm{d}v_\parallel\,\mathrm{d}v_\perp v_\perp J_n^2\left(\frac{k_\perp v_\perp}{\Omega}\right)\delta_+(\omega + n\Omega + k_\parallel v_\parallel)\left[k_\parallel\frac{\partial f}{\partial v_\parallel} + \frac{n\Omega}{v_\perp}\frac{\partial f}{\partial v_\perp}\right]}.$$

The self-field on the particle requires a knowledge of $\boldsymbol{ik}\phi(x_0)$

$$\text{(IV.3.2)}\qquad \boldsymbol{E} = \frac{1}{(2\pi)^4}\int \mathrm{d}^3k\,\mathrm{d}\omega\,\boldsymbol{ik}\phi(\boldsymbol{x}_0) =$$

$$= \frac{\sum\displaystyle\int(-i)^n J_n\left(\frac{k_\perp v_\perp}{\Omega}\right)\exp[in(\varphi-\psi)]\exp\left[i\frac{k_\perp v_\perp}{\Omega}\sin(\varphi-\psi)\right]\boldsymbol{ik}\,\mathrm{d}^3k\,\mathrm{d}\omega}{[k^2 - \omega_0^2 K(n\boldsymbol{E} - k_\parallel v_\parallel, \boldsymbol{k})]}.$$

If into this we place $\boldsymbol{k} = k_\parallel,\ k_\perp\cos\psi,\ k_\perp\sin\psi$, and note $\boldsymbol{v} = (v_\parallel,\ v_\perp\cos\varphi,\ v_\perp\sin\varphi)$ we may integrate over ψ, obtaining

$$\boldsymbol{E}\cdot\hat{\boldsymbol{b}} = \operatorname{Re}\sum\int\frac{ik_\parallel J_n^2(k_\perp v_\perp/\Omega)}{k^2 - \omega_0^2 K(n\Omega - k_\parallel v_\parallel, \boldsymbol{k})},$$

$$\boldsymbol{E}\cdot\hat{\boldsymbol{v}}_\perp = \operatorname{Re}\sum\int\frac{ik_\perp[J_n(J_{n+1} + J_{n-1})]}{k^2 - \omega_0^2 K},$$

$$\boldsymbol{E}\cdot\boldsymbol{b}\times\hat{\boldsymbol{v}}_\perp = -\operatorname{Re}\sum\int\frac{1}{2}k_\perp\frac{J_n(J_{n+1} - J_{n-1})}{k^2 - \omega_0^2 K}.$$

Of particular interest here is the last term, which involves the Hermitian part of $k^2\varepsilon$, and is a force normal to the particle trajectory. For an isotropic distribution function f, $(\partial/\partial v_\perp)\|v_\perp$, and this does not contribute to the F.P.E.

4. – The spectrum.

The power spectrum may be evaluated from (IV.3.2), thus

$$\text{(IV.4.1)}\quad \int \frac{\mathrm{d}^3k\,\mathrm{d}\omega}{(2\pi)^4} P = \int \frac{\mathrm{d}^3k\,\mathrm{d}\omega}{(2\pi)^4} \cdot \int \frac{\mathrm{d}^3k'\mathrm{d}\omega'}{(2\pi)^4} \cdot \Phi(\boldsymbol{k}, \omega), \Phi(\boldsymbol{k}'\omega') =$$

$$= \int \frac{\mathrm{d}^3k\,\mathrm{d}\omega}{(2\pi)^4} \cdot \int \frac{\mathrm{d}^3k'\,\mathrm{d}\omega'}{(2\pi)^4} \cdot (8\pi^2)^2\, e^2 \sum_{i,j} \exp i\boldsymbol{k}\cdot\boldsymbol{x}_1] \exp[i\boldsymbol{k}'\cdot\boldsymbol{x}_i] \cdot$$

$$\cdot \exp[i\boldsymbol{k}\cdot\boldsymbol{x}_i] \sum_{n,m,s,t} \frac{J_n J_m(k_\perp v_\perp^i/\Omega) J_v J_t(k'_\perp v_\perp^j/\Omega)}{k^2\varepsilon(\omega, k) k'^2 \varepsilon(\omega', k')} \delta(\omega + n\Omega + k_\parallel v_\parallel^i) \cdot$$

$$\cdot \delta(\omega' + t\Omega + k'_\parallel v_\parallel^j) \exp[i(m-n)(\varphi^i - \psi) + i(s-t)(\varphi^j - \psi')\,,$$

and on using the random phase approximation on $\sum$, and carrying out the (trivial) integrations over the k', ω', ω and recalling that φ and ψ are the phases of $\boldsymbol{v}_\perp$ and $\boldsymbol{x}_\perp$

$$\text{(IV.4.2)}\quad P(\omega, \boldsymbol{k}) =$$

$$= 32\pi^2 e^2 \int \mathrm{d}^3v \frac{f(\boldsymbol{v}) \sum_{n,m,s,t} J_n J_m J_s J_t \exp[i[(m-n)(\varphi - \psi) + (s-t)(\varphi + \psi)]]}{k^2\varepsilon(\boldsymbol{k}, -n\Omega + k_\parallel v_\parallel) k^2 \varepsilon(-\boldsymbol{k}, k_\parallel v_\parallel - s\Omega)}\,.$$

The diffusion coefficients now take the form

$$\text{(IV.4.3)}\quad \int \frac{\mathrm{d}^3k}{(2\pi)^3} \int \mathrm{d}^3v' \cdot T(\boldsymbol{k}, \boldsymbol{k}) \cdot \sum \frac{J_q J_v J_s J_t(k_\perp v'_\perp/\Omega) f(\boldsymbol{v}')}{k^2\varepsilon[\boldsymbol{k}, -(v\Omega + k_\parallel v_\parallel)] \cdot k^2\varepsilon[-\boldsymbol{k}, k_\parallel v_\parallel - s\Omega]} \cdot$$

$$\cdot \sum J_n J_m \left(\frac{k_\perp v_\perp}{\Omega}\right) 2\pi_i\, \delta_+[(n-r)\Omega + k_\parallel(v_\parallel - v'_\parallel)] \exp[i(n-m)(\varphi - \psi)] \cdot$$

$$\cdot \exp[i[p-q)(\varphi' - \psi) + (s-t))\varphi' + \psi)]]\,,$$

where the $T(\boldsymbol{k}, \boldsymbol{k})$ are tensors constructed from $\boldsymbol{k}$ and $\boldsymbol{b}$.

Some simplification is obtained if we consider the average of the diffusion coefficients over all phase, whereupon they reduce to

$$\text{(IV.4.4)} \qquad \frac{\partial}{\partial \boldsymbol{v}}\frac{\partial}{\partial \boldsymbol{v}}:\boldsymbol{D}f = \frac{\partial^2}{\partial v_\parallel^2}\int k_\parallel^2 P \sum J_n^2\left(\frac{k_\perp v_\perp}{\Omega}\right)\delta_+(0) +$$

$$+\frac{\partial}{\partial v_\parallel}\left(\frac{\partial}{\partial v_\perp}+\frac{1}{v_\perp}\right)\int k_\perp k_\parallel P\frac{1}{2}\sum J_n(J_{n+1}+J_{n-1})\,\delta_+(0) +$$

$$+\frac{1}{2}\left(\frac{\partial}{\partial v_\perp}+\frac{1}{v_\perp}\right)\left[\left(\frac{1}{2}\frac{\partial}{\partial v_\perp}+\frac{1}{v_\perp}\right)\int k_\perp^2 P\sum J_n^2[\delta_+(-1)+\delta_+(1)] +\right.$$

$$\left.+\frac{\partial}{\partial v_\perp}\int k_\perp^2\, P\sum J_n(J_{n+2}+J_{n-2})\,[\delta_+(-1)+\delta_+(1)]\right] +$$

$$+\frac{1}{2}\left(\frac{\partial}{\partial v_\perp}+\frac{1}{v_\perp}\right)\cdot\left(\frac{1}{2}\frac{\partial}{\partial}v_\perp+\frac{1}{v_\perp}\right)\int k_\perp^2 P\sum J_n\,[J_{n-2}+J_{n+2}][\delta_+(-1)-\delta_+(1)]\,,$$

with the following abbreviations

$$\int=\iiiint \mathrm{d}v_\parallel'\,\mathrm{d}v_\perp' v_\perp'\,\mathrm{d}k_\parallel\,\mathrm{d}k_\perp k_\perp\,,$$

$$P = 16\pi^2\,\frac{e^4}{m^2}\cdot\sum_{ps}\frac{J_p^2(k_\perp v_\perp'/\Omega)\,J_s^2(k_\perp v_\perp'/\Omega)}{k^2\varepsilon(\boldsymbol{k},\,-p\Omega-k_\parallel v_\parallel')\,k^2\varepsilon(-\boldsymbol{k},\,k_\parallel v_\parallel'-s\Omega)}\,,$$

$$k^2\varepsilon = k^2-\omega_0^2\int \mathrm{d}v_\parallel\,\mathrm{d}v_\perp v_\perp\sum_n J_n^2 2\pi i\,\delta_+(\omega+n\Omega+k_\parallel v_\parallel)\left[k_\parallel\frac{\partial\hat{f}}{\partial v_\parallel}+\frac{n\Omega}{v_\perp}\frac{\partial\hat{f}}{\partial v_\perp}\right],$$

$$\delta_+(t) = \delta_+[k_\parallel(v_\parallel-v_\parallel')+(n-p+t)\Omega]\,.$$

The functions involved in this expression have been explored to some extent by N. Rostoker, who concludes that at worst the screening parameter is altered, a poor reward for such effort. On the other hand the analysis presented here is incomplete; for in obtaining (IV.4.4) we have assumed f_0 to be phase-independent, whereas, in general one must deal with contributions depending on directions across the B. field. A possible procedure here is to Fourier-expand f in φ, but the reader will be spared this. A more serious defect may be the omission of electromagnetic as opposed to electrostatic interactions, which was rather glibly effected on the first page; for what cooperative effects may be important here is far from clear, and the relative inefficiency of the plasma as a current screen may overweigh the $(v)^2$ in the interparticle force. It is certainly true that in many instabilities, *e.g.* the mirror type, the inductive fields are more important than those produced by charge separation, and similar phenomena may play a rôle in particle interactions. This presentation must, therefore, be considered as a sketch of the process of particle interaction in a magnetic field rather than as a complete account.

V. – Correlation Functions and Scattering of Radiation from a Plasma.

1. – The correlation functions in a plasma.

Since our procedure has given a value for the potential produced by a particle in a plasma, namely (in the absence of a magnetic field)

$$\phi(\boldsymbol{k},\,\omega) = 8\pi^2 e_i\,\frac{\delta(\omega+\boldsymbol{k}\cdot\boldsymbol{v})}{k^2\varepsilon(\boldsymbol{k},\,\omega)}\exp[-i\boldsymbol{k}\cdot\boldsymbol{x}_i]\,, \tag{V.1.1}$$

we may now use the Vlasov equation to calculate the disturbance that this produces in the distribution function, *i.e.*

$$f' = 8\pi^2 e_1\,\frac{\delta(\omega+\boldsymbol{k}\cdot\boldsymbol{v})}{k^2\varepsilon(\boldsymbol{k},\,\omega)}\exp[-i\boldsymbol{k}\cdot\boldsymbol{x}_i]\,\frac{e_2}{m_2}\,\frac{\boldsymbol{k}\cdot(\partial f_0/\partial\boldsymbol{v})}{(\omega+\boldsymbol{k}\cdot\boldsymbol{v})}\,. \tag{V.1.2}$$

Now the probability of finding a particle at $(\boldsymbol{x}_1, \boldsymbol{v}_1, t_1)$ and a particle at $(\boldsymbol{x}_2, \boldsymbol{v}_2, t_2)$ is clearly

$$p(1,2) = f(v_1, x_1, t_1)f(v_2, x_2, 0) + f(x_2, v_2, 0)f(1, 2, x_1, v_1, t_1)\,.$$

However, the last term is just (V.1.2); *i.e.* f' the change in $f(1)$ produced by a particle at 2, hence

$$f(\boldsymbol{x}_1, \boldsymbol{v}_1, t_1, 2) = \frac{e_1 e_2}{m_1}\,\frac{1}{2\pi^2}\int \mathrm{d}^3k\,\mathrm{d}\omega\,\frac{\delta(\omega+\boldsymbol{k}\cdot\boldsymbol{v}_2)\exp[ik\cdot(\boldsymbol{x}_1-\boldsymbol{x}_2)]}{[k^2\varepsilon(\boldsymbol{k},\,\omega)](\omega+\boldsymbol{k}\cdot\boldsymbol{v}_1)}\cdot \cdot\exp[i\omega t]\boldsymbol{k}\cdot\frac{\partial f_0(1)}{\partial\boldsymbol{v}_1}\,. \tag{V.1.3}$$

Having the (space-dependent) perturbation induced in the distribution function by the potential of a charged particle, it is possible to calculate the spatial distribution of electrons and ions in a plasma, in this appoximation. If the zero-order positions of electrons and ions are x_i and X_i, then the plasma potential is

$$(\text{V.1.4})\qquad \phi = 8\pi^2 \sum_i e_- \frac{\delta(\omega + \boldsymbol{k}\cdot\boldsymbol{v}_i)\exp[-i\boldsymbol{k}\cdot\boldsymbol{x}_i] + e_+\delta(\omega+\boldsymbol{k}\cdot\boldsymbol{v}_i)\exp[-i\boldsymbol{k}\cdot\boldsymbol{X}_i]}{k^2\varepsilon(\boldsymbol{k},\omega)},$$

and

$$(\text{V.1.5})\qquad n_-(\boldsymbol{k},\omega) = \sum_i \exp[-i\boldsymbol{k}\cdot\boldsymbol{x}_i]\,\delta(\omega+\boldsymbol{k}\cdot\boldsymbol{v}_i) + \\ + \frac{e_-^2}{m_-}\int \mathrm{d}^3v\, \frac{\delta(\omega+\boldsymbol{k}\cdot\boldsymbol{v}_i)\exp[-i\boldsymbol{k}\cdot\boldsymbol{x}_i]}{k^2\varepsilon(\boldsymbol{k},\omega)(\omega+\boldsymbol{k}\cdot\boldsymbol{v})}\,\boldsymbol{k}\cdot\frac{\partial f_0}{\partial \boldsymbol{v}}.$$

From this, the charge density is

$$(\text{V.1.6})\qquad q_1 = e_-n_-(\boldsymbol{k},\omega) + e_+n_+(\boldsymbol{k},\omega) = \\ = 2\pi\sum\{e_-\exp[-i\boldsymbol{k}\cdot\boldsymbol{x}_i]\,\delta(\omega+\boldsymbol{k}\cdot\boldsymbol{v}_i) + e_+\exp[-i\boldsymbol{k}\cdot\boldsymbol{X}_i]\,\delta(\omega+\boldsymbol{k}\cdot\boldsymbol{V}_i)\}\cdot \\ \cdot\left\{\frac{1+\left[\omega_0^2\int\frac{\mathrm{d}^3v}{(\omega+\boldsymbol{k}\cdot\boldsymbol{v})}\,\boldsymbol{k}\cdot\frac{\partial f_-}{\partial\boldsymbol{v}} + \omega_+^2\int\frac{\mathrm{d}^3v}{(\omega+\boldsymbol{k}\cdot\boldsymbol{v})}\,\boldsymbol{k}\cdot\frac{\partial f_+}{\partial\boldsymbol{v}}\right]}{k^2\varepsilon(\boldsymbol{k},\omega)}\right\}.$$

But

$$k^2\,\varepsilon(\boldsymbol{k},\omega) = k^2 - \omega_0^2\left[\int\frac{\mathrm{d}^3v}{(\omega+\boldsymbol{k}\cdot\boldsymbol{v})}\,\boldsymbol{k}\cdot\frac{\partial f_-}{\partial\boldsymbol{v}} + \frac{\omega_+^2}{\Omega_0^2}\int\frac{\mathrm{d}^3v}{(\omega+\boldsymbol{k}\cdot\boldsymbol{v})}\,\boldsymbol{k}\cdot\frac{\partial f_+}{\partial\boldsymbol{v}}\right],$$

hence

$$4\pi\, q_1 = 8\pi^2\sum\left[\frac{e_-\delta(\omega+\boldsymbol{k}\cdot\boldsymbol{v}_i)\exp[-i\boldsymbol{k}\cdot\boldsymbol{x}_i] + e_+\,\delta(\omega+\boldsymbol{k}\cdot\boldsymbol{V}_i)\exp[-i\boldsymbol{k}\cdot\boldsymbol{X}_i]}{\varepsilon(\boldsymbol{k},\omega)}\right],$$

and the calculation is self-consistent.

2. – Scattering of radiation from a plasma.

It is of interest to note an observable phenomenon which depends on the details of the electron correlation function; this is the scattering of radiation by a plasma. To treat this, we consider a plasma in which the distribution function may be written $f_0(v) + f_1(x, v, t)$ and consider the effect on this of an electric field $E(x, t)$. After Fourier-transforming we obtain for the induced currents

$$(\text{V.2.1})\qquad j_{\text{ind}}(\omega,\boldsymbol{k}) = \\ = -\frac{e^2}{m}\int\mathrm{d}^3v\left[\frac{\boldsymbol{E}(\omega,k)\cdot(\partial f_0/\partial\boldsymbol{v})\boldsymbol{v} + \sum\boldsymbol{E}(\Omega,K)\cdot(\partial/\partial\boldsymbol{v})f(\omega-\Omega,\boldsymbol{K}-\boldsymbol{k})\boldsymbol{v}}{i(\omega+\boldsymbol{k}\cdot\boldsymbol{v})}\right].$$

If the phase velocity is high $\omega/k \gg v_\Theta$; then

$$j_{\text{ind}}(\omega, k) = \frac{1}{4\pi}\frac{\omega_0^2}{i\omega}\left\{E(\omega, k) + \frac{1}{\omega_0^2}\sum \Delta\omega_0^2(\omega - \Omega, k - K)\, E(\Omega, K)\right\}.$$

Maxwell's equations for this field become

$$\nabla^2 \boldsymbol{E} - \frac{1}{c^2}\frac{\partial^2 E}{\partial t^2} = \frac{4\pi}{c^2}\frac{\partial j}{\partial t} + \operatorname{grad}(\operatorname{div}\boldsymbol{E}), \tag{V.2.2}$$

or, on Fourier-transforming, and using the equation of continuity

$$\left(\frac{\omega^2}{c^2} - k^2\right)\boldsymbol{E}(\omega, \boldsymbol{k}) = -\frac{4\pi i\omega}{c^2}\boldsymbol{j} - \boldsymbol{k}(\boldsymbol{k}\cdot\boldsymbol{E}),$$

or

$$\left(\frac{\omega^2 + \omega_0^2}{c^2} - k^2\right)\boldsymbol{E}(\omega, \boldsymbol{k}) = \left[\frac{c^2}{\omega^2} - \boldsymbol{k}\boldsymbol{k}\cdot\right]\sum\frac{\Delta\omega_0^2}{\omega^2}(\omega - \Omega, \boldsymbol{K} - \boldsymbol{k})\,\boldsymbol{E}(\Omega, \boldsymbol{K}).$$

Now, if there is incident on the plasma a field of frequency Ω and wave

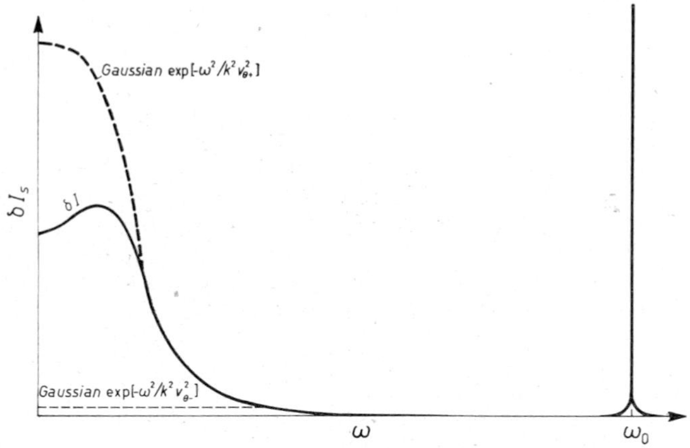

Fig. 5. – Scattering of radiation from a plasma: form of δI_s vs. ω.

number $\boldsymbol{K}$ ($K^2 = \Omega^2/c^2$), a scattered wave will be produced given by

$$E(\omega, k) = \frac{[\omega^2 - c^2\boldsymbol{k}\boldsymbol{k}\cdot]\Delta\omega_0^2(\omega - \Omega, k - K)\,\boldsymbol{E}(\Omega, K)}{[((\omega^2 + \omega)/c^2) - k^2]}. \tag{V.2.3}$$

For scattered and incident waves well above the plasma frequency, this yields an expression for the intensity at large distances in the form

$$\text{(V.2.4)}\qquad \frac{\mathrm{d}^2 I_s(\omega', \boldsymbol{k}')}{\mathrm{d}\omega\,\mathrm{d}\Omega} = I_0(\Omega, \mathbf{K})4\pi\left(\frac{e^2}{mc^2}\right)^2 |\Delta n(\omega, \boldsymbol{k})|^2[1-(\sin\theta\cos\varphi)^2]\,,$$

where θ, φ are the scattering angles, with respect to incident direction and polarization and the last factor $=\frac{1}{2}(1+\cos^2\theta)$ for an unpolarized incident beam. If the only polarization of the electron density is that produced by random fluctuations Δn is given by (V.1.5), *i.e.*

$$\text{(V.2.5)}\qquad \Delta n(\omega, k) = $$

$$= \frac{\sum_i (1+G_-)\exp[-i\boldsymbol{k}\cdot\boldsymbol{x}_i]\,\delta(\omega+\boldsymbol{k}\cdot\boldsymbol{v}_i) + G_+(e_+/e_-)\delta(\omega+\boldsymbol{k}\cdot\boldsymbol{V}_i)\exp[-i\boldsymbol{k}\cdot\boldsymbol{X}_i]}{1-G_- - G_+}$$

where

$$\text{(V.2.6)}\qquad G_\pm = 4\pi\left(\frac{e^2}{m}\right)_\pm \frac{1}{k^2}\int\frac{\boldsymbol{k}\cdot(\partial f_0/\partial\boldsymbol{v})}{(\omega+\boldsymbol{k}\cdot\boldsymbol{v})}\,\mathrm{d}^3v \sim \frac{k_D^2}{k^2}\,.$$

Now, to form $(\Delta n)^2$, the random phase approximation may be invoked whereupon

$$|\sum \exp[-i\boldsymbol{k}\cdot\boldsymbol{x}_i]\,\delta(\omega+\boldsymbol{k}\cdot\boldsymbol{v}_i)|^2 = \int \mathrm{d}^3v\, f(v)\,\delta(\omega+\boldsymbol{k}\cdot\boldsymbol{v})$$

and

$$\text{(V.2.7)}\qquad |\Delta n_-(\omega, k)|^2 = N\left\{\left|\frac{1-G_+}{1-G_- - G_+}\right|^2 \int \mathrm{d}^3v\, f_-(\boldsymbol{v})\,\delta(\omega+\boldsymbol{k}\cdot\boldsymbol{v}) + \right.$$
$$\left. + \left|\frac{G_-}{1-G_- - G_+}\right|^2 \int \mathrm{d}^3v\, f_+(\boldsymbol{v})\,\delta(\omega+\boldsymbol{k}\cdot\boldsymbol{v})\right\}\,.$$

The scattered radiation is then given by (V.2.4) and (V.2.7). If $(k_D/k)^2 \gg 1$ several interesting features appear. Since f_+ is much narrower than f_-, the second term dominates near $\Delta\omega = 0$; but since G_- increases with ω, the scattered wave first increases, then decreases, with a half-width determined by $f_+(\omega/k)$, however, at large frequency shifts the first term dominates. When $\Delta\omega = \omega_0$ the dielectric coefficient $(1 - G_- - G_+)$ becomes small over a narrow region and a sharp narrow peak corresponding to the emission of plasma oscillation appears.

If $T_- \gg T_+$, ion sound waves may appear, and a sharp peak appears at $(\Delta\omega/\Delta k) \sim (KT_-/m_+)$ from the center. If $k_D \ll k$ the G's are small, and only the first terms persist.

The interest in this process lies in the possibility of exploring the correlation function directly; the difficulty lies in the small value of the Thomson cross-section $8\pi(e^2/mc^2)^2 \simeq 10^{-25}\ \text{cm}^2$ which determines the scale of the phenomenon. If because of an instability $\Delta n(\omega, \boldsymbol{k})$ becomes very large for some narrow range of $\omega, \boldsymbol{k}$ a much more spectacular effect would be expected.

BIBLIOGRAPHY

S. Chapman and T. G. Cowling: *The Mathematical Theory of Non-Uniform Gases* (Cambridge, 1939).
W. Marshall: A.E.R.E. Report T/R 2247, 2352, 2419.
M. N. Rosenbluth and A. Kaufman: *Phys. Rev.*, **109**, 1 (1958).
A. Kaufman: *Proc. of Les Houches Summer School* (Paris 1959.
B. B. Robinson and I. B. Bernstein: *Ann. Phys.*, **18**, 110 (1962).
R. Landshoff: *Phys. Rev.*, **82**, 442 (1951).
S. Chandrasekhar: *Rev. Mod. Phys.*, **15**, 2 (1943).
R. S. Cohen, L. Spitzer and P. M. Routley: *Phys. Rev.*, **80**, 230 (1950).
R. Balescu: *Phys. Fluids*, **3**, 52 (1960).
N. Rostoker and M. N. Rosenbluth: *Phys. Fluids*, **3**, 2 (1960).
C. M. Tchen: *Phys. Rev.*, **114**, 394 (1959).
A. Lenard: *Ann. Phys.*, **10**, 390 (1960).
L. J. Landau: *Journ. Phys. U.S.S.R.*, **10**, 25 (1946).
W. Thompson: Risö Report 18 (1960), p. 101 (Danish A. E. A. Risö, Roskilde) Matt. 91 (1961) (Project Matterhorn, Princeton).
D. Pines and D. Bohm: *Phys. Rev.*, **85**, 338 (1952).
W. B. Thompson and J. Hubbard: *Rev. Mod. Phys.*, **32**, 714 (1960).
J. Hubbard: *Proc. Roy. Soc.*, A **260**, 114 (1961); A **261**, 371 (1961).
J. P. Dougherty and D. T. Farley: *Proc. Roy. Soc.*, A **259**, 79 (1960).
J. A. Fejer: *Can. Journ. Phys.*, **38**, 1114 (1960).
E. E. Salpeter: *Phys. Rev.*, **120**, 1528 (1960).
T. Hagfors: *Journ. Geophys. Res.*, **66**, 1699 (1961).

General Stability Theory in Plasma Physics.

RUSSEL KULSRUD

Princeton Plasma Physics Laboratory, Princeton University - Princeton, N. J.

1. – Introduction.

In this course we shall consider three energy principles for the stability of *static* magnetohydrodynamic equilibria. These three energy principles correspond to three different sets of basic equations describing the plasma, each of which applies in various limiting situations. We shall designate these three different approaches by the *fluid* theory, the *adiabatic* theory and the *double adiabatic* theory or Chew-Goldberger-Low theory.

The *fluid* theory corresponds to the strong-collision limit where collisions are so strong that the pressure always remains a scalar, but however, still so weak that the conductivity may be taken as infinite. The *adiabatic* theory corresponds to the limit of no collisions and to the limit where the gyration radius of each type of particle and the Debye length are infinitely small compared with macroscopic quantities. The *double adiabatic* theory is an intermediate theory in which collisions are not strong enough to keep the pressure isotropic but are sufficiently strong to make the heat flow negligible. This latter theory is the theory given in the paper of CHEW, GOLDBERGER and LOW [1].

In each of the three theories first the basic equations are derived, second the equilibrium is discussed, third the linearized equations for small motions about equilibrium are derived, and finally an energy principle is derived which is shown to give a necessary and sufficient condition for the stability of *all* these small motions.

In each case the basic equations are a closed system. In the *fluid* theory the equations are the usual fluid ones with p given by $p = A\varrho^{\gamma}$. In the *adiabatic* theory the pressure is found from the Boltzmann distribution function f but otherwise the equations are essentially the same as in the fluid theory. To find f it is necessary to solve the collisionless Boltzmann or Vlasov equation but the difficulty of solving this equation is reduced by the assumption of small gyration radius. The equation reduces to a one-dimensional Boltzmann equation where the one dimension comprises the position and velocity of a particle along a magnetic line of force. The theory is designated as *adiabatic* because the motion of each particle is governed by the magnetic moment being an adiabatic invariant. In the *double adiabatic* theory a separate equation of state is derived for each of the two independent components of the stress tensor, with each equation similar to the simple equation in the *fluid* theory. Thus, since these equations arise from an adiabatic assumption (adiabatic here means no heat flow) this theory is often called the *double adiabatic* theory. The two uses of the term adiabatic in *adiabatic* theory and *double adiabatic* theory are essentially different.

The theories are developed on generally parallel lines. The arguments for the derivation of an energy principle in each case are formally the same. When the energy principles are derived they will be compared for similar equilibria. It will be shown that if one proves stability for an isotropic equilibrium on the *fluid* theory, the equilibrium will be stable in the other two theories. The comparison of the *double-adiabatic* and *adiabatic* energy principle is not completed as yet but it has been shown that the energy principle of KRUSKAL and OBERMAN [2] which is slightly more pessimistic (more unstable) than the energy principle of the adiabatic theory is definitely more pessimistic than that of the *double adiabatic* theory.

The name *double adiabatic* theory is used rather then the Chew-Goldberger-Low theory since these authors also layed the foundation for the *adiabatic* theory and we shall follow their approach. This latter work is unpublished [3].

The energy principle for the *fluid* theory is the first one to appear in plasma physics. It is the energy principle given in the paper of BERNSTEIN, FRIEMAN, KRUSKAL and KULSRUD [4]. This paper also includes the energy principle of the *double adiabatic* theory. The energy principle of the *adiabatic* theory is a slight generalization of the energy principle developed independently and simultaneously by KRUSKAL and OBERMAN [2] and ROSENBLUTH and ROSTOKER [5].

2. – Fluid theory.

2·1. *Basic equations.* – The basic equations underlying all three theories will be the Fokker-Planck equation [6] for the Boltzmann distribution function of each particle and Maxwell's equations for the electromagnetic field. In each of the three cases we shall make different assumptions of simplicity on these equations and derive three different sets of simple equations which we can handle. It should be emphasized at this point that although the exact structure of the Fokker-Planck equation is still under discussion this will not affect our simpler equations since the uncertainties in the Fokker Planck equation are covered up by our simplifying assumptions.

The fluid equations are

$$\frac{d\varrho}{dt} = -\varrho \nabla \cdot \boldsymbol{V}, \tag{1}$$

$$\varrho \frac{d\boldsymbol{V}}{dt} = -\nabla \cdot p + \boldsymbol{j} \times \boldsymbol{B} + q\boldsymbol{E}, \tag{2}$$

$$\frac{d}{dt}(p/\varrho^{\gamma}) = 0. \tag{3}$$

The Maxwell equations are

$$\nabla \cdot \boldsymbol{B} = 0, \tag{4}$$

$$\nabla \times \boldsymbol{B} = 4\pi \boldsymbol{j} + \frac{1}{c}\frac{\partial \boldsymbol{E}}{\partial t}. \tag{5}$$

$$\frac{\partial \boldsymbol{B}}{\partial t} = -c\nabla \times \boldsymbol{E}, \tag{6}$$

$$\nabla \cdot \boldsymbol{E} = 4\pi q. \tag{7}$$

These equations are coupled by an Ohm's law

$$\boldsymbol{E} + \frac{1}{c}\boldsymbol{V} \times \boldsymbol{B} = \frac{1}{\sigma}\boldsymbol{j}. \tag{8}$$

Here ϱ, $\boldsymbol{V}$, and p are the density, velocity and pressure of the fluid (assumed scalar), $\boldsymbol{j}$, $\boldsymbol{B}$, $\boldsymbol{E}$ and q are the current, magnetic field, electric field and charge density. The units used are Gaussian except $\boldsymbol{j}$ is given in e.m.u.; d/dt indicates $\partial/\partial t + \boldsymbol{V} \cdot \nabla$ as usual. γ, c and σ are the ratio of specific heats, the velocity of light and the conductivity in abamp/abvolt.

Although these equations apply to a number of situations (for instance with $\gamma = \infty$ incompressibility) we have in mind a gas consisting of ions and electrons. ϱ is the total density, $\boldsymbol{V}$ the mass velocity of the combined fluid, and p the total pressure. The assumption is made that collisions are very strong so that the gas is everywhere nearly Maxwellian. The first two fluid equations are derived by taking the first two moments of the Fokker-Planck equations. The next moment involves the heat flow which may be computed by the method of Chapman and Enskog. Similarly a more correct Ohm's law may be derived which gives $\boldsymbol{j}$ with an anisotropic conductivity and heat flow terms. It should also be mentioned that the pressure is a tensor in eq. (2) and the off-diagonal terms give rise to a viscosity term in (2). The correct full equations have been derived by MARSHALL [7] for the ion-electron case and are given in the review article of BERNSTEIN and TREHAN [8].

The condition for the validity of these equations is $\tau \ll T$ where τ is the mean collision time and T is a characteristic macroscopic time. However, if τ is very small all the transport terms (viscosity, heat flow) may be neglected and we get essentially eqs. (1)–(8) (σ may still be anisotropic).

We now wish to consider the limit of infinite conductivity $\sigma \to \infty$, so that (8) is replaced by

$$\boldsymbol{E} + \frac{1}{c} \boldsymbol{V} \times \boldsymbol{B} = 0 . \tag{8}$$

However, in a certain sense $\sigma \to \infty$ corresponds to very large τ which seems to contradict our previous assumption. However, in this case, σ large means $\tau \gg T(L^2\omega_p^2/\varrho^2)^{-1}$ so (1)–(7), and (8a) should be valid if

$$\frac{c^2}{\omega_p^2 L^2} T \ll \tau \ll T ,$$

$\omega_p^2 = 4\pi n e^2/m$ where n, e and m are electron density, charge and mass, and L is a macroscopic length. The first half of this inequality is usually well satisfied so our *fluid* theory is generally valid if the second half is satisfied. This latter is more stringent.

(Of course, there are other conditions that must be satisfied in the fluid theory such as small gyration radius, small Debye length, etc. We do not stress these in the fluid theory.)

Substituting (8′) in (5) and (7) and these in (2) we have

$$\varrho \frac{\mathrm{d}\boldsymbol{V}}{\mathrm{d}t} = -\nabla p + \frac{1}{4\pi} (\boldsymbol{\nabla} \times \boldsymbol{B}) \times \boldsymbol{B} + \frac{1}{4\pi c^2} \frac{\partial}{\partial t} (\boldsymbol{V} \times \boldsymbol{B}) \times \boldsymbol{B} - \frac{1}{4\pi} \frac{\nabla \cdot (\boldsymbol{V} \times \boldsymbol{B}) \boldsymbol{V} \times \boldsymbol{B}}{c^2} .$$

The last two terms are negligible if $\varrho c^2 \gg B^2/4\pi$ so we may neglect them. This corresponds to the neglect of displacement current and the electric force term in the momentum eq. (2). Also we wish at this point to eliminate all the factors 4π which may arise so we set

$$\boldsymbol{B} = \sqrt{4\pi}\, \boldsymbol{B}' ,$$

$$\boldsymbol{J} = \frac{\boldsymbol{J}'}{\sqrt{4\pi}} ,$$

which corresponds to a choice of units about 3 G for B', 3 A for J'.

To summarize our equations are now

Continuity:

$$\frac{d\varrho}{dt} = -\varrho\nabla\cdot\boldsymbol{V}. \tag{9}$$

Momentum:

$$\varrho\frac{d\boldsymbol{V}}{dt} = -\nabla p + \boldsymbol{J}\times\boldsymbol{B}, \tag{10}$$

Energy:

$$\frac{d}{dt}\left(\frac{p}{\varrho^{\gamma}}\right) = 0. \tag{11}$$

Maxwell-Ohm:

$$\nabla\cdot\boldsymbol{B} = 0, \tag{12}$$

$$\frac{\partial\boldsymbol{B}}{\partial t} = \nabla\times(\boldsymbol{V}\times\boldsymbol{B}), \tag{13}$$

$$\nabla\times\boldsymbol{B} = \boldsymbol{J}. \tag{14}$$

The other two equations give q and $\boldsymbol{E}$ which we do not need. Equations (9)–(14) are closed equations for ϱ, $\boldsymbol{V}$, p, and $\boldsymbol{B}$. These are given respectively by (9), (10), (11) and (13), $\boldsymbol{J}$ is given as a definition by (14) and could be eliminated with no trouble. (12) is a side condition; that is always satisfied if it is satisfied at any time. We shall take as *fluid* theory the study of the consequences of eqs. (9)–(14).

Equations (9)–(14) are essentially those given in the paper by BERNSTEIN, FRIEMAN, KRUSKAL and KULSRUD [4]. We parallel the development of the consequences of these equations given in this paper. The energy principle for the *fluid* theory derived in these notes was first given in this paper.

Boundary conditions: There are two types of situations which we might consider: 1) a plasma confined by a rigid infinitely conducting wall on which $\boldsymbol{B}\cdot\boldsymbol{e} = 0$, where $\boldsymbol{e}$ is the normal into the wall (Fig. 1); 2) a plasma adjacent

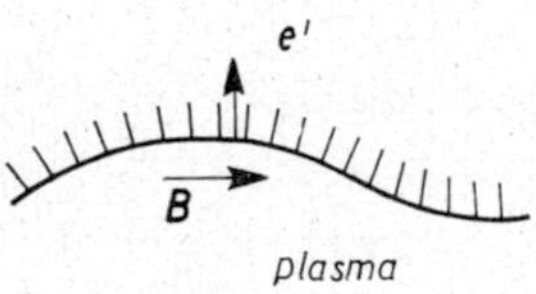

Fig. 1.

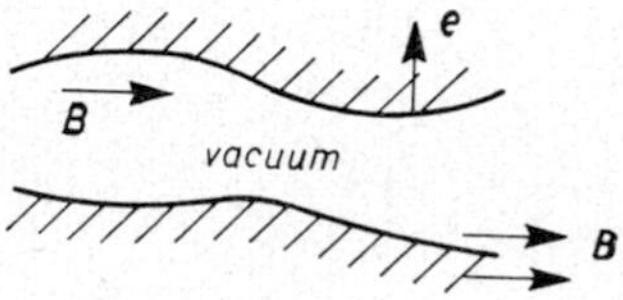

Fig. 2.

to a vacuum adjacent to a rigid wall (Fig. 2); in (2) $\boldsymbol{B}\cdot\boldsymbol{e} = 0$ on the wall and also at the plasma vacuum interface.

For simplicity consider only assumption (1). It is comparatively easy to extend the theory to case 2). We have in the wall

$$\boldsymbol{E} = 0 \, .$$

Also we have the usual boundary conditions (with $\langle A \rangle = A_1 - A_2$)

$$\boldsymbol{V} \cdot \boldsymbol{e} = 0 \, ,$$

$$\boldsymbol{e} \cdot \langle \boldsymbol{B} \rangle = 0 \, ,$$

$$\boldsymbol{e} \times \langle \boldsymbol{E} \rangle = 0 \, ,$$

$$\boldsymbol{e} \times \langle \boldsymbol{B} \rangle = \boldsymbol{J}^* \, ,$$

where $\boldsymbol{J}^*$ is the current flowing in the wall. Thus in the plasma

$$\boldsymbol{e} \times \boldsymbol{E} = 0 \, .$$

Suppose $\boldsymbol{B} = 0$ in the wall at some time; then $\partial \boldsymbol{B}/\partial t = 0$ and $\boldsymbol{B} = 0$ in the wall for all time. Hence since $\langle \boldsymbol{B} \cdot \boldsymbol{e} \rangle = 0$

$$\boldsymbol{B} \cdot \boldsymbol{e} = 0 \quad \text{in the plasma:}$$

so we have case 1) if $\boldsymbol{B} = 0$ in the wall at any time.

It is sufficient in light of (8*a*)–(14) to assume only the boundary conditions

$$\boldsymbol{V} \cdot \boldsymbol{e} = 0 \, , \qquad \boldsymbol{B} \cdot \boldsymbol{e} = 0 \, .$$

Further $\boldsymbol{B} \cdot \boldsymbol{e} = 0$ is a side condition. For,

$$- \boldsymbol{e} \times \boldsymbol{E} = \boldsymbol{e} \times (\boldsymbol{V} \times \boldsymbol{B}) = \boldsymbol{B} \cdot \boldsymbol{e} \boldsymbol{V} - \boldsymbol{V} \cdot \boldsymbol{e} \boldsymbol{B} = 0$$

and therefore

$$\frac{\partial}{\partial t} (\boldsymbol{B} \cdot \boldsymbol{e}) = - c \boldsymbol{e} \cdot \nabla \times \boldsymbol{E} = 0 \, .$$

The last equation follows easily on consideration of the line integral over any closed curve C in the surface

$$0 = \oint \boldsymbol{E} \cdot \mathrm{d}l = \int \boldsymbol{e} \cdot \nabla \times \boldsymbol{E} \, \mathrm{d}S \, .$$

2·2. *Static equilibrium.* – For static equilibrium, $\boldsymbol{V} = 0$, the equations become

$$\nabla p = \boldsymbol{J}\times\boldsymbol{B},$$

$$\nabla\cdot\boldsymbol{B} = 0\,,$$

$$\nabla\times\boldsymbol{B} = \boldsymbol{J}\,.$$

The side condition $\nabla\cdot\boldsymbol{B}=0$ is now a full condition. Note that

$$\boldsymbol{J}\cdot\nabla p = 0\,,$$

$$\boldsymbol{B}\cdot\nabla p = 0\,,$$

so that $\boldsymbol{B}$ and $\boldsymbol{J}$ lie on surfaces of constant p. Thus $\boldsymbol{B}$ has magnetic surfaces *i.e.* surfaces everywhere tangent to $\boldsymbol{B}$. Also p is constant along lines of force. At the rigid wall

$$\boldsymbol{B}\cdot\boldsymbol{e} = 0\,.$$

2·3. *Linearized equations.* – We wish to consider motions of the fluid in the neighborhood of the static equilibrium given above. We write at a fixed point $\boldsymbol{r}$

$$p = p^0 + p'\,,$$

$$\boldsymbol{V} = \boldsymbol{V}'\,,$$

$$\boldsymbol{B} = \boldsymbol{B}^0 + \boldsymbol{B}'\,,$$

$$\varrho = \varrho^0 + \varrho'\,,$$

$$\boldsymbol{J} = \boldsymbol{J}^0 + \boldsymbol{J}'\,.$$

From (14)

$$\boldsymbol{J}' = \nabla\times\boldsymbol{B}'\,.$$

p', $\boldsymbol{V}'$, $\boldsymbol{B}'$ and ϱ' can be given independently initially and then (9)–(14) give their time development.

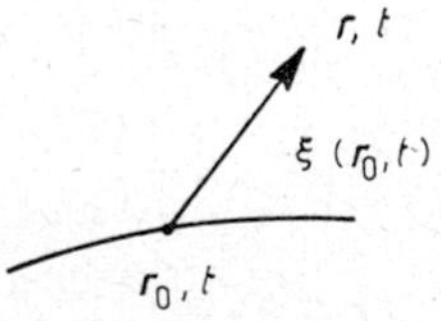

Fig. 3.

Introduce a displacement vector $\boldsymbol{\xi}$, the Lagrangian displacement, (Fig. 3) by

$$\frac{\partial\boldsymbol{\xi}}{\partial t} = \boldsymbol{V}\,.$$

However, it is not necessary to take the Lagrangian point of view. To lowest order we may regard $\boldsymbol{\xi}$ as given at a fixed point and leave it undefined to higher order by $\partial\boldsymbol{\xi}(\boldsymbol{r}, t)/\partial t = \boldsymbol{V}'(\boldsymbol{r}, t)$; $\boldsymbol{\xi}$ will than be an Eulerian variable. We do not consider all displacements p', $\boldsymbol{V}'$, $\boldsymbol{B}'$, ϱ' but only those obtainable in the following manner: assume the system in equilibrium. Then introduce extraneous forces which displace the system away from equilibrium and give it a velocity. During this initial displacement all the equations except (10) hold. Then allow the system to evolve under all equations including (10). Thus initially only $\boldsymbol{\xi}$ and $\boldsymbol{V}$ can be given independently. For instance, this displacement preserves the total mass of the system.

The equations are easily integrated to first order for ϱ', p', $\boldsymbol{B}'$ in terms of $\boldsymbol{\xi}$. Let

$$p^* = p' + \boldsymbol{\xi}\cdot\nabla p\,,$$

$$\varrho^* = \varrho' + \boldsymbol{\xi}\cdot\nabla\varrho\,,$$

(p^* is the perturbed pressure following the displacement). Then

$$\varrho^* = -\,\varrho\,\nabla\cdot\boldsymbol{\xi}\,,$$

$$p^* = -\,\gamma p\,\nabla\cdot\xi\,,$$

$$B' = \nabla\times(\boldsymbol{\xi}\times\boldsymbol{B}) = \boldsymbol{B}\cdot\nabla\boldsymbol{\xi} - \boldsymbol{\xi}\cdot\nabla\boldsymbol{B} - \boldsymbol{B}\nabla\cdot\boldsymbol{\xi}\,,$$

$$B^* = \boldsymbol{B}\cdot\nabla\boldsymbol{\xi} - \boldsymbol{B}\nabla\cdot\boldsymbol{\xi}\,.$$

After the initial displacement $\boldsymbol{\xi}$ is given by

$$\varrho\frac{\partial^2\boldsymbol{\xi}}{\partial t^2} = \varrho\frac{\partial\boldsymbol{V}}{\partial t} = -\,\nabla p' + \boldsymbol{J}'\times\boldsymbol{B} + \boldsymbol{J}\times\boldsymbol{B}'\,,$$

(15) $$\varrho\frac{\partial^2\boldsymbol{\xi}}{\partial t^2} = [\nabla\times(\nabla\times(\boldsymbol{\xi}\times\boldsymbol{B}))]\times\boldsymbol{B} + \boldsymbol{J}\times[\nabla\times(\boldsymbol{\xi}\times\boldsymbol{B})] + \nabla(\boldsymbol{\xi}\cdot\nabla p) + \nabla(\gamma p\cdot\boldsymbol{\xi})\,.$$

Let

$$\boldsymbol{Q} = \nabla\times(\boldsymbol{\xi}\times\boldsymbol{B})\,.$$

Then we have

(16) $$\varrho\frac{\partial^2\boldsymbol{\xi}}{\partial t^2} = \boldsymbol{F}(\boldsymbol{\xi}) = (\nabla\times\boldsymbol{Q})\times\boldsymbol{B} + (\boldsymbol{j}\times\boldsymbol{Q}) + \nabla(\boldsymbol{\xi}\cdot\nabla p) + \nabla(\gamma p\nabla\cdot\boldsymbol{\xi})\,,$$

where $\boldsymbol{F}$ is a linear (differential) operator. The boundary condition on $\boldsymbol{\xi}$ is

(17) $$\boldsymbol{e}\cdot\boldsymbol{\xi} = 0\,.$$

Equation (16) is self-contained giving $\ddot{\boldsymbol{\xi}}$ in terms of $\boldsymbol{\xi}$. (A dot denotes time differentiation). From condition (11), $\boldsymbol{B}'\cdot\boldsymbol{e} = 0$ follows.

(*Note*: one might consider independently perturbations not restricted as above. For instance take $\boldsymbol{\xi} = \dot{\boldsymbol{\xi}} = p' = \boldsymbol{B}' = 0$, ϱ' arbitrary. This produces no time variation. If p' is taken arbitrary and the others zero this corresponds to sound waves or magnetosonic waves. Similarly if $\boldsymbol{B}'$ is independent. Also it is possible to vary the equilibrium state by taking B' and p' together. As will be shown later, none of this extra freedom leads to more generality in the determination of stability.)

2·4. *Stability*. – We now consider the behaviour in time of the perturbations and in particular whether any of them can grow indefinitely. Of course, they must always remain small enough for the linear theory to apply but we may regard $\boldsymbol{\xi}$ as finite and later multiply it by a constant so small as to make the theory valid.

The standard method of studying small oscillations is the normal mode method. One looks for solutions of the form

$$\boldsymbol{\xi} = \boldsymbol{\xi}_n \exp[i\omega_n t],$$

so

$$(18) \qquad -\varrho\omega_n^2\boldsymbol{\xi}_n = \boldsymbol{F}(\boldsymbol{\xi}_n), \qquad \boldsymbol{\xi}_n\cdot\boldsymbol{e} = 0.$$

Assuming the $\boldsymbol{\xi}_n$'s form a complete set we expand any initial perturbation

$$\boldsymbol{\xi}_0 = \sum a_n\boldsymbol{\xi}_n,$$

and then

$$\boldsymbol{\xi} = \sum a_n\boldsymbol{\xi}_n \exp[i\omega_n t]$$

is a solution of (16) which equals $\boldsymbol{\xi}_0$ at $t=0$. (Actually we expand $\dot{\boldsymbol{\xi}}_0 = \sum b_n i\omega_n\boldsymbol{\xi}_n$ also and use the fact that $\boldsymbol{\xi}_n^*$ is a solution to get a more complete solution in an obvious way.) If all ω_n are real $\boldsymbol{\xi}$ is bounded. If any ω is complex ω_m or $-\omega_m$ has a negative imaginary part and $\boldsymbol{\xi}$ is generally unbounded if $a_m \neq 0$, $\boldsymbol{\xi} = \boldsymbol{\xi}_m$ gives an unstable perturbation. Thus the question of stability reduces to examining the ω_n' to see if any is complex.

From this observation we see that no generality is lost in tying ϱ', p', and $\boldsymbol{B}'$ to $\boldsymbol{\xi}$ as we have done above, since if there is an instability, there is one with all quantities growing like $\exp[i\omega_n t]$ (with ω_n complex) and this instability must vanish if continued back to $t=-\infty$. Obviously, for this instability the equations before (15) are satisfied.

The search for a complex ω_n is made easier by the fact that E is self-adjoint *i.e.* for any $\boldsymbol{\xi}$ and $\boldsymbol{\eta}$ satisfying

$$\boldsymbol{\xi}\cdot\boldsymbol{e} = \boldsymbol{\eta}\cdot\boldsymbol{e} = 0,$$

on the the boundary, we have

(19) $$\int \boldsymbol{\xi}\cdot F(\boldsymbol{\eta})\,\mathrm{d}\tau = \int \boldsymbol{\eta}\cdot F(\boldsymbol{\xi})\,\mathrm{d}\tau\,.$$

We show the self-adjointness of F later.
From this we find

a) ω_n^2 is always real.
Proof $\omega_n^{*2}\varrho\boldsymbol{\xi}^* = \boldsymbol{F}(\boldsymbol{\xi}^*)$. This with (18) and (19) gives

$$\omega_n^2\int \varrho\boldsymbol{\xi}^*\cdot\boldsymbol{\xi}\,\mathrm{d}\tau = -\int \boldsymbol{\xi}^*\cdot F(\boldsymbol{\xi})\mathrm{d}\tau = -\int \boldsymbol{\xi}\cdot F(\boldsymbol{\xi}^*)\,\mathrm{d}\tau = \omega_n^{*2}\int \varrho\boldsymbol{\xi}^*\cdot\boldsymbol{\xi}\,\mathrm{d}\tau\,,$$

from which *a*) follows.

Because of *a*) there is no « overstability », that is no growing oscillation can occur. Also if a mode is stable and the equilibrium is varied slightly this mode remains stable (see Fig. 4).

b) The modes are orthogonal

$$\int \varrho\boldsymbol{\xi}_n\cdot\boldsymbol{\xi}_m\,\mathrm{d}\tau = 0 \qquad \text{if} \qquad \omega_n^2 \neq \omega_m^2\,.$$

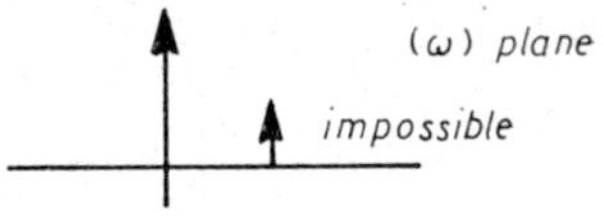

Fig. 4.

Proof

$$(\omega_n^2 - \omega_m^2)\int \varrho\boldsymbol{\xi}_n\cdot\boldsymbol{\xi}_m\,\mathrm{d}\tau = \int (\boldsymbol{\xi}_n\cdot\boldsymbol{F}(\boldsymbol{\xi}_m) - \boldsymbol{\xi}_m\cdot\boldsymbol{F}(\boldsymbol{\xi}_n))\,\mathrm{d}\tau = 0\,.$$

We assume that the $\boldsymbol{\xi}_n$ are chosen orthogonal for equal ω_n's also and normalized so that

(20) $$\int \varrho\boldsymbol{\xi}_n\cdot\boldsymbol{\xi}_m\,\mathrm{d}\tau = \delta_{nm}\,.$$

This property usually goes along with completeness for a set of functions.

c) A variational principle exists for computing ω_n's. Consider

$$\lambda = \frac{-\frac{1}{2}\int \boldsymbol{\xi}\cdot\boldsymbol{F}(\boldsymbol{\xi})\,\mathrm{d}\tau}{\frac{1}{2}\int \varrho\boldsymbol{\xi}^2\,\mathrm{d}\tau}\,,$$

$$\delta\lambda = \frac{\int \delta\boldsymbol{\xi}\cdot(\boldsymbol{F}(\boldsymbol{\xi}) + \lambda\varrho\boldsymbol{\xi})\,\mathrm{d}\tau}{\int \varrho\boldsymbol{\xi}^2\,\mathrm{d}\tau}\,,$$

$\delta\lambda = 0$ for all $\delta\boldsymbol{\xi}$ is equivalent to $\lambda = \omega_n^2$, $\boldsymbol{\xi} = \boldsymbol{\xi}_n$ for some n.

A minimum of λ gives the « most unstable » $\boldsymbol{\xi}_n$. If λ is always positive ω_n^2

is always positive (ω_n is real). Also if λ is ever negative the smallest ω_n^2 (ω_1^2 say) is also negative and the system is unstable. Since the denominator is always positive this means we need only examine the sign of the numerator

$$\delta W = -\tfrac{1}{2}\int \boldsymbol{\xi}\cdot \boldsymbol{F}(\boldsymbol{\xi})\,\mathrm{d}\tau\,, \tag{20}$$

for all $\boldsymbol{\xi}$ with $\boldsymbol{\xi}\cdot\boldsymbol{e}=0$. We have proved $\delta W>0$ for all $\boldsymbol{\xi}$ is a necessary and sufficient condition for stability. Thus

d) There exists an energy principle for stability *i.e.* an expression $\delta W(\boldsymbol{\xi}, \boldsymbol{\xi})$ quadratic in $\boldsymbol{\xi}$ such that stability can be reduced to examining the sign of $\delta W(\boldsymbol{\xi}, \boldsymbol{\xi})$. δW will turn out to be the variation in potential energy of the system.

For a simple example of an energy principle consider the one-dimensional motion of a particle in a potential V:

$$\frac{\partial V}{\partial x} = 0\,,$$

is the condition for equilibrium. A small motion about equilibrium is given by

$$(\delta x)'' = -\frac{\partial^2 V}{\partial x^2}(\delta x)\,.$$

If $\partial^2 V/\partial x^2$ is negative the equilibrium is unstable. The sign of $\partial^2 V/\partial x^2$ may be determined by examining the second variation in ξ

$$V(x+\xi) = \frac{\xi^2}{2}\frac{\partial^2 V}{\partial x^2} = \delta W\,.$$

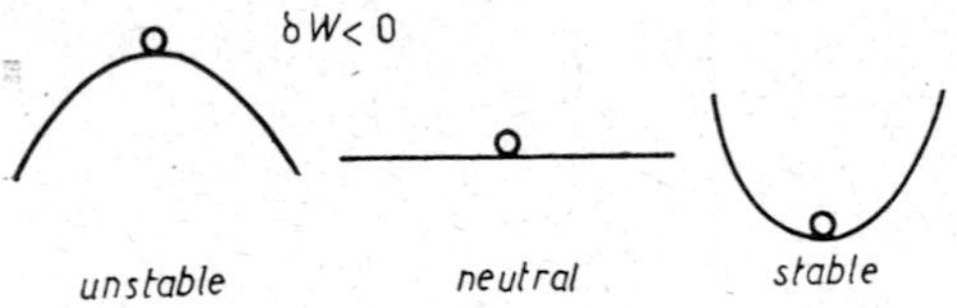

Fig. 5.

If $\delta W<0$ for some ξ then $V_{xx}<0$ and the equilibrium is unstable. Otherwise, it is stable (Fig. 5). The displacement ξ is sometimes called a virtual displacement since it is imagined to be made to test for stability of a real motion.

Similarly for n dimensions

$$\delta W = \sum \frac{\partial^2 V}{\partial x_i \partial x_j}\xi_i \xi_j\,,$$

and one has again an energy principle. The i component of the linearized force is $\sum_j (\partial^2 V/\partial x_i \partial x_j)\xi_j$.

The matrix $\partial^2 V/\partial x_i \partial x_j$ corresponds to our operator $\boldsymbol{F}$ and is obviously self-adjoint, since it is symmetric. Our theory corresponds to a continuum of dimensions but is otherwise algebraically the same.

The proof of *d*) follows the remarks in *c*) with δW given by (20). We make some more remarks on *d*) and give another proof.

The sufficiency of $\delta W > 0$ for stability does not really involve self-adjointness of F if δW is taken *a priori* to be the potential energy rather than defined by (20). For, for an unstable normal mode ω_n is imaginary

$$\delta W \sim \exp[2i\omega_n t];$$

let the kinetic energy be

$$K = \tfrac{1}{2}\int \varrho \dot{\xi}^2 \, d\tau .$$

This is also proportional to $\exp[2i\omega_n t]$. If $\delta W > 0$ we have

$$u = K + \delta W \sim \exp[2i\omega_n t]$$

with a positive coefficient. If ω_n is unstable u must grow indefinitely which is impossible since it is a constant. Therefore ω_n is stable. This argument is essentially due to LIAPUNOFF and is used very skillfully by KRUSKAL and OBERMAN in their paper [2]. It is useful for obtaining weaker energy principles which are only sufficient for stability.

We may identify δW given by (20) with potential energy by considering

$$\frac{\partial}{\partial t}(K + \delta W) = \frac{\partial}{\partial t}\left[\int \varrho \frac{\dot{\boldsymbol{\xi}}^2}{2} d\tau - \frac{1}{2}\int \boldsymbol{\xi}\cdot F(\boldsymbol{\xi})\, d\tau\right] = \int [\varrho \ddot{\boldsymbol{\xi}}\cdot\dot{\boldsymbol{\xi}} - \dot{\boldsymbol{\xi}}\cdot F(\boldsymbol{\xi})]\, d\tau = 0 .$$

Here we use self-adjointness. Since K is kinetic energy δW must be the variation in potential energy.

We give another proof of the energy principle *d*)

Let $\boldsymbol{\xi} = \sum a_n \boldsymbol{\xi}_n$

$$\delta W = -\tfrac{1}{2}\int \boldsymbol{\xi}\cdot F(\boldsymbol{\xi})\, d\tau = -\tfrac{1}{2}\sum_{n,m} a_n a_m \int \boldsymbol{\xi}_n \cdot F(\boldsymbol{\xi}_m) = \tfrac{1}{2}\sum_{n,m} a_n a_m \omega_n^2 \int \boldsymbol{\xi}_n \varrho \boldsymbol{\xi}_m \, d\tau ,$$

$$\delta W = \tfrac{1}{2}\sum_n a_n^2 \omega_n^2 ,$$

by (20). If all $\omega_n^2 > 0$, δW is always positive. Hence $\delta W < 0$ for some $\boldsymbol{\xi}$ implies instability. If $\omega_n^2 < 0$, $\delta W(\xi_n, \xi_n) < 0$ and thus instability $\Rightarrow \delta W < 0$ for some a_n's (or $\boldsymbol{\xi}$). Hence $\delta W > 0$ (for all ξ) $\Leftrightarrow$ stability. We have assumed that the $\boldsymbol{\xi}_n$'s are complete.

We wish an explicit expression for δW. From (16)

$$\delta W = -\tfrac{1}{2}\int \boldsymbol{\xi}\cdot F(\boldsymbol{\xi})\,\mathrm{d}\tau =$$

$$= -\tfrac{1}{2}\int \boldsymbol{\xi}\cdot[(\boldsymbol{\nabla}\times\boldsymbol{Q}\times\boldsymbol{B})] + \boldsymbol{\xi}\cdot\boldsymbol{J}\times\boldsymbol{Q} + \boldsymbol{\xi}\cdot\boldsymbol{\nabla}(\boldsymbol{\xi}\cdot\boldsymbol{\nabla}p) + \boldsymbol{\xi}\cdot\boldsymbol{\nabla}(\gamma p\boldsymbol{\nabla}\cdot\xi)\,.$$

But

$$\boldsymbol{\xi}\cdot(\boldsymbol{\nabla}\times\boldsymbol{Q})\times\boldsymbol{B} = -\boldsymbol{\xi}\times\boldsymbol{B}\cdot\boldsymbol{\nabla}\times\boldsymbol{Q} = \boldsymbol{\nabla}\cdot[(\boldsymbol{\xi}\times\boldsymbol{B})\times\boldsymbol{Q}] - \boldsymbol{\nabla}\times(\boldsymbol{\xi}\times\boldsymbol{B})\cdot\boldsymbol{Q}\,,$$

$$\boldsymbol{\xi}\cdot\boldsymbol{\nabla}(\boldsymbol{\xi}\cdot\nabla p) = \boldsymbol{\nabla}\cdot(\boldsymbol{\xi}\boldsymbol{\xi}\cdot\nabla p) - \boldsymbol{\nabla}\cdot\boldsymbol{\xi}\boldsymbol{\xi}\cdot\boldsymbol{\nabla}p\,,$$

$$\boldsymbol{\xi}\cdot\boldsymbol{\nabla}(\gamma p\,\boldsymbol{\nabla}\cdot\boldsymbol{\xi}) = \boldsymbol{\nabla}\cdot(\boldsymbol{\xi}\gamma p\,\boldsymbol{\nabla}\cdot\boldsymbol{\xi}) - \gamma p(\boldsymbol{\nabla}\cdot\boldsymbol{\xi})^2\,,$$

using Gauss' theorem

$$\delta W = \tfrac{1}{2}\int[\boldsymbol{Q}^2 + \boldsymbol{J}\cdot\boldsymbol{\xi}\times\boldsymbol{Q} + \gamma p(\boldsymbol{\nabla}\cdot\boldsymbol{\xi})^2 + \boldsymbol{\xi}\cdot\nabla p\,\nabla\cdot\boldsymbol{\xi}]\,\mathrm{d}\tau -$$

$$-\int[(\boldsymbol{\xi}\times\boldsymbol{B})\times\boldsymbol{Q} + \boldsymbol{\xi}(\boldsymbol{\xi}\cdot\boldsymbol{\nabla}p) + \boldsymbol{\xi}\cdot\gamma p\boldsymbol{\nabla}\cdot\boldsymbol{\xi}]\cdot\mathrm{d}\boldsymbol{S}\,.$$

The last two terms vanish since $\boldsymbol{\xi}\cdot\boldsymbol{e} = 0$. $(\boldsymbol{\xi}\times\boldsymbol{B})\times\boldsymbol{Q} = \boldsymbol{Q}\cdot\boldsymbol{\xi}\boldsymbol{B} - \boldsymbol{B}\cdot\boldsymbol{Q}\boldsymbol{\xi}$ vanishes also, since $\boldsymbol{B}\cdot\boldsymbol{e} = 0$. $\boldsymbol{B}\cdot\boldsymbol{e} = 0$ follows from (17) by the same argument as used in the nonlinearized case (p. 6).

Thus

$$\delta W = \tfrac{1}{2}\int[\boldsymbol{Q}^2 + \boldsymbol{J}\cdot\boldsymbol{\xi}\times\boldsymbol{Q} + \gamma p(\boldsymbol{\nabla}\cdot\boldsymbol{\xi})^2 + \boldsymbol{\xi}\cdot\boldsymbol{\nabla}p\boldsymbol{\nabla}\cdot\boldsymbol{\xi}]\,\mathrm{d}\tau\,, \tag{21}$$

and we must examine this quadratic functional of $\boldsymbol{\xi}$ for stability. This completes the derivation of the energy principle except for the proof that $\boldsymbol{F}$ is self-adjoint.

We pause to discuss the advantages of the energy principle approach to stability over the normal mode approach.

The most obvious way to check whether $\delta W > 0$ always, is to minimize it over $\boldsymbol{\xi}$. Since it is homogeneous, it is necessary to normalize ξ. If we normalize $\boldsymbol{\xi}$ by

$$\int \varrho\boldsymbol{\xi}^2\,\mathrm{d}\tau = 1\,,$$

we get back the normal mode equations by comment c). However, it is not necessary to choose this particular normalization but only one which makes δW bounded below. There is often a much more convenient normalization.

Sometimes δW is obviously negative by a «trivial» perturbation. In this case the energy principle has the advantage. This is facilitated by the fact that the energy has a physically intuitive significance. The first two terms represent the variation of the magnetic energy and the last two of the plasma energy.

Several important examples of this remarks are

1) Suydam's instabilities [9].

2) Gravitational instability [10].

3) Mercier's generalization of Suydam's criteria [11].

4) Sharp separation of plasma and magnetic field [4].

In each case it is sufficient to exhibit an unstable $\boldsymbol{\xi}$. In cases 1) and 3) the $\boldsymbol{\xi}$ is localized. In case 2) the perturbation is obvious. In case 4) a complete solution is obtained in this way (see Fig. 6). (However, a more general energy principle including gravitation is used.)

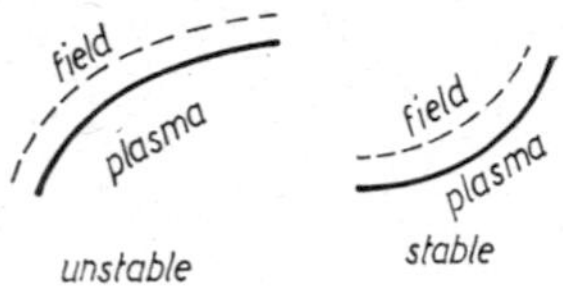

Fig. 6.

It should be remarked that whenever the energy principle is useful a corresponding trick in the normal mode technique can be found. The energy principle simply makes the trick more obvious.

In the energy principle the theorem «$dW > 0 \Rightarrow$ stability» is more natural. The theorem «$\delta W < 0 \Rightarrow$ instability» relies on both completeness and self-adjointness. If $\delta W < 0$ for $\boldsymbol{\xi}$ this need not imply instability in a general case since $\boldsymbol{\xi}$ need not be a normal mode. Although temporarily the system gains kinetic energy at the expense of potential energy it need not remain proportional to $\boldsymbol{\xi}$. $\boldsymbol{\xi}$ may change direction and move into a region of δW and keep $\boldsymbol{\xi}$ finite.

Simple example. – PRENDERGAST [12] has given an example to illustrate a stable situation with $\delta W < 0$.

Consider a 2-dimensional harmonic oscillator with negative spring constant and moving in field $B\perp$ to its plane (see Fig. 7). The potential energy is $-k\xi^2$. Let the oscillator have a charge e. If $\sqrt{k/m} < \frac{1}{2}(eB/mc)$ the system is stable although $\delta W < 0$ always.

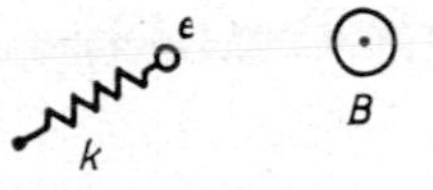

Fig. 7.

Thus it is essential to prove self-adjointness for the energy principle to work both ways.

2·5. *Proof of self-adjointness of F.* – We give two proofs of self-adjointness of F one direct and explicit and the other indirect.

Direct proof. – Let $\boldsymbol{F}=\boldsymbol{F}_1+\boldsymbol{F}_2$

$$\boldsymbol{F}_1(\boldsymbol{\xi}) = [\boldsymbol{\nabla}\times(\boldsymbol{\nabla}\times(\boldsymbol{\xi}\times\boldsymbol{B}))]\times\boldsymbol{B}+\boldsymbol{\nabla}(\gamma p\,\boldsymbol{\nabla}\cdot\boldsymbol{\xi}),$$
$$\boldsymbol{F}_2(\boldsymbol{\xi}) = \boldsymbol{J}\times[\boldsymbol{\nabla}\times(\boldsymbol{\xi}\times\boldsymbol{B})]+\boldsymbol{\nabla}(\boldsymbol{\xi}\cdot\nabla p),$$

$\boldsymbol{F}_1$ and $\boldsymbol{F}_2$ will be separately shown to be self-adjoint.

Let $\boldsymbol{\eta}$ be another vector field with $\boldsymbol{\eta}\cdot\boldsymbol{e}=0$

$$\int\boldsymbol{\eta}\cdot\boldsymbol{F}_1(\boldsymbol{\xi})\,\mathrm{d}\tau=\int\{-(\boldsymbol{\eta}\times\boldsymbol{B})\cdot\big[\boldsymbol{\nabla}\times(\boldsymbol{\nabla}\times(\boldsymbol{\xi}\times\boldsymbol{B}))\big]+\boldsymbol{\eta}\cdot\boldsymbol{\nabla}(\gamma p\boldsymbol{\nabla}\cdot\boldsymbol{\xi})\}\,\mathrm{d}\tau=$$
$$=\int\{\boldsymbol{\nabla}\cdot\big[(\boldsymbol{\eta}\times\boldsymbol{B})\times(\boldsymbol{\nabla}\times(\boldsymbol{\xi}\times\boldsymbol{B}))\big]-[\boldsymbol{\nabla}\times(\boldsymbol{\eta}\times\boldsymbol{B})]\cdot[\boldsymbol{\nabla}\times(\boldsymbol{\xi}\times\boldsymbol{B})]+$$
$$+\boldsymbol{\nabla}\cdot(\gamma p\boldsymbol{\nabla}\cdot\boldsymbol{\xi})-\boldsymbol{\nabla}\cdot\boldsymbol{\eta}\gamma p\boldsymbol{\nabla}\cdot\boldsymbol{\xi}\}.$$

But

$$(\boldsymbol{\eta}\times\boldsymbol{B})\times\boldsymbol{Q}=\boldsymbol{Q}\cdot\boldsymbol{\eta}\boldsymbol{B}-\boldsymbol{Q}\cdot\boldsymbol{B}\boldsymbol{\eta}\qquad\text{and}\qquad\boldsymbol{\eta}\cdot\boldsymbol{e}=0$$

so when the divergence terms are integrated by Gauss' theorem they vanish. The remaining terms are completely symmetric in $\boldsymbol{\xi}$ and $\boldsymbol{\eta}$ and so

$$\int\boldsymbol{\eta}\cdot\boldsymbol{F}_1(\boldsymbol{\xi})\,\mathrm{d}\tau=\int\boldsymbol{\xi}\cdot\boldsymbol{F}_1(\boldsymbol{\eta})\,\mathrm{d}\tau,$$

$\boldsymbol{F}_1$ is self-adjoint. Now consider $\boldsymbol{F}_2$

$$\int[\boldsymbol{\eta}\cdot\boldsymbol{F}_2(\boldsymbol{\xi})-\boldsymbol{\xi}\cdot\boldsymbol{F}_2(\boldsymbol{\eta})]\,\mathrm{d}\tau=\int\{\boldsymbol{\eta}\cdot\boldsymbol{J}\times(\boldsymbol{B}\cdot\boldsymbol{\nabla}\boldsymbol{\xi}-\boldsymbol{\xi}\cdot\boldsymbol{\nabla}\boldsymbol{B}-\boldsymbol{B}\boldsymbol{\nabla}\cdot\boldsymbol{\xi})+$$
$$+\boldsymbol{\eta}\cdot\boldsymbol{\nabla}(\boldsymbol{\xi}\cdot\boldsymbol{\nabla}p)-\boldsymbol{\xi}\cdot\boldsymbol{J}\times[\boldsymbol{B}\cdot\boldsymbol{\nabla}\boldsymbol{\eta}-\boldsymbol{\eta}\cdot\boldsymbol{\nabla}\boldsymbol{B}-\boldsymbol{B}\boldsymbol{\nabla}\cdot\boldsymbol{\eta}]-\boldsymbol{\xi}\cdot\boldsymbol{\nabla}\boldsymbol{\eta}\cdot\boldsymbol{\nabla}p\}\,\mathrm{d}\tau.$$

But

$$-\boldsymbol{\eta}\cdot(\boldsymbol{J}\times\boldsymbol{B}(\boldsymbol{\nabla}\cdot\boldsymbol{\xi}))-\boldsymbol{\xi}\cdot\boldsymbol{\nabla}(\boldsymbol{\eta}\cdot\nabla p)=-\boldsymbol{\eta}\cdot\boldsymbol{\nabla}p\,\boldsymbol{\nabla}\cdot\boldsymbol{\xi}-\boldsymbol{\xi}\cdot\boldsymbol{\nabla}(\boldsymbol{\eta}\cdot\boldsymbol{\nabla}p)=-\boldsymbol{\nabla}\cdot(\boldsymbol{\xi}\boldsymbol{\eta}\cdot\nabla p)$$

which integrates to zero. Similarly

$$+\boldsymbol{\xi}\cdot\boldsymbol{J}\times\boldsymbol{B}(\boldsymbol{\nabla}\cdot\boldsymbol{\xi})+\boldsymbol{\eta}\cdot\boldsymbol{\nabla}(\boldsymbol{\xi}\cdot\boldsymbol{\nabla}p)$$

integrates to zero.

Also

$$\boldsymbol{\eta}\cdot\boldsymbol{j}\times(\boldsymbol{B}\cdot\nabla)\boldsymbol{\xi}=\nabla\cdot[\boldsymbol{B}\boldsymbol{\eta}\cdot\boldsymbol{j}\times\boldsymbol{\xi}]-\nabla\cdot\boldsymbol{B}\boldsymbol{\eta}\cdot\boldsymbol{j}\times\boldsymbol{\xi}-(\boldsymbol{B}\cdot\nabla)\boldsymbol{\eta}\times\boldsymbol{j}\cdot\boldsymbol{\xi}-\boldsymbol{\eta}\times(\boldsymbol{B}\cdot\nabla)\boldsymbol{j}\cdot\boldsymbol{\xi}$$

and the first term integrates out while the second is zero. The third term will cancel the first term in the second bracket in the above expression. Finally

$$0 = \nabla \times (\nabla p) = \nabla(\boldsymbol{j} \times \boldsymbol{B}) = \boldsymbol{B} \cdot \nabla \boldsymbol{j} - \boldsymbol{j} \cdot \nabla \boldsymbol{B}$$

so the last term is

$$\boldsymbol{\eta} \cdot (\boldsymbol{j} \cdot \nabla) \boldsymbol{B} \times \boldsymbol{\xi} .$$

With these remarks we have

$$\int \boldsymbol{\eta} \cdot \boldsymbol{F}_2(\boldsymbol{\xi}) - \boldsymbol{\xi} \cdot \boldsymbol{F}_2(\boldsymbol{\eta}) \, d\tau = \int \boldsymbol{\eta} \cdot [-(\boldsymbol{j} \cdot \nabla) \boldsymbol{B} \times \boldsymbol{\xi} + (\boldsymbol{\xi} \cdot \nabla) \boldsymbol{B} \times \boldsymbol{j} + (\nabla \boldsymbol{B}) \cdot \boldsymbol{\xi} \times \boldsymbol{j}] \, d\tau .$$

To show the bracket vanishes consider

$$[(\boldsymbol{j} \times \boldsymbol{\xi}) \times \nabla] \times \boldsymbol{B} = [\boldsymbol{\xi} \boldsymbol{j} \cdot \nabla - \boldsymbol{j} \boldsymbol{\xi} \cdot \nabla] \times \boldsymbol{B} = \boldsymbol{\xi} \times (\boldsymbol{j} \cdot \nabla) \boldsymbol{B} - \boldsymbol{j} \times (\boldsymbol{\xi} \cdot \nabla) \boldsymbol{B} ,$$

on expanding the triple product in the bracket.

Expanding the triple product with $(\boldsymbol{j} \times \boldsymbol{\xi})$ as a single unit we get

$$(\nabla \boldsymbol{B}) \cdot \boldsymbol{j} \times \boldsymbol{\xi} .$$

Equating these two we find the bracket vanishes. Thus $\boldsymbol{F}_2$ and hence $\boldsymbol{F}$ is self-adjoint.

Indirect proof. – We construct the energy of the system. The energy per unit volume of an adiabatic fluid is $p/\gamma - 1$ since if we let

$$p = A \varrho^\gamma ,$$

$$-\frac{1}{\tau} \int_\infty^\tau p \, d\tau = \varrho \int_0^\varrho A \varrho'^\gamma \frac{d\varrho'}{\varrho'^2} = \varrho \left. \frac{A \varrho^{\gamma-1}}{\gamma - 1} \right|_0^\varrho = \frac{A \varrho^\gamma}{\gamma - 1} = \frac{p}{\gamma - 1} .$$

Therefore the energy U is

$$U = K + W = \int \left(\frac{\varrho \boldsymbol{V}^2}{2} + \frac{\boldsymbol{B}^2}{2} + \frac{p}{\gamma - 1} \right) d\tau ,$$

$\boldsymbol{B}^2/2$ is the energy per unit volume in the magnetic field. All we will need is that U is constant for all motions given by (9)–(14). Explicitly using (9)–(14)

$$\dot{U} = \int \Big[- \nabla\cdot(\varrho\boldsymbol{V})\boldsymbol{V} + \boldsymbol{V}\cdot(-\nabla p + \boldsymbol{j}\times\boldsymbol{B} - \varrho\boldsymbol{V}\cdot\nabla\boldsymbol{V}) + \boldsymbol{B}\cdot\nabla\times(\boldsymbol{V}\times\boldsymbol{B}) - \\ - \frac{\gamma p}{\gamma-1}\nabla\cdot\boldsymbol{V} - \frac{\boldsymbol{V}\cdot\nabla p}{\gamma-1}\Big]\,\mathrm{d}\tau = \\ = \int \Big\{ -\nabla\cdot(\varrho\boldsymbol{V}\boldsymbol{V}) - \frac{\gamma}{\gamma-1}\nabla\cdot(p\boldsymbol{V}) - \nabla\cdot[\boldsymbol{B}\times(\boldsymbol{V}\times\boldsymbol{B})]\Big\}\,\mathrm{d}\tau = 0\,.$$

by Gauss' theorem and (15). ($\boldsymbol{B}\times(\boldsymbol{V}\times\boldsymbol{B}) = B^2\boldsymbol{V} - \boldsymbol{B}\cdot\boldsymbol{V}\boldsymbol{B}$.)

Now we wish to expand U about the equilibrium to second order in $\boldsymbol{\xi}$. To do this we need to define $\boldsymbol{\xi}$ to next order.

We take $\boldsymbol{\xi}$ to be the absolute displacement of a fluid particle, *i. e.* a particle in equilibrium at $\boldsymbol{r}_0$ is at

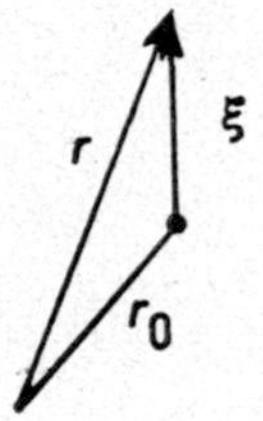

Fig. 8.

$$\boldsymbol{r} = \boldsymbol{r}_0 + \boldsymbol{\xi}(r_0, t) \qquad \text{at time } t\,.$$

(See Fig. 8). To second order

$$\frac{\partial\boldsymbol{\xi}}{\partial t}(\boldsymbol{r}_0, t) = \boldsymbol{V}(\boldsymbol{r}_0 + \boldsymbol{\xi}, t) = \boldsymbol{V}(\boldsymbol{r}_0, t) + \boldsymbol{\xi}\cdot\nabla\boldsymbol{V}(\boldsymbol{r}_0, t)\,.$$

In terms of $\boldsymbol{\xi}$, it is easy to solve for $\boldsymbol{B}$ and p, and ϱ to second order in $\boldsymbol{\xi}$. We have exactly

$$\varrho(\boldsymbol{r}_0 + \boldsymbol{\xi}, t) = \varrho(\boldsymbol{r}_0, 0)\,|\boldsymbol{I} + \nabla_0\boldsymbol{\xi}|^{-1}\,,$$

where $|\boldsymbol{I} + \nabla_0\boldsymbol{\xi}| = \mathrm{Det}\,(\boldsymbol{I} + \nabla_0\boldsymbol{\xi})$

$$\frac{p(\boldsymbol{r}_0 + \boldsymbol{\xi}, t)}{p(\boldsymbol{r}_0, 0)} = \left[\frac{\varrho(\boldsymbol{r}_0 + \boldsymbol{\xi}, t)}{\varrho(\boldsymbol{r}_0, 0)}\right]^{\gamma},$$

$$\frac{\boldsymbol{B}(\boldsymbol{r}_0 + \xi, t)}{\varrho(\boldsymbol{r}_0 + \xi, t)} = \frac{\boldsymbol{B}_0}{\varrho_0} + \frac{\boldsymbol{B}_0}{\varrho_0}\cdot\nabla_0\boldsymbol{\xi}\,,$$

$$\boldsymbol{V} = \frac{\partial\boldsymbol{\xi}}{\partial t}\quad \mathrm{d}\tau = \mathrm{d}\tau_0|\boldsymbol{I} + \nabla_0\boldsymbol{\xi}|\,.$$

We do not make use of these explicit expressions except to show the possibility of expanding U.

Write

$$U = A(\boldsymbol{\xi}) + B(\boldsymbol{\xi}) + K(\dot{\boldsymbol{\xi}}, \dot{\boldsymbol{\xi}}) + M(\dot{\boldsymbol{\xi}}, \boldsymbol{\xi}) + \delta W(\boldsymbol{\xi}, \boldsymbol{\xi})\,,$$

where A and B are linear functionals, K, M, and δW are bilinear functionals and we may take K and δW as symmetric.

What are A and K? To find them we take

$$\boldsymbol{\xi} = 0\ ; \qquad \text{so} \quad \boldsymbol{V} = \dot{\boldsymbol{\xi}}\ , \qquad \delta\boldsymbol{B} = \delta p = \delta\varrho = 0\ , \qquad \mathrm{d}\tau = \mathrm{d}\tau_0$$

and

$$U = \int \frac{\varrho' \dot{\xi}^2}{2}\, \mathrm{d}\tau\ .$$

Thus

$$K = \tfrac{1}{2}\int \varrho\dot{\boldsymbol{\xi}}^2 \qquad \text{and} \qquad A = 0\ .$$

Differentiate (23).

$$\ddot{U} = 0 = B(\dot{\boldsymbol{\xi}}) + O(\boldsymbol{\xi}^2)$$

so $B = 0$, since $\dot{\boldsymbol{\xi}}$ is arbitrary ($\dot{U} = 0$ for all $\boldsymbol{\xi}$ and $\dot{\boldsymbol{\xi}}$ subject to $\boldsymbol{\xi}\cdot\boldsymbol{e} = \dot{\boldsymbol{\xi}}\cdot\boldsymbol{e} = 0$).

Also

$$\dot{U} = 0 = 2\,K[\ddot{\boldsymbol{\xi}}, \dot{\boldsymbol{\xi}}] + M(\dot{\boldsymbol{\xi}}, \dot{\boldsymbol{\xi}}) + M(\boldsymbol{\xi}, \ddot{\boldsymbol{\xi}}) + 2\,\delta W(\dot{\boldsymbol{\xi}}, \boldsymbol{\xi})\ .$$

But by (16)

$$\ddot{\boldsymbol{\xi}} = \boldsymbol{F}(\boldsymbol{\xi})/\varrho$$

so

$$2K\left[\frac{\boldsymbol{F}(\boldsymbol{\xi})}{\varrho}, \dot{\boldsymbol{\xi}}\right] + M(\dot{\boldsymbol{\xi}}, \dot{\boldsymbol{\xi}}) + M\left[\boldsymbol{\xi}, \frac{\boldsymbol{F}(\boldsymbol{\xi})}{\varrho}\right] + 2\,\delta W[\dot{\boldsymbol{\xi}}, \boldsymbol{\xi}] = 0\ ,$$

and this must hold for all $\boldsymbol{\xi}$ and $\dot{\boldsymbol{\xi}}$. Setting here $\boldsymbol{\xi} = 0$ we have

$$M(\dot{\boldsymbol{\xi}}, \dot{\boldsymbol{\xi}}) = 0\ ,$$

the rest of the bilinear functionals vanishing since a bilinear functional vanishes if either argument vanishes. Also from $\dot{\boldsymbol{\xi}} = 0$

$$M\left[\boldsymbol{\xi}, \frac{\boldsymbol{F}(\boldsymbol{\xi})}{\varrho}\right] = 0\ .$$

Thus

$$K\left[\frac{\boldsymbol{F}(\boldsymbol{\xi})}{\varrho}, \dot{\boldsymbol{\xi}}\right] = 2\,\delta W(\dot{\boldsymbol{\xi}}, \boldsymbol{\xi})\ .$$

But δW is symmetrical in $\boldsymbol{\xi}$ and $\dot{\boldsymbol{\xi}}$ so

$$K\left[\frac{\boldsymbol{F}(\boldsymbol{\xi})}{\varrho}, \dot{\boldsymbol{\xi}}\right] = K\left[\frac{\boldsymbol{F}(\dot{\boldsymbol{\xi}})}{\varrho}, \boldsymbol{\xi}\right],$$

and setting $\dot{\boldsymbol{\xi}} = \boldsymbol{\eta}$ and using

$$K = \tfrac{1}{2}\int \varrho \dot{\boldsymbol{\xi}}^2 ,$$

we have again the self-adjointness of F obtained above by a direct method. Let $\dot{\boldsymbol{\xi}} = \boldsymbol{\xi}$

$$\delta W(\boldsymbol{\xi}, \boldsymbol{\xi}) = -K\left[\frac{\boldsymbol{F}(\boldsymbol{\xi})}{\varrho}, \boldsymbol{\xi}\right] = -\frac{1}{2}\int \boldsymbol{\xi}\cdot\boldsymbol{F}(\boldsymbol{\xi})\, d\tau ,$$

and we have shown δW given by (20) is actually the potential energy.

Of the two proofs of self-adjointness the last generalizes the easiest to more complicated cases. *I.e.* (vacuum, adiabatic theory, double adiabatic theory) since it involves the least amount of information on $\boldsymbol{F}$. Notice it is fundamental that $\boldsymbol{\xi}$ and $\dot{\boldsymbol{\xi}}$ be independent as well as that $A = 0$. Also in the establishment of an energy principle we need K to be positive definite.

3. – Adiabatic theory.

3·1. *The Boltzmann equation.* – The *adiabatic* theory corresponds to the limit of no collisions and small gyration radius.

Therefore we start with the Vlasov or collisionless Boltzmann equation for each kind of particle, ions and electrons

$$\frac{\partial f}{\partial t} + \boldsymbol{v}\cdot\nabla f + \frac{e}{m}\left(\boldsymbol{E} + \frac{\boldsymbol{v}\times\boldsymbol{B}}{c}\right)\cdot\nabla_v f = 0 . \tag{1}$$

We pass to the limit of small gyration radius formally by taking very large e (see ref. [5]). We will develop an asymptotic series for all physical quantities in $1/e$. The expansion of the Boltzmann equation follows that of CHEW, GOLDBERGER and LOW [3].

Write

$$f = f^0 + f^1 + \dots ,$$

where f^n is of order $(1/e)^n$. Then to lowest order

$$\left(\boldsymbol{E} + \frac{\boldsymbol{v}\times\boldsymbol{B}}{c}\right)\cdot\nabla_v f = 0 . \tag{2}$$

When necessary f will be distinguished by a subscript i or e. Also the charge and mass. $f(t, \boldsymbol{r}, \boldsymbol{v})$ is the density in velocity space and position space of electrons or ions, ($f\,d^3v\,d^3x$ is the number in $d^3v\,d^3x$).

Write

(3) $$\boldsymbol{E} = -\frac{\boldsymbol{\alpha}\times\boldsymbol{B}}{c} + E_\parallel \boldsymbol{n}\,,$$

where

$$\boldsymbol{n} = \boldsymbol{B}/|B|\,, \qquad \boldsymbol{\alpha}\cdot\boldsymbol{n} = 0\,,$$

(4) $$\boldsymbol{v} = \boldsymbol{\alpha} + \boldsymbol{s} + q\boldsymbol{n}\,, \qquad \boldsymbol{s}\cdot\boldsymbol{n} = 0\,.$$

(2) becomes

$$\frac{\boldsymbol{s}\times\boldsymbol{B}}{c}\cdot\nabla_{\boldsymbol{v}} f + E_\parallel \boldsymbol{n}\cdot\nabla_{\boldsymbol{v}} f = 0\,.$$

From this equation f is constant along a helix in velocity space which extends to infinity if $E_\parallel \neq 0$. Thus f may not approach zero at $q = \infty$ if $E_\parallel = O(1)$.

Hence we take

(5) $$E_\parallel = O(1/e)$$

and regard $\boldsymbol{E}$ and $\boldsymbol{B}$ as known series in $1/e$. Later on we will expand Maxwell's equations in $1/e$ to determine $\boldsymbol{E}$ and $\boldsymbol{B}$.

From

(6) $$\boldsymbol{s}\times\boldsymbol{B}\cdot\nabla_{\boldsymbol{v}} f = 0$$

we get

(7) $$f^0 = F^0(t, \boldsymbol{r}, w, q)\,,$$

where

(8) $$w = s^2/2\,.$$

(7) is the general solution of (6) where F^0 is arbitrary. From (7) F^0 may have any behavior in time. To determine it we write (1) to next order

(9) $$\frac{\partial f^0}{\partial t} + \boldsymbol{v}\cdot\nabla f^0 + \frac{e}{mc}\,\boldsymbol{s}\times\boldsymbol{B}\cdot\nabla_{\boldsymbol{v}} f' + \frac{\varepsilon}{m}\,\boldsymbol{n}\cdot\nabla_{\boldsymbol{v}} f^0 = 0\,,$$

where $E_\parallel = \varepsilon e$. Let φ be the angle between $\boldsymbol{s}$ and some arbitrary vector $\perp$ to $\boldsymbol{B}$, say $\boldsymbol{n}\cdot\nabla\boldsymbol{n}$. Then

$$\boldsymbol{s}\times\boldsymbol{B}\cdot\nabla_{\boldsymbol{v}} f' = |B|\,\frac{\partial f'}{\partial\varphi}\,,$$

and (9) may be solved for f'. However f' must be periodic in φ and this condition is obtained by integrating (9) over φ (for constant t, $\boldsymbol{r}$, w, q)

$$\int_0^{2\pi}\left[\frac{\partial f^0}{\partial t}+\boldsymbol{v}\cdot\nabla f^0+\frac{\varepsilon}{m}\cdot\nabla_v f^0\right]\mathrm{d}\varphi=0\,. \tag{10}$$

Equation (10) gives $\partial F^0/\partial t$. However, in (10) the time derivative $\partial f^0/\partial t$ is at fixed $\boldsymbol{r}$ and $\boldsymbol{v}$.

Let us formally change variables

$$t'=t\,,$$

$$\boldsymbol{r}'=\boldsymbol{r}\,,$$

$$\boldsymbol{s}'=\boldsymbol{v}-\boldsymbol{\alpha}-\boldsymbol{v}\cdot\boldsymbol{n}\boldsymbol{n}\,,$$

$$q'=\boldsymbol{v}\cdot\boldsymbol{n}\,,$$

$$w'=\frac{s'^2}{2}\,,$$

$$\frac{\partial f_0}{\partial t}=\frac{\partial F_0}{\partial t'}+\frac{\partial F_0}{\partial \boldsymbol{r}'}\frac{\partial \boldsymbol{r}'}{\partial t}+\frac{\partial F_0}{\partial w'}\frac{\partial w'}{\partial t}+\frac{\partial F_0}{\partial q'}\frac{\partial q'}{\partial t}\,,$$

$$\frac{\partial t'}{\partial t}=1 \qquad \frac{\partial \boldsymbol{r}'}{\partial t}=0$$

$$\frac{\partial w'}{\partial t}=\boldsymbol{s}'\frac{\partial \boldsymbol{s}'}{\partial t}=\boldsymbol{s}'\cdot\frac{\partial}{\partial t}(-\boldsymbol{\alpha}-\boldsymbol{v}\cdot\boldsymbol{n}\boldsymbol{n})=-\,\boldsymbol{s}'\frac{\partial \boldsymbol{\alpha}}{\partial t}-q'\boldsymbol{s}'\cdot\frac{\partial \boldsymbol{n}}{\partial t}\,,$$

$$\frac{\partial q'}{\partial t}=\boldsymbol{v}\cdot\frac{\partial \boldsymbol{n}}{\partial t}=(\boldsymbol{\alpha}+q\boldsymbol{n}+\boldsymbol{s}')\cdot\frac{\partial \boldsymbol{n}}{\partial t}=(\boldsymbol{\alpha}+\boldsymbol{s}')\cdot\frac{\partial \boldsymbol{n}}{\partial t}\,.$$

Also $\partial\boldsymbol{\alpha}/\partial t$ and $\partial\boldsymbol{n}/\partial t$ are independent of φ while $\oint \boldsymbol{s}'\,\mathrm{d}\varphi=0$. Thus

$$\frac{1}{2\pi}\int\frac{\mathrm{d}f_0}{\partial t}\,\mathrm{d}\varphi=\frac{\partial F_0}{\partial t'}+\boldsymbol{\alpha}\cdot\frac{\partial \boldsymbol{n}}{\partial t}\frac{\partial F^0}{\partial \boldsymbol{q}'}\,.$$

In the same way we find

$$\frac{1}{2\pi}\int\boldsymbol{v}\cdot\nabla f_0\,\mathrm{d}\varphi=(\boldsymbol{\alpha}+q'\boldsymbol{n})\cdot\nabla' F^0-w'(\boldsymbol{\nabla}\cdot\boldsymbol{\alpha}+q'\boldsymbol{\nabla}\cdot\boldsymbol{n}-\boldsymbol{n}\boldsymbol{n}\boldsymbol{:}\boldsymbol{\nabla}\cdot\boldsymbol{\alpha})\frac{\partial F}{\partial w'}+$$
$$+\,(\boldsymbol{\alpha}\boldsymbol{\alpha}\boldsymbol{:}\boldsymbol{\nabla}\boldsymbol{n}+q'\boldsymbol{\alpha}\boldsymbol{n}\boldsymbol{:}\boldsymbol{\nabla}\boldsymbol{n}+w'\boldsymbol{\nabla}\cdot\boldsymbol{n})\frac{\partial F}{\partial \boldsymbol{q}'}\,,$$

$$\frac{1}{2\pi}\int\frac{\varepsilon}{m}\boldsymbol{n}\cdot\boldsymbol{\nabla}_v f_0\,\mathrm{d}\varphi'=\frac{\varepsilon}{m}\frac{\partial F_0}{\partial q'}\,,$$

where we use

$$\int \boldsymbol{ss}\,\mathrm{d}\varphi = w(I - \boldsymbol{nn}) \qquad I \text{ the unit dyadic},$$

in the first equation, and also the notation

$$\boldsymbol{ab}:\boldsymbol{\nabla c} = \boldsymbol{a}\cdot(\boldsymbol{b}\cdot\boldsymbol{\nabla c})\,.$$

Combining these results with (10) we obtain

$$\frac{\partial F^0}{\partial t} + (\boldsymbol{\alpha} + q\boldsymbol{n})\cdot\nabla F^0 - w(\boldsymbol{\nabla}\cdot\boldsymbol{\alpha} + q\boldsymbol{\nabla}\cdot\boldsymbol{n} - \boldsymbol{nn}:\boldsymbol{\nabla\alpha})\frac{\partial F^0}{\partial q} + \\ + \left(\boldsymbol{\alpha\alpha}:\boldsymbol{\nabla n} + q\boldsymbol{\alpha n}:\boldsymbol{\nabla n} + w\boldsymbol{\nabla}\cdot\boldsymbol{n} + \boldsymbol{\alpha}\frac{\partial \boldsymbol{n}}{\partial t} + \frac{\varepsilon}{m}\right)\frac{\partial F^0}{\partial q} = 0\,. \tag{11}$$

We have now dropped the primes.

Equation (11) is the condition on $f^0 = F^0(t, \boldsymbol{r}, w, q)$ that we can solve for f'. Also it gives the time behavior of F^0. F^0 can no longer be taken arbitrarily as a function of t, $\boldsymbol{r}$, w, q as one would conclude from the zeroth order equation, but must be taken to solve (11).

Incidentally we know that

$$\frac{\partial F^0}{\partial t} + \dot{r}\cdot\nabla F^0 + \dot{w}\frac{\partial F^0}{\partial w} + \dot{q}\frac{\partial F^0}{\partial q} = 0\,, \tag{12}$$

so we get by comparison

$$\begin{aligned} \dot{r} &= \alpha + qn\,, \\ \dot{w} &= -\nabla\cdot\alpha - q\nabla\cdot n + nn:\nabla\alpha\,, \\ \dot{q} &= \boldsymbol{\alpha\alpha}:\boldsymbol{\nabla n} + q\boldsymbol{\alpha n}:\boldsymbol{\nabla n} + w\boldsymbol{\nabla}\cdot\boldsymbol{n} + \boldsymbol{\alpha}\frac{\partial\boldsymbol{n}}{\partial t} + \frac{\varepsilon}{m}\,, \end{aligned} \tag{13}$$

to lowest order in $1/e$. The right-hand side of these equations represent the time derivatives averaged over a cyclotron period. These equations suggest an alternative way to obtain the Boltzmann equation to lowest order. Derive (13) directly from the equations of motion and use (12) to derive (11).

3·2. *Maxwell's equations to zero order.* – Equation (11) gives F^0_t in terms of $\boldsymbol{\alpha}$, $\boldsymbol{n}$, ε as functions of $\boldsymbol{r}$ and t, Also the definitions of q and w depend on $\boldsymbol{\alpha}$ and $\boldsymbol{n}$.

To find the behavior of $\boldsymbol{\alpha} = (\boldsymbol{E}\times\boldsymbol{B})/B^2$, $\boldsymbol{n} = \boldsymbol{B}/|B|$ and $\varepsilon = (\boldsymbol{E}\cdot\boldsymbol{n})\boldsymbol{n}$ we must express Maxwell's equation to lowest order. Our development is the same as that sketched by KRUSKAL in the Les Houches notes [13].

Maxwell's equations are

$$\nabla \cdot \boldsymbol{B} = 0\,, \tag{14}$$

$$\frac{\partial \boldsymbol{B}}{\partial t} = -\, c\nabla \times \boldsymbol{E}\,, \tag{15}$$

$$\nabla \times \boldsymbol{B} = \frac{4\pi \boldsymbol{J}}{c} + \frac{1}{c}\frac{\partial \boldsymbol{E}}{\partial t}\,, \tag{16}$$

$$\nabla \cdot \boldsymbol{E} = 4\pi\sigma\,, \tag{17}$$

where σ is the charge density. $\boldsymbol{J}$ and σ are given by

$$\boldsymbol{J} = \sum e \int f \boldsymbol{v}\, \mathrm{d}^3 v\,, \tag{18}$$

$$\sigma = \sum e \int f\, \mathrm{d}^3 v\,, \tag{19}$$

where the sum is over ions and electrons. [We must remember $\boldsymbol{E}$ and $\boldsymbol{B}$ are also expansions in $1/e$ starting with $O(1)$.]

(14) to lowest order is the same, as is (15). In (16) and (17) $\boldsymbol{J}$ and σ are minus first order in $1/e$ by (18) and (19), so (16) and (17) with (18) and (19) become to lowest order

$$\boldsymbol{J}^{-1} = \sum e \int f^0 \boldsymbol{v}\, \mathrm{d}^3 v = 0\,, \tag{20}$$

$$\sigma^{-1} = \sum e \int f^0\, \mathrm{d}^3 v = 0\,. \tag{21}$$

But $f_0 = F^0(t, \boldsymbol{r}, w, q)$. Introduce w and q as integration variables. We get

$$\int f^0 \boldsymbol{v}\, \mathrm{d}^3 v = \int F^0(\boldsymbol{\alpha} + q\boldsymbol{n} + \boldsymbol{s}) 2\pi\, \mathrm{d}w\, \mathrm{d}q\,.$$

The s term vanishes and α is constant so

$$\int f^0 \boldsymbol{v}\, \mathrm{d}^3 v = N^0 \boldsymbol{\alpha} + 2\pi \int F^0 q\, \mathrm{d}q\, \mathrm{d}w\,,$$

where

$$N^0 = 2\pi \int F^0\, \mathrm{d}q\, \mathrm{d}w\,,$$

is the density. Hence by (21) and (20)

$$\boldsymbol{J}^{-1} = \sum n^0 U_\parallel^0 e\boldsymbol{n} = 0$$

or

$$U_\parallel^i = U_\parallel^e ,$$

where

$$N^0 U_\parallel = 2\pi \int qF^0 \, \mathrm{d}q \, \mathrm{d}w .$$

Thus Maxwell's equations to minus first order give

$$\sum e \int F^0 \, \mathrm{d}w \, \mathrm{d}q = 0 , \tag{22}$$

$$\sum e \int F^0 q \, \mathrm{d}w \, \mathrm{d}q = 0 . \tag{23}$$

It is easily shown from (11) that the time derivative of (22) is zero if (23) is satisfied. (This is just $(\partial\sigma^{-1}/\partial t) + \nabla\cdot\boldsymbol{J}^{-1} = 0$.)

Similarly the time derivative of (23) gives

$$\sum \frac{Ne^2}{m} E_\parallel = \sum \frac{e}{m} \boldsymbol{n}\cdot(\nabla\cdot\mathbf{P}^0) , \tag{24}$$

where

$$\mathbf{P}^0 = m \int (\boldsymbol{v} - \boldsymbol{\alpha} - v_\parallel \boldsymbol{n})(\boldsymbol{v} - \alpha - v_\parallel \boldsymbol{n}) f^0 \, \mathrm{d}^3 v ,$$

is the zero order pressure. This may be obtained more simply by taking a moment of the unexpanded Boltzmann equation for ions and electrons dotting with $\boldsymbol{n}$ and subtracting (see ref. [13]). It is easily shown that

$$\mathbf{P}^0 = p_\perp (I - \boldsymbol{nn}) + p_\parallel \boldsymbol{nn} , \tag{25}$$

where

$$p_\perp = m \int w F^0 2\pi \, \mathrm{d}w \, \mathrm{d}q , \tag{26}$$

$$p_\parallel = m \int q^2 F^0 2\pi \, \mathrm{d}w \, \mathrm{d}q . \tag{27}$$

Thus, Maxwell's equations to lowest order (together with (11)) give $E_\parallel$ and $\partial\boldsymbol{B}/\partial t$ but not $\boldsymbol{E}_\perp$. (22) and (23) are side conditions which if once satisfied are always satisfied. (14) is also a side condition.

$\partial \boldsymbol{E}/\partial t$ is given by (16) to zeroth order. Hence we must proceed to zeroth order in (16) to find $\partial \boldsymbol{E}/\partial t$ (strictly speaking (16) is three equations). The component parallel to $\boldsymbol{n}$ is minus first order, which we used to find $E_\parallel$. The part perpendicular to $\boldsymbol{n}$ is zero order which we use to find $E_\perp$).

To proceed to zero order in (16) we need $\boldsymbol{J}$ to zero order which by (18) involves f'. Because of (11) eq. (9) may be solved for f' but not uniquely; only up to a function of t, $\boldsymbol{r}$, w and q as in the case of the solution for f^0 (at first). However such a function gives no component of $\boldsymbol{J}$ perpendicular to $\boldsymbol{B}$ except for the convected current $\sigma^0\alpha$, so $J_\perp$ is uniquely determined in terms of F^0 (and E^0). We find $\boldsymbol{j}$ by multiplying (19) by $m\boldsymbol{v}$ and integrating. Let $\boldsymbol{U}=\boldsymbol{\alpha}+\boldsymbol{U}_\parallel \boldsymbol{n}$. The result is

$$\varrho^0 \frac{\mathrm{d}\boldsymbol{U}^0}{\mathrm{d}t} = \varrho^0\left(\frac{\partial \boldsymbol{U}^0}{\partial t} + \boldsymbol{U}^0\cdot\nabla \boldsymbol{U}^0\right) = \boldsymbol{J}^0\times\boldsymbol{B}^0 - \nabla\cdot\mathbf{P}^0 + \sigma^0\boldsymbol{E}^0 . \tag{28}$$

Thus the perpendicular part of (28) gives $\boldsymbol{J}_\perp$. The parallel part of (28) is already satisfied in virtue of (11). Remember $\boldsymbol{E}^0\cdot\boldsymbol{n}=0$, $\varrho=\sum mN$. $\mathbf{P}$ is here the total pressure. (28) is obviously the equation of motion. σ^0 is found from (17) to zero order. Solving (28) for $\boldsymbol{J}_\perp$ we may substitute in (16) to find $\partial \boldsymbol{E}_\perp/\partial t$.

This would complete our system of equations for zeroth order quantities. This system is (11), (15), $(16)_\perp$ and (24) with side conditions (14), (22) and (23). [$(16)_\perp$ means the perpendicular part of (16), the parallel part is actually (24)] (11) is an equation for F^0, (15) for $\boldsymbol{B}$, $(16)_\perp$ for $\boldsymbol{E}_\perp$ and (24) for $\boldsymbol{E}_\parallel$. In (11), $\boldsymbol{\alpha}$, $\boldsymbol{n}$, and ε occur. These are defined by (3), $\boldsymbol{B}/|\boldsymbol{B}|$ and $eE_\parallel$ respectively. In $(16)_\perp$, $\boldsymbol{j}_\perp$ occurs which is defined by $(28)_\perp$. In (24) $\mathbf{P}$, and N occur which are given by (25)–(27) and $2\pi\int F^0\mathrm{d}q\,\mathrm{d}w$ respectively. σ^0 in eq. (28) is given by (17). These equations which form the system include all of the Boltzmann and Maxwell equations each to their lowest order. $(16)_\perp$ and $(16)_\parallel$ are of different orders of course. (17) is taken to minus first and zeroth order but in zeroth order it determines σ^0 which gives a condition on f' which we do not use. It is clear that this procedure may be iterated to obtain a set of equations valid to first order [*i.e.* order $1/e$]. All those conditions on zeroth-order quantities which are necessary conditions for the solutions of the first order equations have been obtained. Thus, we have attained our goal of determining a system of equations for the zeroth-order quantities.

In these notes we shall work only with lowest order quantities. In general, these equations are valid if 1) the gyration radius and Debye length of both ions and electrons are small compared to macroscopic lengths and 2) both gyration frequencies and plasma frequencies are large compared to macroscopic frequencies. It is not always the case that these conditions are sufficient for finding stability from the lowest order equations. For instance, if ω/Ω is of order $(a/R)^2$ where ω and R are a characteristic macroscopic frequency and length, and Ω and a are the gyration frequency and radius, then it is necessary to proceed to next order in the expansion to determine stability. (See the paper of ROSENBLUTH, KRALL and ROSTOKER [14].)

3.3. *Summary of equations for the adiabatic theory.* – For convenience we collect in one place the zero order system of equations. However, we note first that instead of solving (28) for $\boldsymbol{J}$, and substituting in (17) we may solve (17) for $\boldsymbol{J}_\perp$ and substitute in (28). The result is

$$\varrho \frac{\mathrm{d}\boldsymbol{U}}{\mathrm{d}t} = -\nabla\cdot\mathbf{P} + \frac{1}{4\pi}(\nabla\times\boldsymbol{B})\times\boldsymbol{B} - \frac{1}{c}\frac{\partial\boldsymbol{E}}{\partial t}\times\frac{\boldsymbol{B}}{4\pi} + \frac{\nabla\cdot\boldsymbol{E}\boldsymbol{E}}{4\pi}. \tag{29}$$

Since $\boldsymbol{E}^0 = -\boldsymbol{\alpha}\times\boldsymbol{B}/c$, the last two terms are small by the factor $B^2/\varrho c^2$ when compared with the left-hand side. We may thus write (29) as

$$\varrho \frac{\mathrm{d}\boldsymbol{U}}{\mathrm{d}t} = -\nabla\cdot\mathbf{P} + \boldsymbol{J}\times\boldsymbol{B}, \tag{30}$$

where now $4\pi\boldsymbol{J} = \nabla\times\boldsymbol{B}$.

Of course, only the perpendicular part of (30) is independent of (11). To make our scheme closer to the fluid theory we derive the continuity equation from (11). Our equations may now be written:

Continuity:

$$\frac{\partial\varrho}{\partial t} + \nabla\cdot(\varrho\boldsymbol{U}) = 0. \tag{31}$$

Momentum:

$$\varrho \frac{\mathrm{d}\boldsymbol{U}}{\mathrm{d}t} = \boldsymbol{J}\times\boldsymbol{B} - \nabla\cdot\mathbf{P}. \tag{32}$$

Energy:

$$\mathbf{P} = p_\parallel \boldsymbol{nn} + p_\perp(I - \boldsymbol{nn}), \tag{33}$$

$$p_\parallel = \sum m \int F^0 (q - U_\parallel)^2 2\pi \,\mathrm{d}q\,\mathrm{d}w, \tag{34}$$

$$p_\perp = \sum m \int F^0 w 2\pi \,\mathrm{d}q\,\mathrm{d}w, \tag{35}$$

$$\frac{\partial F^0}{\partial t} + (\boldsymbol{\alpha} + q\boldsymbol{n})\cdot\nabla f_0 + w(\boldsymbol{nn}:\nabla\boldsymbol{\alpha} - \nabla\cdot\boldsymbol{\alpha} - q\nabla\cdot\boldsymbol{n})\frac{\partial F_0}{\partial w} + \left(\boldsymbol{\alpha}\cdot\frac{\partial\boldsymbol{n}}{\partial t} + \boldsymbol{\alpha\alpha}:\nabla\boldsymbol{n} + q\boldsymbol{\alpha n}:\nabla\boldsymbol{n} + w\nabla\cdot\boldsymbol{n} + \frac{eE_\parallel}{m}\right)\frac{\partial F_0}{\partial q} = 0, \tag{36}$$

$$\boldsymbol{\alpha} = \boldsymbol{U} - \boldsymbol{U}\cdot\boldsymbol{nn} \tag{37}$$

Maxwell's equations:

$$\nabla\cdot\boldsymbol{B}=0\,, \tag{38}$$

$$\nabla\times\boldsymbol{B}=\boldsymbol{J}\,, \tag{39}$$

$$\frac{\partial\boldsymbol{B}}{\partial t}=\nabla\times(\boldsymbol{\alpha}\times\boldsymbol{B})\,, \tag{40}$$

$$\sum e\int F^0\,\mathrm{d}q\,\mathrm{d}w=0\,, \tag{41}$$

$$\sum e\int F^0 q\,\mathrm{d}q\,\mathrm{d}w=0\,. \tag{42}$$

Ohm's law:

$$E_{\parallel}=\frac{\sum(e/m)\boldsymbol{n}\cdot(\nabla\cdot\mathbf{P})}{\sum Ne^2/M}\,. \tag{43}$$

Definitions (besides $\mathbf{P}$ and $\boldsymbol{\alpha}$):

$$\boldsymbol{n}=\frac{\boldsymbol{B}}{|B|}\,, \tag{44}$$

$$\frac{\mathrm{d}\boldsymbol{U}}{\mathrm{d}t}=\frac{\mathrm{d}\boldsymbol{U}}{\partial t}+\boldsymbol{U}\cdot\nabla\boldsymbol{U}\,, \tag{45}$$

$$\textstyle\sum \text{ means sum over } i \text{ and } e\,, \tag{46}$$

$$N=\int F\,\mathrm{d}^3v=\int F2\pi\,\mathrm{d}q\,\mathrm{d}w\,. \tag{47}$$

The names are in general heuristic. The equations are complete and independent. New dependent variables have been taken to replace $\boldsymbol{E}$, $\boldsymbol{B}$ and F^0. For instance,

$$\varrho=\sum m\int f\mathrm{d}^3v\,, \tag{48}$$

originally, but this definition can be replaced by a differential equation (31). (48) is now a side condition. Similarly we have the side condition

$$\varrho\,\boldsymbol{U}\cdot\boldsymbol{n}=\sum m\int F^0 q2\pi\,\mathrm{d}w\,\mathrm{d}q\,. \tag{49}$$

$\boldsymbol{E}_{\perp}$ has been eliminated. The Boltzmann equation now plays only the role of giving the equation of state. (38), (41) and (42) are side conditions. Rationalized units are used for $\boldsymbol{J}$ and $\boldsymbol{B}$ as in the fluid theory.

The system is (31)–(49). The differential equations are (31), (32), (36), (40), for ϱ, $\boldsymbol{u}$, F^0, $\boldsymbol{B}$. The definitions are (33)–(35), (37), (39), (43)–(47). The side conditions (initial conditions) are (38), (41), (42), (48), (49).

This system of equations is for all quantities to zero order. It is in principle easy to develop the system of equations to next order or to any order in $1/e$ by iteration.

It is possible to simplify the Boltzmann equation somewhat by introducing the magnetic moment $\nu = w/\beta$ for a particle in place of w, where $\beta = |B|$. Also it is possible to write

$$E_{\parallel} = -\boldsymbol{n}\cdot\nabla\psi \,, \tag{50}$$

where this equation of course only expresses part of E' as a gradient. Then let

$$\varepsilon = w + \frac{q^2}{2} + \frac{e}{m}\psi \,, \tag{51}$$

so

$$q = \sqrt{2\left(\varepsilon - \nu\beta - \frac{e}{m}\psi\right)} \,. \tag{52}$$

$$w = \nu\beta \,. \tag{53}$$

We may transform the Boltzmann equation to these variables, t, $\boldsymbol{v}$, ν and ε just as we did from $\boldsymbol{trv}$ to $\boldsymbol{tr}wq$. The notation is the same as that of Kruskal and Oberman [2]. The introduction of ψ is due to NEWCOMB [15].

The equation becomes after some algebra

$$\frac{\partial f}{\partial t} + (\boldsymbol{\alpha} + q\boldsymbol{n})\cdot\nabla f + \left[\nu\beta\boldsymbol{nn}\cdot\nabla\boldsymbol{\alpha} - \nu\beta\nabla\cdot\boldsymbol{\alpha} - q^2\boldsymbol{nn}:\nabla\boldsymbol{\alpha} + {} + q\boldsymbol{n}\cdot\nabla\frac{\alpha^2}{2} + \frac{e}{m}\frac{D\psi}{Dt}\right]\frac{\partial f}{\partial q} = 0 \,, \tag{54}$$

where

$$\frac{D}{Dt} = \frac{\partial}{\partial t} + \boldsymbol{\alpha}\cdot\nabla \,.$$

In deriving (54) we had to use (40) and (41) to show

$$\boldsymbol{\alpha}\cdot\frac{D\boldsymbol{n}}{Dt} = \boldsymbol{\alpha n}:\nabla\boldsymbol{\alpha} = \frac{1}{2}\boldsymbol{n}\cdot\nabla\frac{\alpha^2}{2} \,.$$

Also we have set

$$f(t, \boldsymbol{r}, \nu, \varepsilon) = 2\pi F^0(t, \boldsymbol{r}, w, q) \tag{55}$$

and in (54) q is given by (52) in terms of t, $\boldsymbol{r}$, ν, ε. f is a different function than $f(t, \boldsymbol{r}, \boldsymbol{v})$.

We note that there is no coefficient of $\partial f/\partial \nu$ which corresponds to the fact that the zero-order ν is a constant of the motion. Also the coefficient of $\partial f/\partial t$ vanishes if α and $\partial \psi/\partial t$ vanish which corresponds to the fact that for $\alpha = 0$ $E^0 = 0$) and $E_\parallel$ constant, ε is a constant of the motion. These facts may be derived directly from eq. (13).

In computing moments we also need to know the Jacobian of the transformation to $\nu\varepsilon$ which if β/q $(2\pi(\beta/q)\,\mathrm{d}\nu\,\mathrm{d}\varepsilon = \mathrm{d}^3 V)$. Hence

$$\left\{\begin{aligned} N &= \sum\nolimits_\pm \int \frac{\beta}{q}\,\mathrm{d}\nu\,\mathrm{d}\varepsilon f\,, \\ U_\parallel &= \sum\nolimits_\pm \int \frac{\beta}{q}\,\mathrm{d}\nu\,\mathrm{d}\varepsilon f q\,, \\ p_\perp &= \sum\nolimits_\pm m \int \frac{\beta}{q}\,\mathrm{d}\nu\,\mathrm{d}\varepsilon f \nu\beta\,, \\ p_\parallel &= \sum\nolimits_\pm m \int \frac{\beta}{q}\,\mathrm{d}\nu\,\mathrm{d}\varepsilon\, f q^2\,, \end{aligned}\right. \tag{56}$$

the factor 2π having been absorbed into f. Since two values of q correspond to one of f we must sum the integrals over the $+$ and $-$ values. Thus the $\sum_\pm$.

Note also for toroidal situations ψ is in general multivalued which is permissible but awkward. We shall, henceforth, restrict ourselves to situations where ψ may be taken singlevalued. Such a situation is a mirror machine where all the particles are «trapped», turned around by the magnetic field as in a mirror machine.

Boundary conditions. – We will assume the simplest boundary conditions namely $\boldsymbol{\alpha}\cdot\boldsymbol{n} = 0$ on all boundaries. The magnetic lines of force of the mirror machine are assumed to enter a rigid infinitely conducting wall (see Fig. 9).

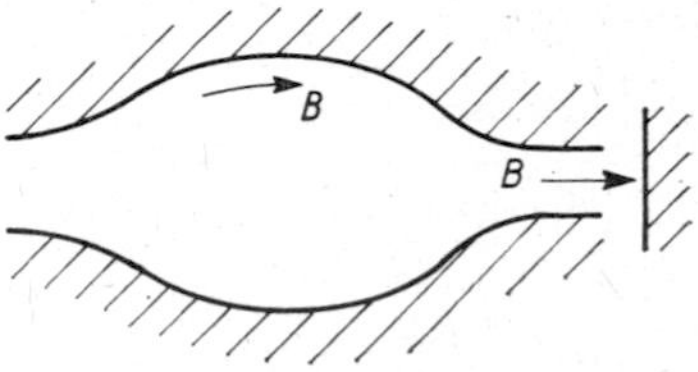

Fig. 9.

For particles which are turning around we have

$$f^+(t, \boldsymbol{r}, \nu, \varepsilon) = f^-(t, \boldsymbol{r}, \nu, \varepsilon) \qquad \text{when } q = 0. \tag{57}$$

Where the plus and minus refer to the different signs of q as one approaches the point where $q = 0$.

We now consider the consequences of the equations given in the summary with (36) replaced by (54) and where appropriate integrals over w and q are replaced by integrals over ε and ν (and summed over $\pm$) with the the corresponding factor of β/q.

3·4. *Static equilibrium.* – We first consider the case of a static equilibrium satisfying the equations of the adiabatic theory given in Section **3**·3. The side conditions now become full equations. Denoting the equilibrium f by g we find from the Boltzmann equation with $\boldsymbol{\alpha}=0$ ($\boldsymbol{u}=0$)

$$q\,\boldsymbol{n}\cdot\nabla g = 0 \tag{58}$$

or g is constant along each line of force L

$$g = g(L, \nu, \varepsilon)\,. \tag{59}$$

Also from (42) and $u_{\parallel}=0$

$$0 = \int \frac{\beta}{|q|}\,\mathrm{d}\nu\,\mathrm{d}\varepsilon f q g\,. \tag{60}$$

The other non-trivial equations are

$$\nabla\cdot\mathbf{P} = \boldsymbol{J}\times\boldsymbol{B}\,, \tag{61}$$

$$\boldsymbol{J} = \nabla\times\boldsymbol{B}\,, \tag{62}$$

$$\nabla\cdot\boldsymbol{B} = 0\,, \tag{63}$$

$$\sum_{\pm i,e} e\int \frac{\beta}{|q|}\,g\,\mathrm{d}\nu\,\mathrm{d}\varepsilon = 0\,. \tag{64}$$

The other equations including (43) become redundant. The simplicity of (58) explains the choice of ε as an independent variable.

3·5. *Linearized theory.* – As in the fluid theory we do not consider any arbitrary perturbation of the equations of Section **3**·3 from the state given in Section **3**·4 but restrict ourselves to those satisfying constraints which do not involve loss in generality with regard to stability.

Let

$$\frac{\mathrm{d}\boldsymbol{\xi}}{\mathrm{d}t} = \boldsymbol{\alpha}\,. \tag{65}$$

Then since $\boldsymbol{B}\cdot\boldsymbol{\alpha}=0$

$$\frac{\partial}{\partial t}(\boldsymbol{\xi}\cdot\boldsymbol{B}) = \boldsymbol{B}\cdot\boldsymbol{\alpha} + \boldsymbol{\xi}\cdot\frac{\partial\boldsymbol{B}}{\partial t} = O(\xi^2)\,. \tag{66}$$

Thus we may assume

$$\boldsymbol{\xi}\cdot\boldsymbol{B} = 0\,.$$

We take

$$\boldsymbol{B}' = \nabla\times(\boldsymbol{\xi}\times\boldsymbol{B})\,, \tag{67}$$

which is a solution of the linearized equations. (Remember an arbitrary constant in time can be added to (67) in general, so (67) involves a constraint on the perturbation.)

The linearized momentum equation is

$$\varrho\,\frac{\mathrm{d}^2\boldsymbol{\xi}}{\partial t^2} + \nabla\cdot\mathbf{P}' = (\nabla\times\boldsymbol{B}')\times\boldsymbol{B} + \boldsymbol{J}\times\boldsymbol{B}'\,. \tag{68}$$

(We remind the reader $\boldsymbol{B}'$ is the perturbed $\boldsymbol{B}$ at a fixed point; $\boldsymbol{B}^* = \boldsymbol{B}' + \boldsymbol{\xi}\cdot\nabla\boldsymbol{B}$ is the perturbed quantity following $\boldsymbol{\xi}$.)

Now

$$\mathbf{P}' = \mathbf{P}^* - \boldsymbol{\xi}\cdot\nabla\mathbf{P}^0\,,$$

and

$$\mathbf{P}^* = p_\perp^*(I - \boldsymbol{nn}) + (p_\parallel - p_\perp)(nn)^* + p_\parallel^*\,\boldsymbol{nn}\,.$$

$p_\perp^*$ and $p_\parallel^*$ are found from (56)

In evaluating $p_\perp^*$ care must be taken to vary the factor β/q. ((q is given by (52)). To do this note that $(\partial q/\partial\varepsilon)_{t,r,v} = 1/q$ so, for instance,

$$\left[\int\frac{\beta}{q}\,\mathrm{d}v\,\mathrm{d}\varepsilon v\beta f\right]^* = -\left[\int\beta q\,\mathrm{d}v\,\mathrm{d}\varepsilon v\beta f_\varepsilon\right]^* = -\int(\beta^2 q)^*\,\mathrm{d}v\,\mathrm{d}\varepsilon v g_\varepsilon + \int\frac{\beta}{q}\,\mathrm{d}v\,\mathrm{d}\varepsilon v\beta f^*\,.$$

In this way the expression for $\mathbf{P}^*$ is found to be

$$\begin{aligned}\mathbf{P}^* = {}&(c + 2\beta_\perp)(nn:\nabla\xi - \nabla\cdot\xi)(I - nn) + \\ &+ (p_\parallel - p_\perp)[nn\cdot\nabla\xi + n\cdot\nabla\xi n - nn(nn\cdot\nabla\xi + \nabla\cdot\xi)] + \\ &+ \sum_{i,e}\pm m\int\frac{\beta}{|q|}\,\mathrm{d}v\,\mathrm{d}\varepsilon[v\beta I + q^2 - v\beta nn]\left(f^* + \frac{e}{m}\,g_\varepsilon\psi^*\right),\end{aligned} \tag{69}$$

where

$$c = \sum_{i,e}\pm m\int\frac{\beta}{|q|}\,\mathrm{d}v\,\mathrm{d}\varepsilon v^2\beta g_\varepsilon\,. \tag{70}$$

To complete the linearized equations we need equations for f^* and one for ψ^*.

The equation for f^* is from (54)

$$\frac{\mathrm{d}f^*}{\mathrm{d}t} + q\boldsymbol{n}\cdot\nabla f^* - \frac{\mathrm{d}\zeta}{\mathrm{d}t}\,g_\varepsilon = 0\,, \tag{71}$$

where

$$\zeta = q^2 nn:\nabla\xi + \nu\beta(\nabla\cdot\xi - nn:\nabla\xi) - \frac{e}{m}\,\psi^*\,, \tag{72}$$

and the equation for ψ^* is (from (41))

$$0 = \sum_{i,e}\pm\, e\int\frac{\beta}{|q|}\,\mathrm{d}\nu\,\mathrm{d}\varepsilon(f^* - \zeta g_\varepsilon)\,. \tag{73}$$

Equation (42) is easily shown to follow automatically.

From boundary conditions (57) and (71) we can show that

$$\sum\pm\int\frac{\beta}{|q|}\,\mathrm{d}l(f_t - \zeta_t g_\varepsilon) = 0\,, \tag{74}$$

where the line integral is between the points where $q = 0$. We restrict ourselves to perturbations satisfying

$$\sum\pm\int\frac{\beta}{|q|}\,\mathrm{d}l(f - \zeta g_\varepsilon) = 0\,, \tag{75}$$

which once satisfied is always satisfied by eq. (71). Hence, no loss in generality is involved by restriction (75).

3·6. *Stability theory.* – The stability theory follows closely the *fluid theory* of stability. The details are given in the author's paper « On the Necessity of the Energy Principle of Kruskal and Oberman », *Physics of Fluids*, **2**, 192 (1962) [16].

Instead of a single equation for $\boldsymbol{\xi}$ as in the fluid theory we have another equation for f which is first order. To arrive at two second-order equations we make use of a suggestion of NEWCOMB [15] and set

$$\bar{f}(t, r, \nu, \varepsilon) = f^*_+(t, r, \nu, \varepsilon) + f^*_-(t, r, \nu, \varepsilon)\,,$$

$$h(t, r, \nu, \varepsilon) = f^*_+(t, r, \nu, \varepsilon) - f^*_-(t, r, \nu, \varepsilon)\,,$$

$\bar{f}$ is the total number of particles with given ε regardless of the sign of q, and h represents the asymmetry of the particles between the two signs of q. Writing

(71) for both f_+^* and f_-^* and adding and subtracting we obtain for $\bar{f}$ and h

$$(76)\qquad \begin{cases} \dfrac{\partial \bar{f}}{\partial t} + |q|\boldsymbol{n}\cdot\nabla h - 2\dot{\zeta} g_\varepsilon = 0\,, \\[2ex] \dfrac{\partial h}{\partial t} + |q|\boldsymbol{n}\cdot\nabla \bar{f} = 0\,. \end{cases}$$

Eliminating h from these equations we have

$$(77)\qquad \frac{\partial^2 \bar{f}}{\partial t^2} - q\boldsymbol{n}\cdot\nabla(q\boldsymbol{n}\cdot\nabla\bar{f}) - 2\ddot{\zeta} g_\varepsilon = 0\,,$$

a second order equation for $\bar{f}$. Since only $\bar{f}$ enters into $\mathbf{P}^*$ and (73) which gives ψ^* we may regard (68), (73) and (77) as a closed system of equations. Regarding (73) as a definition for ψ^* we may write the two second-order equations as

$$(78)\qquad \begin{cases} \ddot{\boldsymbol{\xi}} = \boldsymbol{F}(\xi, \bar{f})\,, \\ \dot{\bar{f}} = J(\xi, \bar{f})\,, \end{cases}$$

and derive the energy principle from these equations. For this purpose it is necessary that $\boldsymbol{F}$ and J together satisfy a self-adjointness. This self-adjointness was first discovered by NEWCOMB [15] who gave a direct proof of it. In these notes we follow the indirect method of the fluid theory in finding and demonstrating this self-adjointness.

We consider an arbitrary perturbation away from the equilibrium described in Section 3'4. Ordinarily one could specify $\boldsymbol{\xi}_t$, and f^* while f_t^* is determined by (71). These are subject to the constraints $\boldsymbol{\xi}\cdot\boldsymbol{n} = \boldsymbol{\xi}_t\cdot\boldsymbol{n} = 0$ and (75). However, it is possible to assign both $\bar{f}$ and $\bar{f}_t$ arbitrarily since $\bar{f}$ satisfies a second-order differential equation. Of course, $\bar{f}$ must satisfy restriction (75), and one must be able to determine h from eq. (76). It is easily seen that the latter requires that (75) be satisfied by $\bar{f}_t$. In summary the stability problem is reduced to examining all solutions of (78) subject to the constraints (on the initial values of the perturbations) $\boldsymbol{\xi}\cdot\boldsymbol{n} = \boldsymbol{\xi}_t\cdot\boldsymbol{n} = 0$ and $\bar{f}$ and $\bar{f}_t$ satisfy (75).

In order to define our perturbations to higher order (as in the fluid theory) we define

$$(79)\qquad \frac{\partial \boldsymbol{\xi}}{\partial t} = \boldsymbol{\alpha}(\boldsymbol{r} + \boldsymbol{\xi}, t)\,,$$

so $\boldsymbol{\xi}$ is the true Lagrangian variable. We generalize (75) by noting that it is a linearized version of a nonlinear constraint given in the paper of KRUSKAL and OBERMAN [2]. Namely, introduce a labeling (*i.e.*, two parameters) of the lines of force, L, which gives the original position of a line of force passing through $\boldsymbol{r}$ at time t. Then for any function G

$$\mu_G = \int G(v, f, L)\, d^3v\, d\tau \,, \tag{80}$$

is a constant. We consider only those perturbation which have the same value of μ as the equilibrium for any choice of G. It is easy to show that to first order in the displacement this is just constraint (75). Because (80) is nonlinear f^{**}, the second-order perturbation of f, is not completely independent but may be related to f^{*2} and therefore it is possible to express the energy to second order in terms of $\boldsymbol{\xi}$ and $\bar{f}$ alone.

After these remarks we now follow the argument given in the fluid theory to establish the self-adjointness of the operators in (78). For convenience, we write (78) as one equation

$$\ddot{\Lambda} = H(\Lambda) \,, \tag{81}$$

where Λ represents $\boldsymbol{\xi}$ and $\bar{f}$ together. The energy U is

$$U = \sum m \int \frac{\beta}{q}\, dv\, d\varepsilon f \varepsilon\, d\tau + \int \frac{B^2}{2}\, d\tau \,,$$

which may be shown to be constant by means of the equations of Section 3·3 Expanding it to second order in Λ, we get

$$U = A(\Lambda) + B(\dot{\Lambda}) + K(\dot{\Lambda}, \dot{\Lambda}) + M(\Lambda, \dot{\Lambda}) + W(\Lambda, \Lambda)$$

which must be constant for all Λ and $\dot{\Lambda}$ subject to the constraints. We find A and K by setting $\Lambda = \bar{f}$, $\boldsymbol{\xi} = 0$

$$U = \frac{1}{2} \int \varrho \dot{\boldsymbol{\xi}}^2 + \sum m \int \frac{\beta}{q}\, dv\, d\varepsilon \left(f^{**} + \frac{e}{m} \psi^{**} \right) \varepsilon \,.$$

(We may ignore the change in β/q produced by $\boldsymbol{\xi}$, since $\boldsymbol{\xi} = 0$ and the change produced by ψ^{**} is simply $(e/m)\beta/q\psi^{**}$.) On the other hand the nonlinear constraint (80) to second order gives

$$\int \frac{\beta}{q} \left[G_f \left(f^{**} + \frac{e}{m} \psi^{**} \right) + G_{ff} \frac{f^{*2}}{2} \right] dv\, d\varepsilon = 0 \,.$$

Taking $G_f = \varepsilon$ (for $f = g$) we have $G_{ff} = -1/g_\varepsilon$ and

$$U = \frac{1}{2}\int \varrho \dot{\boldsymbol{\xi}}^2 - \sum \frac{m}{8}\int \frac{\beta}{q}\, \mathrm{d}v\, \mathrm{d}\varepsilon\, \frac{1}{g_\varepsilon}\, h^2 = K(\dot{\Lambda}, \dot{\Lambda}) \,, \tag{82}$$

and $A = 0$. We must assume $g_\varepsilon < 0$ since it must vanish nowhere, and this makes K positive definite. h is to be expressed in terms of $\bar{f}_t$ by means of (76). Now from the same formal argument as in the fluid theory one obtains

$$K[H(\Lambda), \Lambda'] = K[H(\Lambda'), \Lambda] \,, \tag{83}$$

where Λ and Λ' are any two Λ's satisfying the constraints. Also

$$W(\Lambda, \Lambda) = - K[H(\Lambda), \Lambda] \tag{84}$$

and $W(\Lambda, \Lambda)$ being positive for all Λ gives a necessary and sufficient condition for stability as can be shown by the last proof of condition d) in Section **2·4**. The analogous equation here is

$$W(\Lambda, \Lambda) = \sum a_n^2\, K(\Lambda_n, \Lambda_n)\omega_n^2$$

and K is positive definite.

To obtain an explicit energy principle it is necessary to find W by means of eq. (84). Let $h(\Lambda) = h(f, \boldsymbol{\xi})$ represent an abbreviation for the solution of

$$q\, \boldsymbol{n}\cdot\nabla h = f + 2\zeta g_\varepsilon \tag{85}$$

then

$$W(\Lambda, \Lambda) = -\frac{1}{2}\int \varrho \boldsymbol{\xi}\cdot \boldsymbol{F}(\boldsymbol{\xi}, f)\, \mathrm{d}\tau + \sum \frac{m}{4}\int \frac{\beta}{q}\frac{1}{g_\varepsilon}\, h(\Lambda) h[H(\Lambda)]\, \mathrm{d}\tau\, \mathrm{d}\varepsilon\, \mathrm{d}v \,. \tag{86}$$

But is is easily seen from (71) that

$$h[H(\Lambda)] = -|q|\, \boldsymbol{n}\cdot\nabla \bar{f} \,. \tag{87}$$

Using (87) in (86), integrating the last term by parts and using (85) we have

$$W(\Lambda, \Lambda) = -\frac{1}{2}\int \varrho \boldsymbol{\xi}\cdot \boldsymbol{F}(\boldsymbol{\xi}, f) - \sum m \int \frac{\beta}{q}\frac{1}{g_\varepsilon}\, (\bar{f} - \zeta g_\varepsilon)\bar{f}\, \mathrm{d}v\, \mathrm{d}\varepsilon\, \mathrm{d}\tau \,.$$

The remaining integrations by parts (on $\boldsymbol{F}$) are standard and one obtains after some algebra

(88) $$\delta W_A=\frac{1}{2}\int\{\boldsymbol{Q}^2+\boldsymbol{J}\cdot\boldsymbol{\xi}\times\boldsymbol{Q}+\boldsymbol{\xi}\cdot\nabla p_\perp\nabla\cdot\boldsymbol{\xi}+ + 2p_\perp(\boldsymbol{nn}:\nabla\boldsymbol{\xi}-\nabla\cdot\boldsymbol{\xi})^2+(p_\parallel-p_\perp)[-\boldsymbol{nn}:(\boldsymbol{\xi}\cdot\nabla\nabla\boldsymbol{\xi})+ + (\boldsymbol{nn}:\nabla\boldsymbol{\xi})^2-(\boldsymbol{n}\cdot\nabla\boldsymbol{\xi})^2-(\boldsymbol{n}\cdot\nabla\boldsymbol{\xi})(\nabla\boldsymbol{\xi}\cdot\boldsymbol{n})]+ + c(\boldsymbol{nn}:\nabla\boldsymbol{\xi}-\nabla\cdot\boldsymbol{\xi})^2\}\mathrm{d}\tau-\frac{1}{4}\sum_{i,e}m\int\frac{1}{g_\varepsilon}f^{A^2}\frac{\beta}{q}\,\mathrm{d}\nu\,\mathrm{d}\varepsilon\,\mathrm{d}\tau= = W_1-\frac{1}{4}\sum m\int\frac{f^{A^2}}{g_\varepsilon}\frac{\beta}{q}\,\mathrm{d}\nu\,\mathrm{d}\varepsilon\,\mathrm{d}\tau\,,$$

where

$$f^A=\bar{f}+\frac{2e}{m}\,\psi^* g_\varepsilon\,,$$

f^A is the value of f^* given by KRUSKAL and OBERMAN and differs from ours due to a difference in definition of ε. Their ε is

$$\varepsilon_{\mathrm{KO}}=\nu\beta+q^2/2$$

with no ψ.

Expression (88) is identical with that given by KRUSKAL and OBERMAN after one sets $f_+=f_-$ in their expression (a trivial minimization). Thus their δW gives a necessary and sufficient condition for stability.

To be useful this expression should be minimized over f^A (or $\bar{f}^-$) subject to constraints (73) and (75). This has not yet been done but a good sufficient condition (for stability) is obtained by minimization subject only to (75). The resulting expression is

(89) $$\delta W_{\mathrm{KO}}=W_1-\sum m\int\frac{\beta}{q}\,\lambda^2 g_\varepsilon\,\mathrm{d}\nu\,\mathrm{d}\varepsilon\,\mathrm{d}\tau\,,$$

where

(90) $$\lambda=-\frac{\int[(q^2-\nu\beta)\boldsymbol{nn}:\nabla\boldsymbol{\xi}+\nu\beta\nabla\cdot\boldsymbol{\xi}]\,\mathrm{d}l/|q|}{\int\mathrm{d}l/|q|}\,.$$

3·7. *Comparison theorems.* – To compare δW_A given by (88) with δW_F of fluid theory we must consider the case of an isotropic g. We know that $\delta W_A > \delta W_{\mathrm{KO}}$ of (89) and δW_{KO}, simplifies considerably for isotropic g,

(91) $$\delta W_{\mathrm{KO}}=\frac{1}{2}\int[Q^2+j\cdot\xi\times Q+\xi\cdot\nabla p\nabla\cdot\xi]-\sum m\int\frac{\beta}{q}\,g_\varepsilon\lambda^2\,\mathrm{d}\nu\,\mathrm{d}\varepsilon\,\mathrm{d}\tau\,.$$

Write

$$(92)\quad \lambda = \frac{\varepsilon\int[2-3y\beta)\boldsymbol{nn}:\nabla\boldsymbol{\xi}+y\beta\,\nabla\boldsymbol{\xi}]\,\mathrm{d}l/\sqrt{1-y\beta}}{\int\mathrm{d}l/\sqrt{1-y\beta}} = \frac{J}{K};\quad K=\int\frac{[\mathrm{d}l]}{\sqrt{1-y\beta}};\quad y=\frac{\nu}{\varepsilon}.$$

Then the last term of (91) is

$$(93)\qquad W_2 = \frac{15}{4}\int\mathrm{d}\psi\int\mathrm{d}y\int\mathrm{d}l\,\frac{p}{(1-y\beta)^{\frac{3}{2}}}\,\frac{J^2}{K^2},$$

where $\mathrm{d}\psi$ is a flux element and we have used

$$p=\frac{8\sqrt{2}}{15}\sum m\int\varepsilon^{5/2}g_\varepsilon\,\mathrm{d}\varepsilon\,.$$

But by Schwarz's inequality

$$\int\frac{J^2}{K}\,\mathrm{d}y=\int\mathrm{d}y\left(\frac{J}{K}\right)^2K\geqslant\frac{[\int\mathrm{d}yJ]^2}{\int\mathrm{d}yK},$$

so

$$W_2\geqslant\frac{15}{4}\int\mathrm{d}\psi p\,\frac{[\int\mathrm{d}yJ]^2}{\int\mathrm{d}yK}$$

Now

$$\int\mathrm{d}yK=\frac{2}{\beta},$$

$$\int\mathrm{d}yJ=\frac{2}{3}\int\mathrm{d}l\,\frac{\nabla\cdot\boldsymbol{\xi}}{\beta},$$

so

$$W_2\geqslant\frac{1}{2}\frac{5}{3}\int\mathrm{d}\tau p\,\langle\nabla\cdot\boldsymbol{\xi}\rangle^2,$$

where

$$\langle\nabla\cdot\boldsymbol{\xi}\rangle=\frac{\int(\nabla\cdot\boldsymbol{\xi})\,\mathrm{d}l/\beta}{\int\mathrm{d}l/\beta}\,.$$

However we must remember that in the adiabatic theory $\boldsymbol{\xi}\cdot\boldsymbol{n}=0$ while in the fluid theory $\boldsymbol{\xi}\cdot\boldsymbol{n}\neq0$. Minimizing δW_F over $\boldsymbol{\xi}\cdot\boldsymbol{n}$ is equivalent to making $\boldsymbol{n}\cdot\nabla(\nabla\cdot\boldsymbol{\xi})=0$ or $\nabla\cdot\boldsymbol{\xi}=\langle\nabla\cdot\boldsymbol{\xi}\rangle$. Call δW_F minimized over $\boldsymbol{\xi}\cdot\boldsymbol{n}$, $\delta W_F'$. Hence we obtain the important result

$$\delta W_A>\delta W_{\mathrm{KO}}>\delta W_F'\,.$$

But $\delta W'_F$ gives a necessary and sufficient result for stability on the fluid theory. Hence if the fluid theory gives stability so must the adiabatic theory. Thus, we see the fluid theory gives reliable results even in cases where one would suppose it was unreliable. The comparison theorem and its proof is given in both references [2] and [5].

4. – Double adiabatic theory.

4·1. *Basic equations.* – We now consider the situations intermediate between those of weak collisions, adiabatic theory, and strong collisions, fluid theory. That is the situations in which collisions are not sufficiently strong to keep the pressure a tensor but sufficiently strong to prevent heat flow and other transport processes. The basic equations are

$$\frac{\mathrm{d}\varrho}{\mathrm{d}t} = -\varrho\nabla\cdot\boldsymbol{V}, \tag{1}$$

$$\varrho\frac{\mathrm{d}\boldsymbol{V}}{\mathrm{d}t} = -\nabla\cdot\mathbf{P} + \boldsymbol{J}\times\boldsymbol{B}, \tag{2}$$

$$\mathbf{P} = p_\perp(I - \boldsymbol{nn}) + p_\parallel\,\boldsymbol{nn}, \tag{3}$$

$$\frac{\mathrm{d}}{\mathrm{d}t}\left(\frac{p_\perp}{\varrho\beta}\right) = 0 \qquad \frac{\mathrm{d}}{\mathrm{d}t}\left(\frac{p_\parallel\beta^2}{\varrho^3}\right) = 0, \tag{4}$$

$$\frac{\partial\boldsymbol{B}}{\partial t} = \nabla\times(\boldsymbol{V}\times\boldsymbol{B}), \tag{5}$$

$$\nabla\cdot\boldsymbol{B} = 0, \tag{6}$$

$$\boldsymbol{J} = \nabla\times\boldsymbol{B}. \tag{7}$$

The only difference between these equations and the fluid eqs. (9)–(14) of Section **2** is that in (10) the pressure is replaced by a tensor whose two independent components are determined by the two adiabatic equations of state (4) instead of the single equation of state (11) of the fluid theory.

One derives (3) and (4) as follows: Starting with the Boltzmann equation (without collisions)

$$\frac{\partial f}{\partial t} + \boldsymbol{V}\cdot\nabla f + e\left(\boldsymbol{E} + \frac{\boldsymbol{V}\times\boldsymbol{B}}{c}\right)\cdot\nabla_v f = 0, \tag{8}$$

and introducing

$$N = \int f\,\mathrm{d}^3v, \tag{9}$$

(10) $$NU = \int Vf \, d^3v \, ,$$

(11) $$\mathbf{P} = m \int (\boldsymbol{v} - \boldsymbol{U})(\boldsymbol{v} - \boldsymbol{U}) f \, d^3v \, ,$$

(12) $$Q = m \int (\boldsymbol{v} - \boldsymbol{U})(\boldsymbol{v} - \boldsymbol{U})(\boldsymbol{v} - \boldsymbol{U}) f \, d^3v \, ,$$

one finds after multiplying (8) by $(\boldsymbol{v} - \boldsymbol{U})(\boldsymbol{v} - \boldsymbol{U})$ and integrating by parts

(13) $$\frac{d\mathbf{P}}{dt} + \nabla \cdot Q + \mathbf{P}(\nabla \cdot \boldsymbol{U}) + \mathbf{P} \cdot \nabla \boldsymbol{U} + (\mathbf{P} \cdot \nabla \boldsymbol{U})^{tr} + \frac{e}{mc}(\boldsymbol{B} \times \mathbf{P} - \mathbf{P} \times \boldsymbol{B}) = 0 \, ,$$

where $(\mathbf{P} \cdot \nabla \boldsymbol{U})^{tr}$ is the transposed dyadic of $\mathbf{P} \cdot \nabla \boldsymbol{U}$.

Let e (or B) be large and expand $\mathbf{P}$

$$\mathbf{P} = \mathbf{P}^0 + \mathbf{P}' + \dots \, ,$$

then from (13)

$$\boldsymbol{B} \times \mathbf{P}^0 = \mathbf{P}^0 \times \boldsymbol{B}$$

or

(14) $$\mathbf{P} = p_\perp (I - \boldsymbol{nn}) + p_\parallel \boldsymbol{nn} \, .$$

Equation (13) to first order involves $\mathbf{P}'$ but first taking the trace and second double-dotting with $\boldsymbol{nn}$ eliminates $\mathbf{P}$, and gives

(15) $$\frac{d}{dt}(2p_\perp + p_\parallel) + (2p_\perp + p_\parallel)\nabla \cdot \boldsymbol{U} + 2p_\perp(\nabla \cdot \boldsymbol{U} - \boldsymbol{nn} : \nabla \boldsymbol{U}) + \\ + 2p_\parallel(\boldsymbol{nn} : \nabla \boldsymbol{U} - \nabla \cdot \boldsymbol{U}) + 2p_\parallel \nabla \cdot \boldsymbol{U} = 0 \, ,$$

and

(16) $$\frac{dp_\parallel}{dt} + p_\parallel(\nabla \cdot \boldsymbol{U}_0) + 2p_\parallel \boldsymbol{nn} : \nabla \boldsymbol{U} = 0 \, ,$$

respectively. Now from (1) and (5)

(17) $$\nabla \cdot U = -\frac{1}{\varrho}\frac{d\varrho}{dt} \, ,$$

(18) $$\boldsymbol{nn} : \nabla \boldsymbol{U} - \nabla \cdot \boldsymbol{U} = \frac{1}{\beta}\frac{d\beta}{dt} \, ,$$

and combining these with (16) gives the second half of (4). Combining these with (15) and the second half of (4) now gives the first half of (4). The derivation is essentially given in reference [8].

4·2. *Static equilibrium.* – The equations for static equilibrium are

$$J \times B = \nabla \cdot \mathbf{P}\,, \tag{19}$$

$$J = \nabla \times B\,, \tag{20}$$

$$\nabla \cdot B = 0\,. \tag{21}$$

These equations are the same as those of the adiabatic theory except $p_\perp$ and $p_\parallel$ are free and not determined by f.

4·3. *Linearized equations.* – Introduce ξ as before

$$\frac{\partial \xi}{\partial t} = U\,.$$

Then the basic linearized equation is

$$\varrho \frac{d^2\xi}{dt^2} = -\nabla \cdot \mathbf{P}^* + \nabla \cdot (\xi \cdot \nabla \mathbf{P}) + J' \times Q + (\nabla \times Q) \times B\,, \tag{22}$$

where:

From (1), (3), (4) (and (5) for n^*),

$$\mathbf{P}^* = p_\perp(nn : \nabla\xi - \nabla \cdot \xi) I - p_\perp(3\,nn : \nabla\xi - \nabla \cdot \xi) nn + \\ + (p_\parallel - p_\perp)[n \cdot \nabla\xi\, n + nn \cdot \nabla\xi - nn(4\,nn : \nabla\xi + \nabla \cdot \xi)]\,. \tag{23}$$

From (22) and (23) we have

$$\varrho \frac{d^2\xi}{dt^2} = F_{DA}(\xi)\,, \tag{24}$$

where $F_{DA}(\xi)$ is a linear operator in ξ corresponding to the F of the fluid theory.

4·4. *Stability.* – The treatment of stability follows the indirect proof of section 2·4. It is easy to show from (1)–(7) of this section that the energy

$$U = \int \left(\frac{\varrho}{2} V^2 + \frac{B^2}{2} + p_\perp + \frac{p_\parallel}{2} \right) d\tau\,, \tag{25}$$

is constant. One expands U in ξ to second order

$$U = A(\xi) + B(\dot{\xi}) + K(\dot{\xi}, \dot{\xi}) + M(\xi, \dot{\xi}) + W(\xi, \xi) \tag{26}$$

and A and B are easily shown to be zero and again

$$K(\dot{\boldsymbol{\xi}}, \dot{\boldsymbol{\xi}}) = \frac{1}{2}\int \varrho\dot{\boldsymbol{\xi}}^2 \,\mathrm{d}\tau \,. \tag{27}$$

As before from $\dot{U}=0$ for all $\boldsymbol{\xi}$ and $\dot{\boldsymbol{\xi}}$ one has

$$K\left(\frac{\boldsymbol{F}_{\mathrm{DA}}(\boldsymbol{\xi})}{\varrho}, \dot{\boldsymbol{\xi}}\right) = K\left[\frac{\boldsymbol{F}_{\mathrm{DA}}(\boldsymbol{\xi}')}{\varrho}, \boldsymbol{\xi}\right], \tag{28}$$

or

$$\int \boldsymbol{F}_{\mathrm{DA}}(\boldsymbol{\xi}), \boldsymbol{\xi}' \,\mathrm{d}\tau = \int \boldsymbol{F}_{\mathrm{DA}}(\boldsymbol{\xi}')\cdot\boldsymbol{\xi}\,\mathrm{d}\tau \,. \tag{29}$$

From this ω^2 is real for normal modes and an energy principle exists, *i.e.*

$$\delta W_{\mathrm{DA}} = -\int \boldsymbol{\xi}\cdot\boldsymbol{F}_{\mathrm{DA}}(\boldsymbol{\xi})\,\mathrm{d}\tau \,, \tag{30}$$

gives a necessary and sufficient condition for the stability of the solutions of the linearized double adiabatic solutions of Section **3**.3. By some integrations by parts one finds

$$\begin{aligned}\delta W_{\mathrm{DA}} = \frac{1}{2}\int & Q^2 + \boldsymbol{J}\cdot\boldsymbol{\xi}\times\boldsymbol{Q} + \boldsymbol{\xi}\cdot\nabla p_\perp(\boldsymbol{\nabla}\cdot\boldsymbol{\xi}) + \frac{5}{3}(\boldsymbol{\nabla}\cdot\boldsymbol{\xi})^2 p_\perp + \\ & + \frac{1}{3}p_\perp(\boldsymbol{\nabla}\cdot\boldsymbol{\xi} - \boldsymbol{nn}\boldsymbol{\nabla}\boldsymbol{\xi})^2 + (p_\parallel - p_\perp)[-(\boldsymbol{n}\cdot\boldsymbol{\nabla}\boldsymbol{\xi})^2 - \\ & - (\boldsymbol{n}\cdot\boldsymbol{\nabla}\boldsymbol{\xi})\cdot(\boldsymbol{\nabla}\boldsymbol{\xi}\cdot\boldsymbol{n}) + 4(\boldsymbol{nn}:\boldsymbol{\nabla}\boldsymbol{\xi})^2 + \boldsymbol{nn}\cdot\boldsymbol{\nabla}\boldsymbol{\xi}\,\boldsymbol{\nabla}\cdot\boldsymbol{\xi} - \boldsymbol{nn}:\boldsymbol{\nabla}\cdot(\boldsymbol{\xi}\cdot\boldsymbol{\nabla}\boldsymbol{\xi})] \,.\end{aligned} \tag{31}$$

From (31) and eq. (21) of Section **2** one sees that if $\mathbf{P}$ is isotropic ($p_\perp = p_\perp$) and we set $\gamma = \frac{5}{3}$,

$$\delta W_{\mathrm{DA}} > \delta W_F \,. \tag{32}$$

Thus, if an equilibrium is found to be stable on the fluid theory it will be stable on the double adiabatic theory. (It will also be an equilibrium).

To find the comparison between the adiabatic and double adiabatic theories we find from (31) of this section, and eq. (88) and (89) of Section **3**, that

$$\begin{aligned}\delta W_{\mathrm{KO}} = \delta W_{\mathrm{DA}} + I - \frac{1}{2}\int \mathrm{d}\tau\boldsymbol{nn}:\boldsymbol{\nabla}\boldsymbol{\xi}[p_\perp(2\boldsymbol{\nabla}\cdot\boldsymbol{\xi} + \boldsymbol{nn}:\boldsymbol{\nabla}\boldsymbol{\xi}) + \\ + 3(p_\parallel - p_\perp)(\boldsymbol{nn}:\boldsymbol{\nabla}\boldsymbol{\xi})^2] \,,\end{aligned} \tag{33}$$

where

$$I = -\sum \frac{m}{2} \int \frac{\beta}{q} \, d\nu \, d\varepsilon \, d\tau g_\varepsilon [\lambda^2 - \nu^2 \beta^2 (\boldsymbol{nn} : \nabla \boldsymbol{\xi} - \nabla \cdot \boldsymbol{\xi})^2] , \tag{34}$$

and λ is given by (90) of Section **3**.

But by the Schwarz inequality

$$\lambda^2 < \frac{\int (dl/q) \, [q^2 \boldsymbol{nn} : \nabla \boldsymbol{\xi} + \nu\beta(\nabla \cdot \boldsymbol{\xi} - \boldsymbol{nn} \cdot \nabla \boldsymbol{\xi})]^2}{\int dl/q} , \tag{35}$$

(Hint: write

$$\lambda^2 = \left(\int ab \, dl \right)^2 \Big/ \int \frac{dl}{q} < \int a^2 \, dl \int b^2 \, dl \Big/ \left(\int \frac{dl}{q} \right)^2 ,$$

where

$$a = \frac{1}{\sqrt{q}} , \quad b = \frac{1}{q} [\;] \Bigg) .$$

Use of (34) and (35) gives

$$I < \frac{1}{2} \int d\tau \, [p_\perp (2 \nabla \cdot \boldsymbol{\xi} - \boldsymbol{nn} : \nabla \boldsymbol{\xi}) + 3 p_\parallel \boldsymbol{nn} : \nabla \boldsymbol{\xi}] \boldsymbol{nn} : \nabla \boldsymbol{\xi} . \tag{36}$$

Therefore from (33) and (36)

$$\delta W_{\text{KO}} < \delta W_{\text{DA}} . \tag{37}$$

We thus have the inequalities

$$\begin{cases} \delta W'_F < \delta W_{\text{KO}} < \delta W_{\text{DA}} , \\ \delta W_{\text{KO}} < \delta W_A . \end{cases} \tag{38}$$

The relationship between δW_{DA} and δW_A has not yet been worked out.

From inequalities (38) we get the theorems

Stability	on	F.T. ⇒	Stability	on	K.O.T.
»	»	F.T. ⇒	»	»	A.T.
»	»	F.T. ⇒	»	»	D.A.T.
»	»	K.O. ⇒	»	»	A.T.
»	»	K.O. ⇒	»	»	D.A.

where we use the abbreviations:

F.T. $\equiv$ Fluid theory,

A.T. $\equiv$ Adiabatic theory,

D.A.T. $\equiv$ Double adiabatic theory,

K.O.T. $\equiv$ Kruskal Oberman theory,

$\Rightarrow$ $\equiv$ implies.

Again the derivation of the comparison theorems follows that given in both references [2] and [5].

* * *

I should like to thank DIETRICH VOSLAMBER for critically reading these notes and making useful comments on them.

Part of this work was sponsored by the US Atomic Energy Commission.

REFERENCES

[1] G. F. CHEW, M. L. GOLDBERGER and F. E. LOW: *Proc. Roy. Soc.*, A **236**, 112 (1956).
[2] M. D. KRUSKAL and C. R. OBERMAN: *Phys. Fluids*, **1**, 275 (1958).
[3] G. F. CHEW, M. L. GOLDBERGER and F. E. LOW: *Los Alamos Lecture Notes on Physics of Ionized Gases*, LA-2055 (1955).
[4] I. B. BERNSTEIN, E. A. FRIEMAN, M. D. KRUSKAL and R. M. KULSRUD: *Proc. Roy. Soc.*, A **244**, 17 (1958).
[5] M. N. ROSENBLUTH and N. ROSTOKER: *Phys. Fluids*, **2**, 23 (1959).
[6] M. N. ROSENBLUT, W. M. MCDONALD and D. L. JUDD: *Phys. Rev.*, **107**, 1 (1957).
[7] W. MARSHALL: Report No. A.E.R.E. T/R 2419.
[8] I. B. BERNSTEIN and S. K. TREHAN: *Nucl. Fusion*, **1**, 3 (1960).
[9] B. R. SUYDAM: *Proc. of the Second International Conference on the Peaceful Uses of Atomic Energy*, **31**, 157 (1958).
[10] M. N. ROSENBLUTH and C. L. LONGMUIR: *Ann. Phys.*, **1**, 120 (1957).
[11] C. MERCIER: *Proc. of the Conference on Plasma Physics and Controlled Nuclear Fusion Research*, paper no. 95 (Salzburg, 1961).
[12] K. PRENDERGAST: private communication.
[13] M. D. KRUSKAL: *Chapter in La theorie des Gaz Neutres et Ionisés*, edited by C. DEWITT and J. F. DETOEUF (New York, 1960).
[14] M. N. ROSENBLUTH, N. A. KRALL and N. ROSTOKER: *Conference on Plasma Physics and Controlled Nuclear Fusion Research*, paper CN-10/170 (Salzburg, 1961).
[15] W. A. NEWCOMB: to be published.
[16] R. M. KULSRUD: *Phys. Fluids*, **2**, 192 (1962).

Gas Discharge Theory.

G. ECKER
Bonn University - Bonn

Introduction.

In this course on « advanced plasma theory » the topic « theory of gas discharges » is quite different from all the others. Therefore I feel that a few introductory remarks may be helpful to permit a better judgement of the following analysis.

A plasma was defined by Langmuir to be a partially ionized gas in which the Debye length is small compared with other lengths of interest. Consequently in a course on the theory of the plasma you will expect that one model, namely an unbounded many-particle system consisting of an electron, ion and neutral particle component, is treated in various mathematical approximations. Therefore the different lectures here are essentially distinguished by the approximation procedure, but not by the model.

This is different with the lectures on gas discharge theory where not so much a special approximation but rather the extension of the model characterizes the course.

In a gas discharge we again consider electrons, ions and neutral particles. But already the number of particle components may be larger. Multiply charged positive and negative ions can influence the behaviour of our system. More important we consider the plasma under the actual experimental conditions, that means the qualities of the circuit and the boundary conditions of the electrodes and the container walls enter our calculations.

In other words: whereas the other lectures develop solutions of certain typical differential equations paying little attention to the initial and boundary conditions, our analysis—using these solutions—will be governed just by these initial and boundary conditions.

This produces two consequences:

first, since we have an infinite number of possible initial and boundary conditions corresponding to the experimental set-ups it will be necessary to select certain discharge types which we choose to consider in this course; secondly, under the influence of the boundary conditions the *region of interest* will present itself not as a uniform region, but as composed of various different *model regions.* Since each model region is described by a number of physical laws the gas discharge problem is much more complex than the unbounded plasma problem.

We can therefore anticipate that in the unbounded uniform plasma the evaluation may be carried to a higher and more satisfactory degree of accuracy than will be possible in the case of a gas discharge where we have several interacting model regions governed by a large number of essentially different physical laws. Just for this reason most calculations in gas discharge theory are extremely cumbersone. Therefore in this lecture I will restrict myself to describe the mathematical formulations but leave out the details of the evaluation which in general are less ingenious than tiresome.

1. – Region of interest.

From the large number of possible discharges we want to select an object which is at the same time of practical importance and theoretical interest. To this end we consider a simple experimental set-up of two planar electrodes in a cylindrical discharge vessel. With the help of an external resistance we change the current and observe the V-I-characteristic shown in Fig. 1.

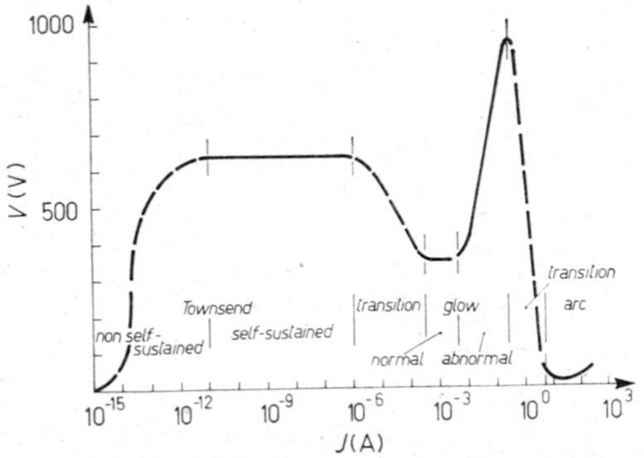

Fig. 1. – Schematical voltage-current characteristic.

For very low currents of the order 10^{-15} A we have spontaneous current bursts which are due to photoelectric emission from the cathode. The electron emitted passes through the gas producing an electron avalanche.

With increasing current the secondary processes of the initial electron are able to produce a new electron from the cathode, for instance via photons which cause photoelectric emission. As soon as we reach this point we have a self-sustained discharge. In this region the voltage is practically constant.

New features occur at 10^{-6} A when space charge becomes of importance. In the corona-discharge where we have inhomogeneous fields due to space-charge build-up, more economic particle-production processes are available. Also the discharge becomes visible.

With further increase in the current the carrier distribution in the discharge volume is rearranged and at the low-voltage end of the transition region we have the so-called subnormal glow discharge which consists of a cylindrical plasma column and the typical cathodic parts of the glow. In these cathodic parts there is a mutual production of electrons at the cathode surface by impinging ions and ions in the negative glow by electrons accelerated through the cathode-fall region. In the column particles are produced by electron collisions and lost by diffusion and wall recombination. Characteristic for the subnormal region is that the diffusion is not ambipolar since the particle densities are too small to build up the ambipolar field.

When with current increase the particle densities in the column are finally sufficient to provide the full ambipolar field, then we are in the region of the normal glow discharge. Here with increasing current the voltage does not change, but the area covered by the discharge increases with the current, keeping the current density constant.

If the total cathode area is covered so that the current density is forced to increase, then we reach the region of the abnormal glow in which the potential rises again. With the increasing potential and the increasing current gradually more and more energy is given to the neutral gas component and temperature effects and volume recombination become important.

This is the point where the transition to the arc discharge occurs. The voltage drops rapidly. The concept of local quasi-equilibrium becomes applicable and the production of the carriers is mainly due to thermal ionization. There is a complete change in the cathode mechanism.

We see that already for this simple geometry we have a wide variety of discharge types. The two most important ones are the glow and the arc discharge, each of them composed of a column and two electrode components.

We talked about these discharges in general terms without giving a clear-cut definition. It is indeed very difficult to assign a group of necessary and sufficient characteristics to the glow and the arc since there are so many different types. Also there are mixed phenomena which have for instance the electrode components of a glow discharge and the column of an arc.

In an attempt to give the essential characteristics let us propose the following definitions:

a *glow discharge* is a self-sustained conducting gas region between two electrodes. In the column the electrons have a much higher temperature than the ions and the neutrals. The carrier production in the column is due to electrons collisions. Particle destruction is dominated by diffusion and mobility losses to the wall and not by volume effects. For the cathode mechanism the mutual production of electrons by ion bombardment and photoeffect at the cathode and of ions by electron collisions in the negative glow is of vital importance;

an *arc discharge* is a self-sustained conducting gas zone between two electrodes. The carriers in the column are predominantly produced by thermal means. Particle destruction is dominated by volume recombination and not by drift and mobility losses. The electrode mechanism is distinguished from the cathode mechanism of the glow by very large current densities and a very small cathode-voltage drop. Temperature, field and individual electron emission at the cathode surface, thermal ionization and field ionization in the cathode spot are vital for the cathode mechanism of the arc.

In addition we further define:

the *electrode components* include the entire region of the arc, respectively, the entire region of the glow in so far as it is influenced by the presence of the electrodes. That means the entire region within which alterations are produced by an infinitesimal displacement of the electrodes.

With these definitions we can now state the regions of interest treated in these lectures:

first, we will consider the *electrode components of the arc.* This subject is interesting because it is typical for the problems and difficulties which occur in gas discharge theory. Also the electrode components of the arc show experimentally a number of striking phenomena which were not understood until very recently. We will show that a rather crude mathematical approach is able to explain most of the observations;

secondly, we will deal with the *column of the glow.* This region of interest allows a detailed analysis since it has the advantage of certain symmetry properties and relatively simple boundary conditions. Also it is of strong experimental interest in the studies of elementary processes, for instance of the diffusion in a magnetic field.

2. – List of symbols.

The following symbols with their subsequent meanings are chosen for the description of the observations and the formulation of the laws. For technical reasons different symbols had to be used in Sections **3** and **4**.

Section **3**:

$\boldsymbol{B}$ Magnetic induction.
c_0 Velocity of light.
d Extension of model region.

D	Diffusion coefficient.
e	Elementary charge.
E	Energy.
h	Plank's constant.
$\boldsymbol{i}$	Saturation current per unit area.
$\boldsymbol{I}$	Saturation current.
$\boldsymbol{j}$	Current per unit area.
$\boldsymbol{J}$	Current.
k	Boltzmann constant.
$\boldsymbol{k}$	Direction vector.
L	Energy loss of the electrode.
m	Mass of particles.
n	Number of particles per unit volume.
O	Point of existence.
p	Pressure.
q	Ratio of electron and ion current density.
r	Radial distance.
R	End contraction in front of the electrode.
T	Temperature.
V	Electric potential.
$\boldsymbol{v}$	Average mass velocity.
$\mathbf{X}$	Electric field.
β	Field enhancement factor.
γ	Coefficient of particle liberation.
δ	Dirac function.
φ	Work function.
μ	Mobility.
ω	Supply function.
Π	Current of energy.

Section **4**: $\boldsymbol{B}$, c_0, D, e, h, k, $\boldsymbol{k}$, m, n, p, r, T, V, $\mathbf{X}$, μ, have the same meaning as in Section **3**:

$\boldsymbol{v}$	Particle velocity.
$\boldsymbol{c}$	Relative particle velocity.
$\boldsymbol{v}_d$	Drift velocity.
R	Radius of the column.
I	Discharge current.
$\mathbf{E}$	Longitudinal electric field component.
U	Radial potential distribution.
α	Ionization coefficient for electron-neutral collisions.
β	Attachment coefficient.
δ	Detachment coefficient.
σ	Recombination coefficient for positive and negative ions.
Δ	General net production term.
$\mathbf{\Gamma}$	Particle current density.
ν	Collision frequency.
ω	Cyclotron frequency.
$\bar{\omega}$	Mode frequency.
η	Resistivity of a fully ionized gas.

$\sigma_\pm$, σ_0 Interaction parameters of positive ions with electrons.
ϱ Scattering function.
χ Scattering angle.
Ω Solid angle of the scattering parameters.
$\overleftrightarrow{\theta}$ Identity tensor.
$\bar{\alpha}$ Instability parameter.

3. – Electrode components of the arc discharge.

Fifty years of intensive experimental research have produced a vast amount of experimental details about the behaviour of the electrode components of an arc discharge. The material seems not always consistent. We summarize first those results which are sufficiently secured and distinct. These results we can expect to describe by a general theory.

The following features are characteristic:

the column of the arc constricts in front of both electrodes. In front of the cathode this constriction process causes an increase in the current density up to $(10^6 \div 10^7)$ A/cm². In general the constriction in front of the anode is smaller than that in front of the cathode;

the voltage requirement of the electric components of the arc is small of the order of 10 V. It is much smaller than the voltage requirement of the cathode region of the glow discharge.

There exist certain « critical values » for the pressure (p_k) and the current (I_k). With decreasing pressure the extension of the cathode spot of the arc increases gradually. When we reach the critical pressure p_k then the increase stops and discontinuously the onset changes to a highly constricted mode which moves erratically on the cathode surface. With increasing current of the arc the cathode spot also widens. When the critical current I_k is reached again the arc changes discontinuously to a completely different mode, the so-called arc without a cathode spot.

With very clean electrodes it is difficult to ignite an arc.

Finally, in a transverse magnetic field the cathodic part of the arc shows the unexpected phenomenon of « retrograde motion ». Depending on the pressure and the current, the onset at the cathode moves in a direction opposite to Ampère's rule.

From this experimental knowledge we formulate our theoretical problem in the following set of questions:

1) As is obvious from Fig. 2 a very large production of electrons and ions is required in the small cathode onset area. How are these particles

produced which provide the current continuity in front of the cathode and anode?

2) How is it possible that in spite of space-charge effects a current of $(10^6 \div 10^7)$ A/cm² can be transported with a small voltage like 10 V?

3) Why does the arc discharge contract at all in front of the electrodes?

4) We know that we have arcs with extreme constriction and with very little constriction. What are the parameters which define the degree of constriction?

5) Why is the contraction in front of the anode in general smaller than that at the cathode? Is there any theoretical indication for the micro-spot observed at pure carbon arcs with increased current density?

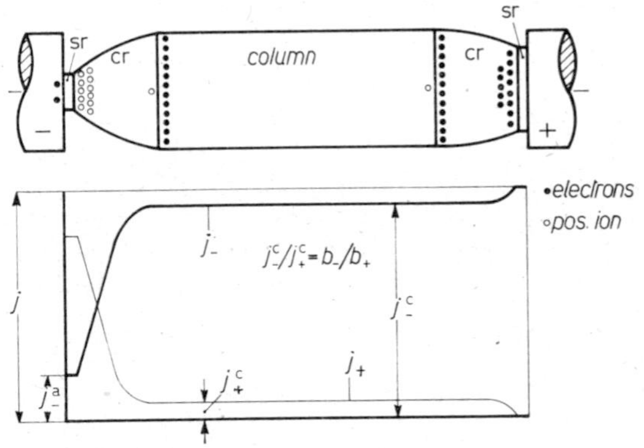

Fig. 2. – The current densities j_- and j_+ are given schematically as a function of position in the arc.

6) Is there a theoretical explanation of the existence of the critical pressure p_k? Can the discontinuity of the transition between the two arc types be understood?

7) Is there an explanation for the existence of the critical current I_k? Can the discontinuity of the change be explained here too?

8) Are there grounds for believing that the arc cannot exist under certain experimental conditions or can exist only if certain favourable surface conditions are present?

9) Is there a reason why the arc exhibits retrograde motion in a transverse magnetic field?

3·1. *Current continuity in front of the electrodes.* – The problem of current continuity in front of the cathode is the problem most widely studied in the theory of the arc cathode. Only two types of carriers really come under consideration for the current transport:

electrons freed from the cathode or ions produced in the gas area. The liberation of ions from the electrode is hardly of practical interest.

Electrons can be freed from the metal thermally (T-emission), by an external electric field (F-emission), by the combined effect of the above (T-F-emission), by the γ effect of single or multiply charged ions (γ_+-emission), by the γ effect of excited atoms (γ_m-emission) or by the photoeffect (γ_p-emission). Ions can be produced by electrons emitted from the cathode and accelerated in the space-charge zone (F-production of ions) or thermally (T-production of ions).

All these possibilities have been taken into consideration.

One can show that of all these effects only the T-F-emission and the T-production of ions play an important role.

The T-F-emission can be calculated from the equation

$$j_e = e\int_{E_{min}}^{\infty} P_e(X_c, E)\cdot\omega(E)\,\mathrm{d}E\,, \tag{3.1}$$

where P_e denotes the emission probability of an electron of energy E, $\omega(E)$ the supply function. Both functions are given in the work of MURPHY and GOOD (‡). There is no general analytical formulation of this emission current as a function of the electric field (X_c) but only approximate analytical formulae for very high or very small electric fields.

To judge whether the T-F-mechanism is able to explain the production of charge carriers in the cathode region one plots the current density j_e as given by Good and Murphy's formula *vs.* the cathodic field X_c multiplied by a factor β which accounts for the surface roughness. The result is shown in Fig. 3 for various values of cathode temperature and working potential.

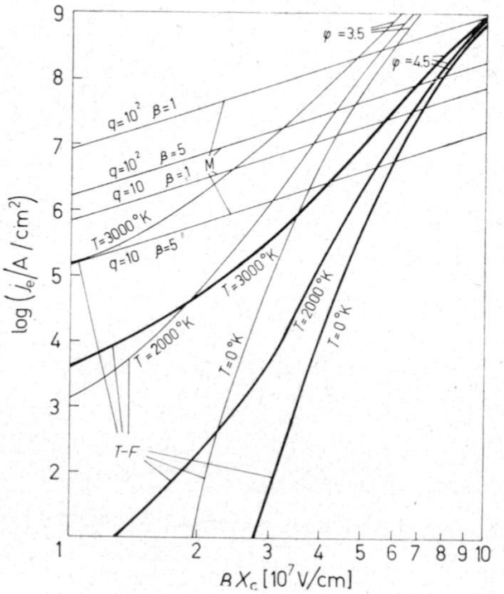

Fig. 3. – Electron emission density j_e of the cathode as a function of the temperature T and the effective field, $X_{ef}=\beta X_c$, (F-T-curves). The M curves show Mac Keown's equation as calculated for Hg for different values of $q=j_e/j_+$ and the field enhancement factor β.

(‡) G. ECKER: *Ergeb. d. Exakt. Naturwiss.*, **33**, 1 (1961). This comprehensive article also contains an extensive list of the literature and to simplify the references in the following the asterisk refers either to the article or the reference list.

Further one invokes the so-called MacKeown equation of bipolar space charge motion

(3.2)
$$X_c^2 = 7.6 \cdot 10^5 \cdot V_s^{\frac{1}{2}} \left\{ j_+ \left(\frac{m_+}{m_-} \right)^{\frac{1}{2}} - j_e \right\} [V^2/\text{cm}^2] \,,$$

where j_e is the emission current density at the cathode and V_s the potential across the space-charge region. V_s, X_c and j_+ have to be measured in V, V/cm and A/cm², respectively. MacKeowns formula is based on the assumption that electrons and ions move inertia-limited under the influence of their own space-charge field. The uncertainty included in the application of this formula to the cathode region of the arc is small since only the cathode drop itself which is sufficiently known enters the formula. MacKeown's law is also plotted in Fig. 3 assuming certain values of the parameters $q = j_e/j_+$ and β. The intersection points of the T-F- and M-curves define the value βX_c and the electron emission current density j_e for which the T-F-mechanism can operate.

As one sees from the figure the result depends strongly on the value of the parameter q and also on the roughness factor β. The value of q is limited by the assumption of a pure T-F-mechanism, since within the frame of this mechanism the electrons must produce the ions coming back to the cathode surface. Considering the energy balance and the ionization ability of the electrons emitted from the cathode it is easy to show that q must fulfil the condition $q \gg 1$.

Under these circumstances high current densities are required for the operation of the T-F-mechanism even if one assumes favourable values of the roughness factor β.

We conclude, that the T-F-emission can give a satisfactory explanation of the current transport in cases with extreme current densities at the cathode. But there are undoubtedly cases with less favourable conditions which cannot be understood on the basis of the T-F-mechanism. Some authors have therefore advocated the assumption that an important feature of electron emission is still not understood.

We ourselves in investigating this problem arrived at a new concept, the so-called « emission by individual field components » (I-F-emission).

On the cathode surface the field is produced by individual charge carriers. Due to this fact the field at a single point of the cathode surface shows statistical fluctuations with a mean value which can be calculated from Poisson's law.

If we wish to calculate the mean cathode emission density $\bar{j}_e$ we must have the emission current density as a function of the field X_c and after multiplying with the field probability $P(X_c)$ we average over all possible field values.

Thus we get

$$\bar{j}_e = \iiint P(X_c)\, j_e(X_c)\, \mathrm{d}X_c\,. \tag{3.3}$$

In general this $\bar{j}_e$ is not identical with $j_e(X_0)$ where X_0 is the mean value calculated from Poisson's law. The latter would be true only if the relation $j_e(X_c)$ could be approximated in a linear form which is certainly not possible for the Nordheim-Fowler law or the Murphy-Good relation.

The problem of the calculation of the probability distribution of the electric field at the cathode surface is a special case of the problem of « random flight ». The solution can therefore be given for large numbers of particles by the Markoff method as

$$P(\mathbf{X}_c) = \delta(X_x)\cdot\delta(X_y)\cdot\frac{1}{\pi X_0}\cdot\int_0^\infty \cos\,(vy + v^{\frac{3}{2}})\cdot\exp\,(-\,v^{\frac{3}{2}})\,\mathrm{d}v = \delta(X_x)\cdot\delta(X_y)\cdot\frac{1}{X_0}\cdot\phi(y). \tag{3.4}$$

in which the abbreviations

$$y = \frac{X_z - X_0}{X_0}\,, \qquad X_0 = e\pi\left(\frac{16}{15}\,n\right)^{\frac{2}{3}}, \tag{3.5}$$

have been used. These integrals can be traced to the solution of the confluent

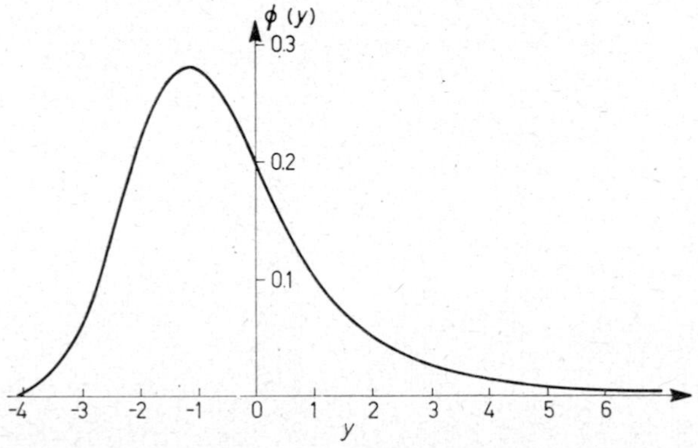

Fig. 4. – $\phi(y)$ is proportional to the probability distribution $P(X_c)$ of the field X_c at the cathode surface.

hyper-geometric equation and have been determined with the help of a computer, producing the results shown in Fig. 4.

From this field-probability distribution one evaluates the average emission density $\bar{j}_e$ which we call the *I-F*-emission density. The effect calculated in this way resembles somewhat the influence of surface roughness (β), however in distinction to β it is a completely fundamental and general effect and does not depend on special experimental conditions.

Using a procedure similar to that demonstrated in Fig. 3, we can relate the current density j_e to the value of q required for the *F*-, *T-F*- or *I-F*-mechanism. The result is shown in Fig. 5. Since—as we have already stated—the parameter q must fulfil the condition $q \gg 1$, we see that the *I-F*-mechanism is applicable much further down to lower values of the current density j than the *T-F* or the *F*-mechanism.

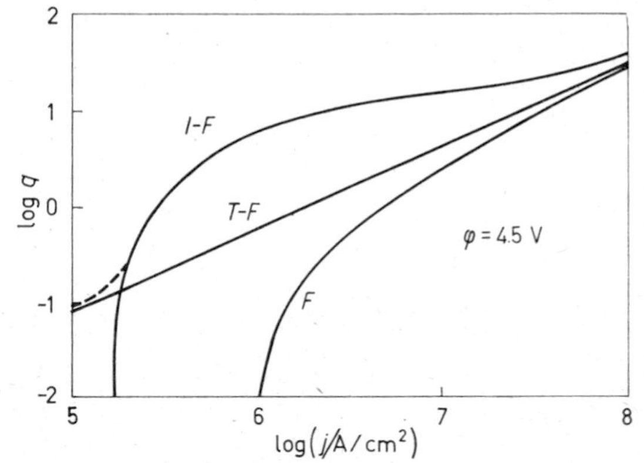

Fig. 5. – $q = j_e/j_+$ plotted *vs.* the current density j calculated for the *F*, *T-F* or *I-F* mechanism.

In the preceding we have already used the concept of the *space-charge zone*. This is one of the model regions of our problem. The others are the

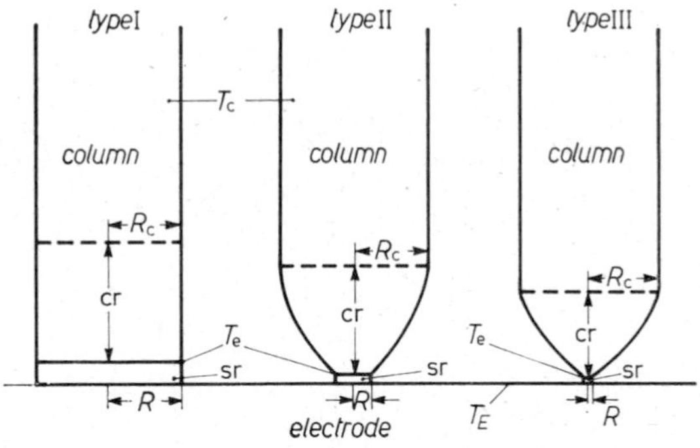

Fig. 6. – This figure shows schematically the following model zones: column, cr-contraction zone, sr-inertia-limited region. It also indicates the meaning of the symbols R_0, R, T_0, T_s, T_c. The sketch is given for three values of end contraction R (types I-III) which play a part in the theoretical evaluations.

contraction region (cr) which is essential for the *T*-production of ions and the *inertia-limited zone* (sr).

In Fig. 6 we show schematically these model regions for three typical cases.

The ion current to the cathode surface (I_+) can be estimated from Langmuir's saturation current

$$I_{+} = e \cdot \pi \cdot R^{2} \cdot \frac{n_{+} \cdot \bar{c}_{+}}{4} \tag{3.6}$$

calculated at the boundary of the contraction region (s). To find the particle density $(n_{+})_{s}$ and the average velocity $(c_{+})_{s}$ it is necessary to know the temperature close to the boundary (s) of the contraction region. This in turn requires the evaluation of the energy balance for the whole cathode region of the arc.

This is an extremely difficult problem, the details of which are given elsewhere (‡). We sketch here only the basic assumptions and equations.

The calculation of the contraction zone follows in principle the procedures of the column theory of Elenbaas and Heller (*) and like the latter it assumes quasi-thermal equilibrium and quasi-neutrality. With the basic equations of electrodynamics, the energy balance, the equations of carrier motion, the Saha equation and the radiation laws we have under the assumptions outlined above the following relations

$$\left\{\begin{array}{l} \boldsymbol{\nabla} \times \boldsymbol{X} = 0, \quad (\boldsymbol{\nabla} \cdot \boldsymbol{X}) = 0, \quad (\boldsymbol{\nabla} \cdot \boldsymbol{j}) = 0, \quad \boldsymbol{j} = \boldsymbol{j}_{-} + \boldsymbol{j}_{+}, \quad n_{-} \simeq n_{+} = n\,, \\ \boldsymbol{j} \simeq \boldsymbol{j}_{-} = e\mu_{-} n \boldsymbol{X} + eD_{-} \boldsymbol{\nabla} n, \qquad \boldsymbol{X} \cdot \boldsymbol{j} = S - \boldsymbol{\nabla}(\lambda\, \boldsymbol{\nabla} T) - V_{i} \cdot (\boldsymbol{\nabla} \cdot \boldsymbol{j}_{-})\,, \\ \dfrac{n_{-}^{2}}{n_{0}^{2} - n_{-}^{2}} = \dfrac{k^{\frac{5}{2}}}{h^{3}} (2\pi m)^{\frac{3}{2}} \cdot \dfrac{T^{\frac{5}{2}}}{p} \cdot \exp(-\, eV_{i}/kT)\,, \\ S \propto \exp(-\, eV/kT) \qquad \text{resp.} \qquad S \propto T^{4}\,. \end{array}\right. \tag{3.7}$$

The laws of free fall and space-charge limitation govern the energy balance of the inertia-limited zone (ILZ).

The conditions in the electrode are defined by the laws of energy conservation and heat conduction.

Energy is exchanged between the contraction region and the electrode via the inertia-limited region.

The solution of this problem produces the temperature at the end of the contraction region (T_{s}) and the temperature in the cathode onset (T_{c}) as a function of the contraction parameter R_{0}/R, where R_{0} is the radius of the column. Figures 7a) and b) show the ion saturation current I_{+} and the temperature T_{s} for various currents and pressures of a representative high pressure mercury discharge assuming a constant cathode temperature T_{c}. The essential phenomenon is that the ion saturation current shows a maximum and disappears for very low contractions and very high contractions. We will rec-

(*) W. Finkelnburg and H. Maecker: *Handb. der Phys.*, Vol. **22** (Berlin-Göttingen-Heidelberg, 1956), p. 2.

ognize the importance of this result in the following section on the existence diagram.

As may be seen from Fig. 2 the ratio of the number of carriers to be

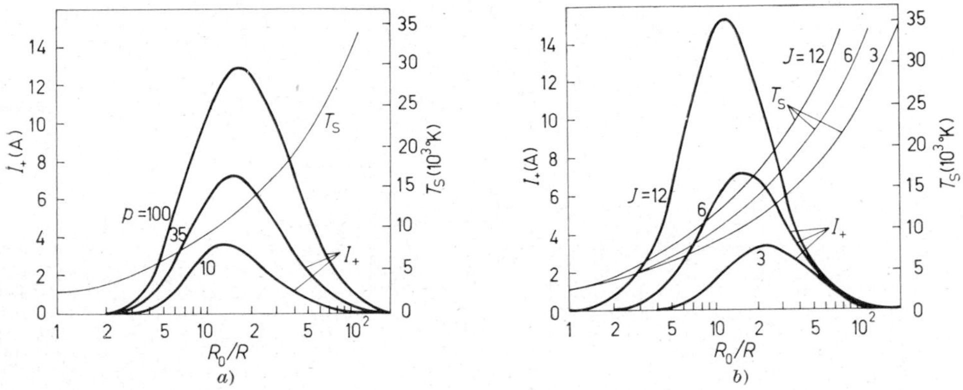

Fig. 7. – Ion saturation current I_+ for the representative Hg discharge plotted *vs.* R_0/R for various pressures (10, 35, 100 atm) and various currents (3,6 and 12 A). Also the temperature T_s is given.

produced in the anode region to those to be produced in the cathode region is μ_+/μ_-. Therefore in front of the anode the question of current continuity is less problematic and interesting. Further a considerable simplification arises from the lack of any emission process at the anode if we disregard, for the time being, the Beck-arc. Thus the sections corresponding to the *F*-, *T*-*F*- and *I*-*F*-emission at the cathode have no equivalent at the anode.

F-ionization is of little importance in front of the anode (‡).

The problem of *T*-carrier production in front of the anode is basically the same as in front of the cathode. Again we have the question of calculating the saturation current, (now I_-), flowing from the gas to the anode. Exactly the same laws hold for the contraction zone. Complete symmetry is broken only by the different masses and the charge sign of the particles carrying the saturation currents I_+ or I_- respectively. Since for quasi-neutrality the particle densities of the electrons and ions are the same and since the average thermal velocities vary proportionally to the square root of the mass we have $(I_+\sqrt{m_+})_{\text{cath}} = (I_-\sqrt{m_-})_{\text{an}}$. Therefore we can apply the curves of Fig. 7 and 8 directly to the anode if we multiply the ordinates by the factor $\sqrt{m_+/m_-}$.

3·2. *Voltage requirement of the electrode components.* – The voltage drop of the electrode components of the arc is composed of two essentially different parts.

The one (V_{0s}) belongs to the contraction region. It is caused by the energy losses to the electrode on the one hand and the increased current density on the other hand. It depends strongly on the value of the end contraction and is found automatically from the solution of the problem of the contraction region described in the preceding paragraph. The result is shown in Fig. 8.

The other contribution comes from the sheath immediately in front of the electrode where the concept of quasi-neutrality and the diffusion and mobility laws do not hold. Our discussion of this region starts by distinguishing four separate cases according to whether or not a one-dimensional or collision-free description is possible.

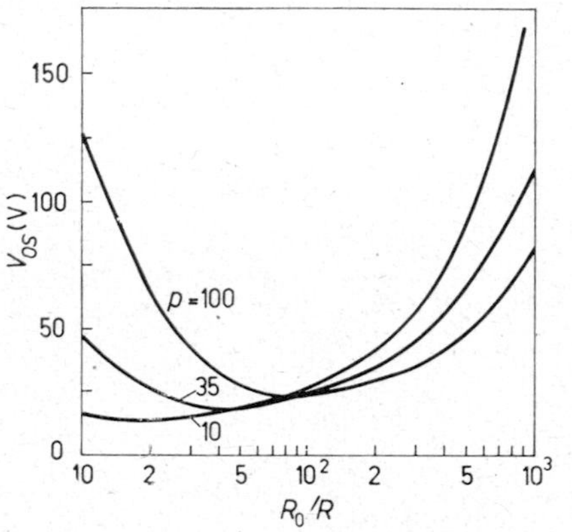

Fig. 8. – Voltage requirement V_{0s} of the contraction zone of the representative Hg discharge.

The cases are:

($a\alpha$) one-dimensional collision-free motion;

($a\beta$) one-dimensional motion with collision;

($b\alpha$) multi-dimensional problem with collisions;

($b\beta$) multi-dimensional problem with collision-free motion.

($a\alpha$) is by far the most interesting case. It has been investigated almost exclusively and only for this case is there a satisfactory theory. ($a\alpha$) dominates the entire experimental area except for extremely large current densities with small pressures or very small current densities. For the other three cases only qualitative pictures have been developed (‡).

The problem ($a\alpha$) is governed by the laws of bipolar space-charge movement. The general formulation can be set down in the equations

$$(3.8)\qquad \begin{cases} j_- = h_1^2(1, \gamma_1)\cdot \dfrac{1}{9\pi}\sqrt{\dfrac{2e}{m_-}}\cdot\dfrac{V_s^{\frac{3}{2}}}{d_s^2}\,, \\[2ex] j_+ = \gamma_1 h_1^2(1, \gamma_1)\cdot\dfrac{1}{9\pi}\sqrt{\dfrac{2e}{m_+}}\cdot\dfrac{V_s^{\frac{3}{2}}}{d_s^2}\,, \qquad\qquad \gamma_1 = \sqrt{m_+/m_-}\cdot q\,, \\[2ex] h_1(1, \gamma_1) = \dfrac{3}{4}\displaystyle\int_0^1 \dfrac{dy}{\sqrt{\sqrt{y} + \gamma_1(\sqrt{1-y} - 1)}}\,. \end{cases}$$

If, in particular, $j_-/j_+ \ll m_+/m_-$—and this is almost always true—then the general law reduces to the simple Langmuir $-(\frac{3}{2})$-law.

With the help of the eqs. (3.8) we can calculate the potential drop across the inertia-limited zone (ILZ) using the experimental knowledge of the current density and of the extension of the ILZ. The results show deviations from the experimental data which are outside of the uncertainties of the parameters and the experiments.

To remove these difficulties we have argued that electrons enter the inertia-limited zone not only from the cathode but also from the contraction zone. Although these latter electrons do not contribute to the current transport, they do advance towards the electrode and in this way compensate a part of the space-charge of the ions (counter field diffusion).

To assess this effect it is important to remember that the electrons themselves alter the potential distribution. Otherwise one might come to the inaccurate conclusion that because of their small thermal energy (order of magnitude 1 eV) the electrons could not approach the cathode to any appreciable extent.

The differential equation describing the potential distribution in the inertia-limited region is

$$\frac{d^2V}{dx^2} = -4\pi\frac{j_+}{\sqrt{kT_s/m_+}}\left\{\frac{1}{\sqrt{1-2eV/kT_s}} - \exp[eV/kT_s]\right\}, \tag{3.9}$$

or after integrating twice

$$d_s = -\int\limits_{V=0}^{V=V_s} \frac{dV}{\sqrt{(8\pi j_+\sqrt{kT_s m_+}/e\,\{(\sqrt{1-2eV/kT_s}-1)+(\exp[eV/kT_s]-1)\}+X_0^2}}, \tag{3.10}$$

where X_0 is the field at the entrance of the ILZ. The evaluation of (3.10) has been done graphically.

The counter field diffusion of the electrons reduces the voltage requirement of the inertia-limited region very much. For instance for our representative high-pressure discharge we require 6 instead of 19 V according to Langmuir's theory. For other experimental conditions a reduction results of as much as from 70 to 12 V. What is more, the result now becomes practically independent of the extension of the ILZ in contrast to the Langmuir theory, where the voltage varies nearly proportionally to the extension. Therefore the new data are more reliable.

Of course our result means that the extension of the space-charge region is not identical with the extension of the inertia-limited zone. There is an effective space-charge region which is not influenced by changes in the extension of the inertia-limited zone.

3·3. *The existence diagram and its consequences.* – Applying the knowledge accumulated so far, we are able to answer the questions 3) to 8) regarding the contraction of the arc discharge in front of the electrodes.

We claim that the contraction itself and the extension of the contraction follows simply from the application of the physical laws of electrodynamics, statistical mechanics and quantum mechanics to our problem.

To prove this assertion we make use of our preceding calculations for the cathode, the inertia-limited zone and the contraction region. As a result we have already the electron current J_e and the saturation current I_+ (resp. I_-).

One of the important laws of electrodynamics is of course the law of charge conservation which in our special case requires that the total current be the sum of the saturation current and the emission current

$$J = I_+ + J_e \,. \tag{3.11}$$

For a possible stationary state of the arc this condition definitely must be fulfilled. So our concept is to plot the saturation current and the electrode emission current as a function of the contraction parameter R_0/R to see whether this condition is met for all values of end contraction, or only for a limited number, or for no values of the end contraction at all. This is the scheme of the existence diagram (E-diagram).

Unfortunately our preceding calculations are not yet general enough since they underlied the assumption of an independent electrode temperature T_c. For the following more precise evaluations we have to define the temperature at the end of the contraction region T_s *and* the temperature T_c of the cathode from the energy balance. This can be done as follows.

If Π_{sc} is the effective energy supply from the end of the contraction zone to the electrode then this is a function of T_s, T_c and R; *e.g.* $\Pi_{sc}(T_s, T_c, R)$. This energy supply must be compensated by a corresponding loss L of the electrode. Therefore we have the condition

$$\Pi_{sc}(T_s, T_c, R) = L(p_\nu) \tag{3.12}$$

where p_ν are the parameters governing the electrode energy losses.

A second relation between T_s and T_c results from the requirement of the continuity of energy flow at the boundary (s), contraction zone—inertia-limited zone. In accordance with the laws of the conduction of gases a certain energy current flows to this boundary from inside of the contraction zone (Π_{0s}). This energy current must be transported to the electrode by collision-free energy transport, Π_{sc}, in the ILZ. This condition gives the second relation

$$\Pi_{0s}(T_s, R) = \Pi_{sc}(T_s, T_c, R) \tag{3.13}$$

and from (3.12) and (3.13) we may determine T_s and T_c as a function of R

and p_ν. With this one obtains

$$(3.14) \qquad I_+ = \pi R^2 i_+ , \qquad I_- = \pi R^2 i_- , \qquad J_e = \pi R^2 j_e .$$

From (3.7), (3.6) and Fig. 3 one can now construct the existence diagram.

However clear this general plan may be, it turns out to be extremely difficult to carry it through, since so many laws of a very difficult nature enter into the balances (3.12) and (3.13). We succeeded in solving approximately the problem for the case of the typical high-pressure mercury discharge making certain assumptions about the shape and the quality of the electrodes and the loss processes (‡). We content ourselves to present here the results.

The general shape of the curves $T_s(R)$ corresponds to the curves given in Fig. (7) and (8). The electrode temperature $T_c(R)$ is shown in Fig. 9 for various currents J and gas temperatures T_s. The dependence of the temperature T_c on the parameter R_0/R is open to a physical interpretation on the basis of heat-conduction and electron cooling effects. Having so calculated T_c, T_s, we find $I_+(R)$ and $J_e(R)$ and are now in a position to study the question of charge conservation. For this purpose we plot the diagram showing on the one hand the ion saturation current of the gas, $I_+(R)$ (resp. $I_-(R)$), and on the other the defect current, $J - J_e(R)$ as a function of the contraction parameter R.

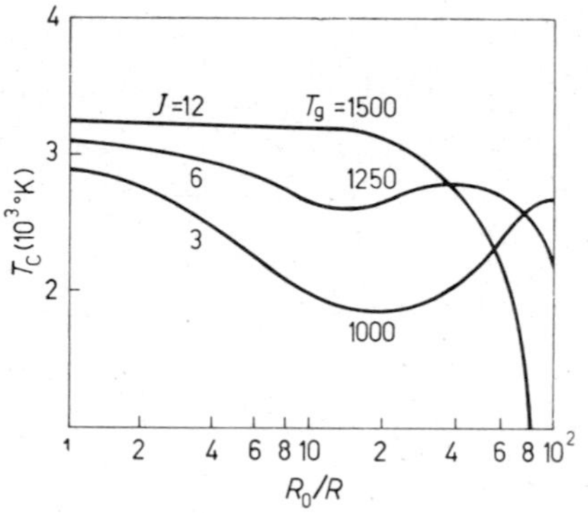

Fig. 9. – Cathode temperature, T_c, of the representative Hg discharge as a function of the end contraction R_0/R. T_g is the gas temperature in the neighbourhood of the cathode outside the contact area.

As was already stated we call this whole arrangement the « diagram of existence », that is the « E-diagram ». We choose this term because the points of intersection of I_+ and $J - J_e$, the points of existence or the E-points, belong to those values of R for which the arc can exist in a stationary state.

The E-diagram is extremely revealing and provides the answers to the questions 3)–8):

Answ. 3) The arc must contract in front of the cathode since otherwise the laws of electrodynamics, statistical mechanics and quantum mechanicscannot be fulfilled simultaneously. This is quite clear from Fig. 10. If we have already taken into account all laws except that of charge conservation, than this last requirement can be met only for definite values of the end contraction, R. Therefore the contraction is not the result of a minimum principle, but an essential physical necessity to satisfy the laws of nature.

Generally the E-diagram can show up to four E-points. Which one of the several points is stable depends on the growth of small deviations in the external and internal parameters. We have judged the question of stability using Steenbeck's principle of minimum voltage requirement which (within certain limits) Maecker and Peters traced back to the requirement of minimum entropy production of irreversible thermodynamics.

Answ. 4) Our I_+ and $J-J_e$ curves depend on the experimental parameters. Consequently the position of the E-points varies with the experimental conditions. Moreover it depends on the voltage requirement which of the E-points is stable. So the stable point may be one with an extreme end contraction or with very little end contraction. One can easily conceive that all possible current densities have been measured with different experimental arrangements. Different measurement results are not necessarily contradictory, but can exist side by side and can be explained on the basis of our theory.

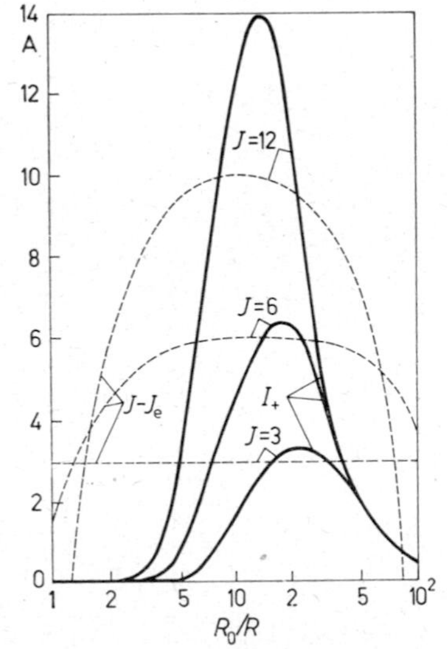

Fig. 10. – E-diagram of the representative Hg discharge for various currents J.

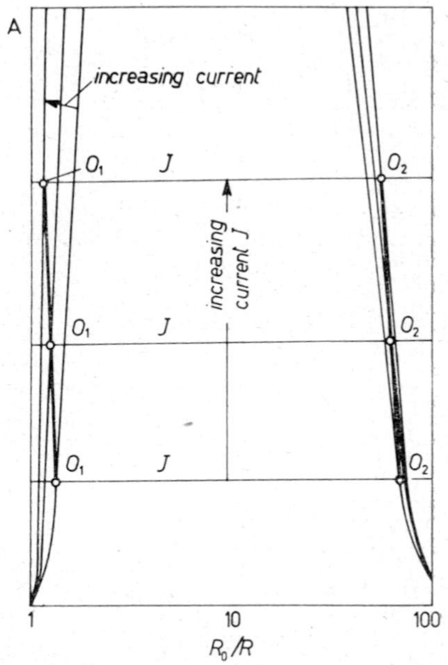

Fig. 11. – Schematical E-diagram of the anode.

Answ. 5) We argued above that for the ion saturation current in front of the cathode (I_+) and the electron saturation current in front of the anode (I_-) the relation

$$I_+ \cdot \sqrt{m_+} = I_- \cdot \sqrt{m_-} \tag{3.15}$$

holds. Consequently the E-diagram at the anode is demonstrated in Fig. 11. Because of the large scale it is not possible to reproduce the whole curves. However one can imagine the curves as extended in accordance with Fig. 10. The essential feature is, that in any case a point of intersection occurs in the range of weak end contraction ($R_0/R \approx 1$). In addition we have only one other E-point which belongs to extreme contraction. The point with very little contraction (O_1) is generally favoured by the minimum principle. Thus the weaker contraction in front of the anode can be understood without additional assumption, simply from the different mass of the carriers of the saturation current. Further conclusions can be drawn from the fact that certain circumstances may arise in which the conditions for the voltage may be otherwise. This is of interest for the phenomenon of the microspot at the anode. If for example the arc is compelled because of radial limitation to increase its current density then the voltage requirement will have to increase causing vaporization and with that a further increase in voltage requirement. Therefore with increasing current the E-point with weak contraction (O_1) might well require more voltage than the E-point with strong contraction (O_2). Under these circumstances, then, the arc will change to the strongly contracted « microspot mode » (O_2). It is true, that in the microspot too there is considerable vaporization, but the microspot is able to avoid this vapour jet because of its small radial extension. It joins the anode at the side of the origin of the jet. It is particularly satisfying to know that the conclusions drawn from the E-diagram are in agreement with the experimental results for both the cathode as well as the anode.

Answ. 6) The existence of a critical pressure p_k necessarily follows from the E-diagram of Fig. 12. If, for example we start from point (O_1) and lower the pressure, than we move from (O_1) to (O_2). Simultaneously and in accordance with experiment the contraction, R_0/R, decreases. When we reach the point (O_2) the possibility of existence near contractions of the order (O_2) lapses

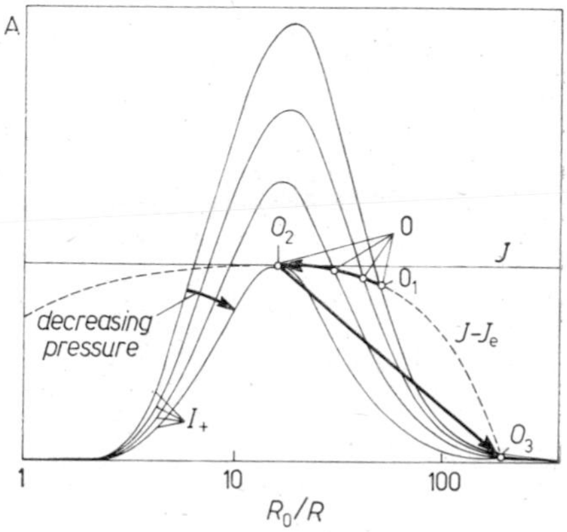

Fig. 12. – E-diagram which explains the existence of the critical pressure p_k.

with the lowering of the pressure and the arc must of necessity change discontinuously to the E-point (O_3). It is then a field arc with much stronger contraction. This theoretical result is in complete accord with the experimental findings.

Answ. 7) The discontinuous change from an arc with spot to an arc without spot at a given current J_k is also an obvious result of our E-diagram. If the phenomenon of the arc without cathode spot had not already been observed, we would have to infer it from the calculations. To indicate the process quite clearly we repeat in Fig. 13 the curves of Fig. 10 which are essential for our purpose. According to Fig. 13 there suddenly appears with increasing current an additional E-point at $R_0/R \simeq 1$,—that is, without contraction. It is caused by the high-temperature thermal electron emission from the cathode. Thus with an increasing current the arc first goes from O_1 to O_2 and then changes discontinuously to O_3. The change occurs because O_3 is favoured by the minimum-principle.

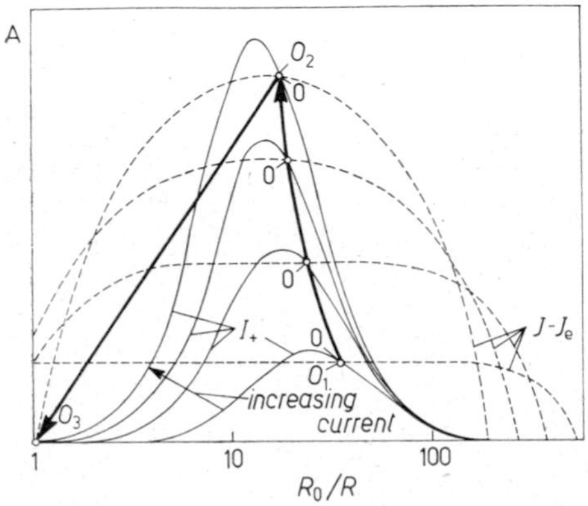

Fig. 13. – E-diagram which explains the critical current J_k.

Answ. 8) One can imagine experimental conditions for which the E-curves would show the form given in Fig. 14. In this case there is only one « E-point » in the region with extreme contraction. However this E-point is only possible, if certain assumptions are fulfilled. We must remember that our calculation of I_+ was based on thermal ionization, which is negligible if R is of the order of the mean free path of the electrons λ_-: This limitation $R \gg \lambda_-$ is indicated in Fig. 14 by the subdivision in the zone III a, III b. In III b the thermal ion component disappears. If an arc is to exist in the zone III b one must first show that the field ionization can produce a sufficient quantity of ions for the build-up of

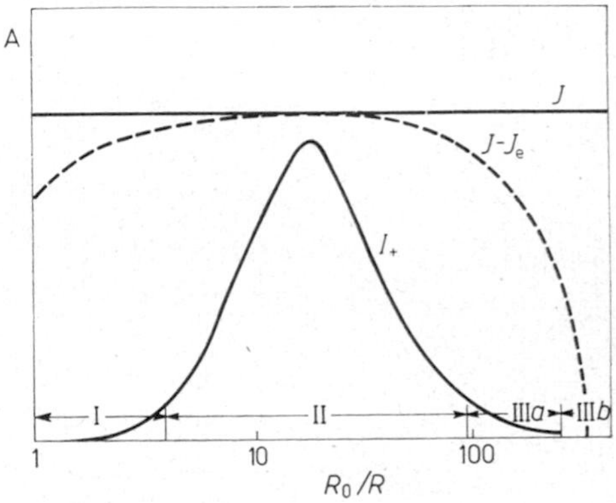

Fig. 14. – E-diagram demonstrating the ignition difficulties of arcs with very pure electrodes.

the cathode-field. As we know this requirement is tied up with the condition $q \gg 1$. However, under this condition the mechanism of T-F or I-F-field electron emission leads to difficulties. Only if the conditions of the surface favour the process of emission is there any hope of igniting the arc with extreme contraction. If these favourable influences are not present then it may be impossible to ignite a stationary arc discharge.

This last discussion leads us to our final *question* 9) related to the « problem of retrograde motion ». The reason is that a critical discussion (‡) of all possible contraction types shows that retrograde motion can be expected only in an arc of the type III b, that means an arc where the condition $R \simeq \lambda_-$ is met.

Under these circumstances we find in the space-charge region conditions of the type $(b\beta)$. Because of the thermal motion, ions and electrons leave the discharge zone. In a stationary state the loss of ions and electrons must be equal. For this a radial potential fall V_R is required. (eV_R) must be at least of the magnitude of the thermal energy of the electrons. Thereby the electrons are held back and can still compensate an appreciable portion of the space charge of the ions.

If this were not the case, then either the cathode fall would have to be of the order of several thousand volts or there would have to be potentials of the same magnitude over the cross-section of the discharge, as can be seen from the evaluation of the space-charge problem.

We consequently get the following picture for the cathode onset III b:

the space charge region extends beyond the inertia-limited zone. An ion stream of high density enters the ILZ from the end of the contraction zone. The space charge of the ions is not compensated. A certain portion of the field lines leaves the discharge channel radially and in this way forms a potential tube. The depth of the potential tube must be at least several electronvolts but cannot be predicted more accurately. The electrons emitted at the cathode with little energy must necessarily follow this potential tube, independently of the direction of the emission. It is assumed that because of the importance of the space-charge and the ion bombardment the center of electron emission lies in the contact area of the ion tube with the cathode.

On the basis of this model let us try to understand the influence of a transverse magnetic field. For a description of the ion motion we use the Lorentz equations

$$(3.16)\qquad \begin{cases} \dfrac{\partial n^+}{\partial t} + \boldsymbol{\nabla}(n_+ \boldsymbol{v}_+) = 0\,, \\[2ex] m_+ \cdot \left(\dfrac{\partial}{\partial t} + \boldsymbol{v}_+ \cdot \boldsymbol{\nabla}\right) \boldsymbol{v}_+ = e\left\{\boldsymbol{X} + \dfrac{\boldsymbol{v}_+ \times \boldsymbol{B}}{c_0}\right\} + \boldsymbol{K}_-\,, \end{cases}$$

and the Poisson equation

$$\nabla \cdot \boldsymbol{X} = 4\pi(n_+ - n_-)e \,. \tag{3.17}$$

c_0 is the velocity of light.

The term K_- accounts for the radial coupling of the ions and electrons through the potential tube. In other words the electrons transfer their Lorentz force to the ions. From this we get

$$K_- = \frac{n_-}{n_+} \cdot e \cdot \frac{\boldsymbol{v}_- \times \boldsymbol{B}}{c_0} \,. \tag{3.18}$$

Using

$$q = j_-/j_+ \,, \tag{3.19}$$

in (3.16) and (3.18) we find

$$m^+ \frac{\mathrm{d}\boldsymbol{v}_+}{\mathrm{d}t} = e \left\{ \boldsymbol{X} + \frac{\boldsymbol{v} \times \boldsymbol{B}}{c_0} (1 + q) \right\} . \tag{3.20}$$

In a certain approximation we have solved the problem represented by the differential equations (3.16)-(3.20) and found the path of the ions in the inertia-limited region.

Within the frame of the above picture of the cathode region the electrons travel back from the cathode through the potential tube of the ions and enter the gas on the « retrograde side » of the cathode spot. Here they produce a new ion cloud by ionization and these ions move back to the cathode being deflected in the correct Lorentz direction. Again electrons come from the cathode, are forced by the potential tube to travel in the direction opposite to the Lorentz direction and create a new ion cloud on the « retrograde side » of the cathode spot. By this process the cathode spot can move in the « counter-Lorentz direction ».

Before we show the experimental results in comparison to the theory let us once more state clearly the physically important fact. If we had electrons or ions moving independently to the cathode surface, the deflection would be always too small to explain the experimental data. The reason is that due to their high velocities the electrons undergo a strong Lorentz force, however just for this reason they stay in the inertia-limited zone only for a very short time. On the other hand the ions stay in the inertia-limited zone for a longer time because they have a smaller velocity. However, since they have a smaller velocity they undergo only a small Lorentz force. Therefore the coupling between the electrons and ions due to the radial potential of the space-charge tube is decisive. It transfers the strong Lorentz force experienced by the electrons to the slowly moving ions and in that way creates a large deflection.

The Fig. 15 and 16 compare the theoretical results with the experimental data. The first two Fig. 15 *a*) and *b*) show the retrograde velocity as a function of the magnetic field and pressure for different values of the current. A number of parameters which are not adequately known from experiments are included in the theoretical description. To avoid the uncertainty of these parameters we have adjusted the theoretical curves by setting them equal with the experimental curves at the point indicated in the figures by a circle. Obviously the theoretical dependence is in satisfactory agreement with the experimental data.

The point at which the normal velocity turns back to the retrograde velocity, that means the point where the spot is at rest, is called the critical point and is indicated by the index (0). Figures 16 *a*) and *b*) give the relation between critical magnetic field, current and pressure and at the same time the theoretical data. Again there is very satisfactory agreement between experiment and theory.

With that we have answered on the basis of one theoretical concept all the questions which we used to characterize the arc electrode problem. It was essential that we restricted

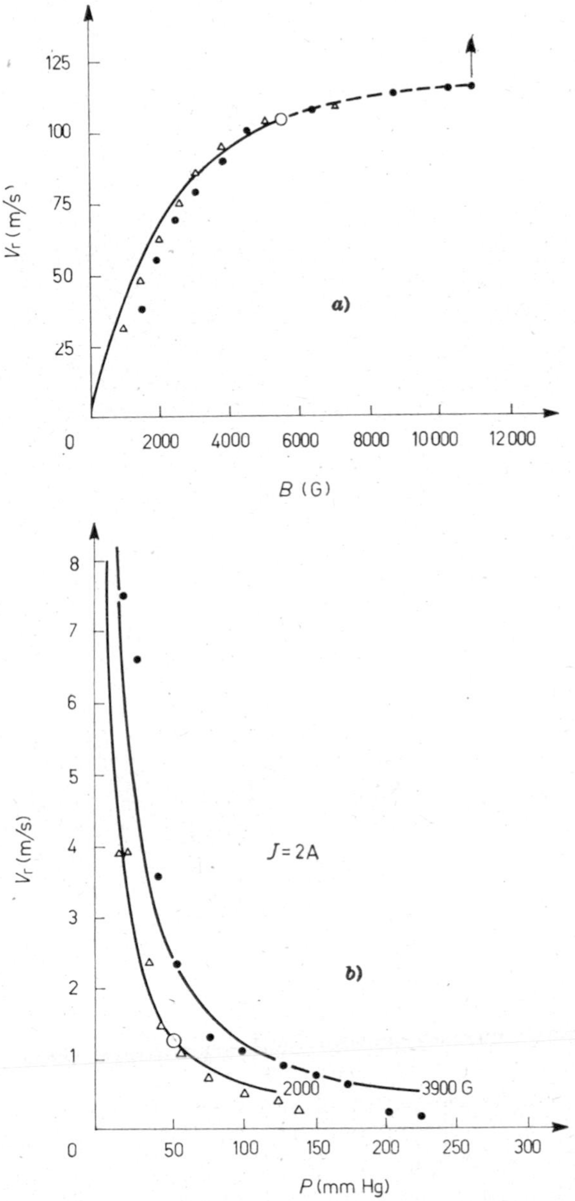

Fig. 15. – Retrograde velocity V_r as a function of magnetic induction B. The points give typical measurements, the solid lines are the results of the theory: ● $J = 2.0$ A; △ $J = 5.5$ A.

our questions to features sufficiently distinct to be explainable by a theory of the accuracy possible with a problem so complicated like a gas discharge.

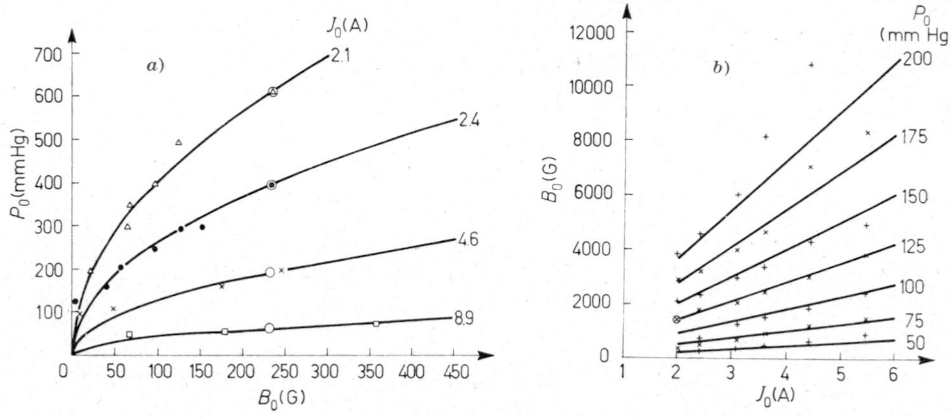

Fig. 16. – The mutual dependence of the critical data of the reversal point (0) as taken from measurements (points) and as calculated from theory (solid lines).

One more concluding remark. We believe that many of the seeming discrepancies in the experimental work can be easily explained and understood taking into account the different experimental conditions.

4. – The collision-dominated cylindrical plasma column.

The collision-dominated cylindrical plasma column occurs in many gas discharges. It is also of importance for the studies of elementary processes in gaseous electronics. Just in this latter application the cylindrical plasma column has been recently used frequently to investigate the problem of « enhanced diffusion » in an external magnetic field.

The problem of the plasma column, principally, is different from that of the « electrode components of the arc ». In the latter case we have a large number of model regions in our region of interest. For the column we have practically only one model region, since the sheath close to the wall of the container enters our considerations only as a boundary condition. The fact that we have only one model simplifies the situation. On the other hand for the applications of the column theory a higher degree of accuracy is required and the set of differential equations describing the physical situation together with the boundary conditions confronts us with a difficult mathematical problem.

The collision-dominated cylindrical plasma column may exhibit many different features and qualities depending on the experimental conditions. The phenomena change with increasing energy input from the « nonthermal glow column » to the « arc column ». They also depend decisively on the mechanisms of particle production and destruction and on the boundary conditions. Different production mechanisms are for instance present in the normal glow column (electron collisions) and in the contact-ionized cesium discharge and the hollow cathode discharge (external production). Different destruction processes occur, *e.g.*, in the normal glow column, in gases with a negative ion component and in the arc column (volume recombination).

We decide to consider the three following problems:

1) The weakly ionized, nonthermal, four-component system (including negative ions) without a magnetic field in the whole normal and subnormal region.

2) The three-component, quasi-neutral system in a magnetic field including temperature variation and external particle production.

3) The instability of a weakly ionized, quasi-neutral three-component system in a magnetic field.

All these problems are described by the same set of differential equations and it seems therefore reasonable to begin with the description of these equations.

As usual our mathematical treatment is based on the transport equations, which for an arbitrary quantity $V(v)$ read

$$(4.1)\qquad \frac{\partial}{\partial t}(n\overline{V}) + \boldsymbol{\nabla}(n\overline{\boldsymbol{v}V}) - \frac{e}{m}\,n\overline{\boldsymbol{X}\cdot\boldsymbol{\nabla}_v V} + \frac{e}{m}\,n\overline{\boldsymbol{B}\times\boldsymbol{v}\cdot\boldsymbol{\nabla}_v V} = \sum_i \int_{(8)} (V'-V) f_i(\boldsymbol{v}_i)\cdot f(\boldsymbol{v})\, c_i \varrho_i \, \mathrm{d}v^3\, \mathrm{d}v_i^3\, \mathrm{d}^2\Omega\,,$$

where f is the distribution function, v the velocity, c the relative velocity and ϱ the scattering function. The quantities without index refer to the particle under consideration, the index (i) to the particle components with which the test particle collides. The prime $(')$ characterizes quantities after collision. The bar indicates the velocity average (v).

As is well known, the law of particle conservation follows from eq. (4.1) by substituting $V=1$, where ionizing and recombining collisions may be taken into account by using $V'=2$ and $V'=0$ resp. The continuity equation reads

$$(4.2)\qquad \operatorname{div}\Gamma + \frac{\partial n}{\partial t} = (\nu_{\text{ion}} - \nu_{\text{rec}})\,n = \Delta$$

where the coefficient Δ gives the net particle production and Γ the particle current density.

The law of conservation of momentum follows from eq. (4.1) if we identify V with $m\vec{v}$. We have

$$\frac{\partial \mathbf{\Gamma}}{\partial t} + \mathbf{\nabla}_{\gamma}(n\overline{\overleftarrow{v}\vec{v}}) - \frac{e}{m}\mathbf{X}n - \mathbf{\Gamma}\times\mathbf{B}\frac{e}{m} = \sum_i \int_{(8)} (\vec{v}' - \vec{v})c_i\varrho_i(c_i)f_i(v_i)f(v)\,\mathrm{d}v_i^3\,\mathrm{d}v^3\,\mathrm{d}^2\Omega \tag{4.3}$$

where the tensor $\overline{nm\overleftarrow{v}\vec{v}}$ is related to the normal pressure tensor $\overleftrightarrow{p}$ by

$$\overline{nm\overleftarrow{v}\vec{v}} = \mathbf{P} = \overleftrightarrow{p} + \overleftarrow{v}_d\,\vec{v}_d\,m\,. \tag{4.4}$$

A difficulty arises in the evaluation of the collision terms on the right-hand side. This is shown by a simple transformation

$$\int_{(3)}\int_{(3)}\int_{(2)} (\mathbf{V}' - \mathbf{v})c_i\varrho_i(c_i)f_i(v_i)f(v)\,\mathrm{d}^3v_i\,\mathrm{d}^3v\,\mathrm{d}^2\Omega = \int_{(3)}\int_{(3)} (v_i - v)\frac{\nu_i(c_i)}{n_i}f_i(v_i)f(v)\,\mathrm{d}^3v_i\,\mathrm{d}^3v, \tag{4.5}$$

introducing the collision frequency

$$\nu_i(c_i) = \int_{(2)} n_i\varrho_i(1 - \cos\chi)\,\mathrm{d}^2\Omega\,, \tag{4.6}$$

where χ is the deflection angle in the laboratory system. We see that for the collision integral simple expressions in terms of the current density occur only if we can neglect the dependence of the collision frequency ν_i on the relative velocity c_i. This can be justified to some extent for the interaction of ions with neutrals (Maxwell r^{-5} interaction) or for very light particles colliding with heavy particles. In all other cases however the assumption of a collision frequency independent of the relative velocity c_i is—in spite of its wide application—nothing more than a convenient approximation which contributes to the limitations of this theory.

With this assumption we find

$$\frac{\partial \mathbf{\Gamma}}{\partial t} + \mathbf{\nabla}_r(n\overleftarrow{v}\vec{v}) - \frac{e}{m}n\mathbf{X} - \mathbf{\Gamma}\times\mathbf{B}\cdot\frac{e}{m} = \sum_i\left(\frac{n}{n_i}\Gamma_i - \Gamma\right)\nu_i\,, \tag{4.7}$$

where it is useful to remember that the ν_i stand for average values and that the collision frequencies of electrons with ions ν_{-+} and ions with electrons ν_{+-} are related to the resistivity of a fully ionized quasi-neutral plasma by

$$\eta = \frac{m_+\nu_{+-}}{e^2n} = \frac{m_-\nu_{-+}}{e^2n}\,. \tag{4.8}$$

Finally, the substitution of $V=mv^2$ in eq. (4.1) yields the law governing the energy flux in the discharge. We have

$$\frac{\partial}{\partial t}\overline{mv^2}+\nabla\cdot\overline{n\boldsymbol{v}v^2}-\frac{2e}{m}n\boldsymbol{X}\cdot\boldsymbol{v}_d=\sum_i\int_{(8)}(v'^2-v^2)f_i(v_i)f(v)\varrho_i(c_i)\cdot c_i\,\mathrm{d}^3v_i\,\mathrm{d}^3v\,\mathrm{d}^2\Omega\,. \tag{4.9}$$

In the general case the scattering function ϱ_i must be considered as a sum

$$\varrho_i=\varrho_{0i}+\sum_x\varrho_{xi}\,, \tag{4.10}$$

where ϱ_{0i} accounts for the elastic and ϱ_{xi} for the inelastic collisions.

For the elastic collisions we can evaluate the integrals and the collision terms again, assuming that we may consider the collision frequencies as being approximately independent of the relative velocity. Under these circumstances it is easy to show that we have

$$\int_{(8)}[(v')^2-v']f_i(v_i)f(v)c_i\varrho_{i0}(c_i)\,\mathrm{d}^2\Omega\,\mathrm{d}^3v_i\,\mathrm{d}^3v=\frac{2\nu_i}{m_i+m}n\,\{m_i\overline{v_i^2}-\overline{mv^2}\}\,. \tag{4.11}$$

For inelastic collisions we use

$$\int_{(8)}[(v')^2-v^2]f_i(v_i)f(v)c_i\sum_x\varrho_{ix}\,\mathrm{d}^2\Omega\,\mathrm{d}^3v_i\,\mathrm{d}^3v=-\,n\sum_x\nu_{ix}V_{ix}\cdot\frac{2e}{m}\,, \tag{4.12}$$

with

$$\nu_{ix}=n_i\int_{(2)}\varrho_{ix}c_i\,\mathrm{d}^2\Omega\,. \tag{4.13}$$

Introducing eq. (4.12) and eq. (4.11) into eq. (4.9) we find

$$\frac{\partial\overline{mv^2}}{\partial t}+\nabla\overline{nvv^2}-\frac{2e}{m}Xn\cdot\boldsymbol{v}_d=\sum_i\frac{2\nu_i n}{m_i+m}(m_i\overline{v_i^2}-\overline{mv^2})-\sum_{i,x}\frac{n\cdot\nu_{ix}\cdot V_{ix}\cdot 2\cdot e}{m}\,. \tag{4.14}$$

From eq. (4.14) the field X may be eliminated with the help of the momentum balance equation.

It is quite clear that the eqs. (4.2), (4.7) and (4.14) are not a complete set which can be solved without further approximation, because the momenta

$$\overleftrightarrow{P}=\overline{nm\overleftrightarrow{v}\overleftrightarrow{v}}\,;\qquad \boldsymbol{W}=\overline{n\cdot m\boldsymbol{v}\cdot v^2} \tag{4.15}$$

are unknown. Since we do not want to introduce higher order momentum equations we use the following approximations

$$\overleftrightarrow{P} = \overleftrightarrow{p} + m\vec{v}_a\vec{v}_a = \overleftrightarrow{\theta} nkT + m\vec{v}_a\vec{v}_a\,, \qquad \boldsymbol{W} = \overline{mn\boldsymbol{v}v^2} = \boldsymbol{\Gamma}\cdot\overline{mv^2}\,, \tag{4.16}$$

where $\overleftrightarrow{\theta}$ is the identity tensor.

4.1. *The weakly ionized four-component system without magnetic field.* – The theory which we present here includes as a special case the well-known Schottky theory for the quasi-neutral three-component positive column. It also uses one of the basic assumptions of this theory, namely the assumption that the temperatures are constant in a weakly ionized system, where the energy gain in the electric field and the collision loss with the neutral particles are the decisive terms in the energy balance. Under these circumstances the energy law (4.14) has a simple solution which provides a relation between the electron temperature and the $X_z = E$-field of the column. Otherwise the relations (4.14) may be disregarded.

Introducing further the abbreviations

$$\mu = \frac{e}{m\nu}\,, \qquad D = \mu\frac{kT}{e}\,, \tag{4.17}$$

and remembering that due to the weak ionization only collisions with neutral particles are of importance, one obtains the particle and momentum conservation laws for a stationary state of the charged components according to eqs. (4.2) and (4.7)

$$\operatorname{div}\boldsymbol{\Gamma}_i = \Delta_i \tag{4.18}$$

and

$$\boldsymbol{\Gamma}_i = -D_i \operatorname{grad} n_i + \frac{l_i}{|l_i|}\mu_i n_i \boldsymbol{X}\,. \tag{4.19}$$

The interesting features are the terms of net particle production Δ_i.

The electron production can be given in the form

$$\Delta_e = \alpha n_e + \delta n_e n_- - \beta n_e\,, \tag{4.20}$$

where the first term describes electron production by electron collisions with neutral particles, the second electron detachment from negative ions and the third electron loss due to attachment of electrons.

For the negative ions we have

$$\Delta_- = \beta n_e - \sigma n_+ n_- - \delta n_e n_- \,, \tag{4.21}$$

where the first term is due to electron attachment, the second due to recombination of positive with negative ions and the third due to detachment processes.

Finally we have

$$\Delta_+ = \alpha n_e - \sigma n_+ n_- \,, \tag{4.22}$$

where the first term describes production by electron collisions with neutrals and the second loss due to recombination with negative ions.

Since we have a weakly ionized plasma the neutral gas can and has been assumed to be uniformly distributed.

The boundary conditions for this set of differential equations are—apart from the usual regularity conditions—given by the requirements that the current continuity at the sheath edge is satisfied and the total electric current to the isolated wall is zero.

The first requirement means for a system with two carrier components that we have

$$(\mathrm{d}\ln n_+/\mathrm{d}r)_s = \alpha(\bar{v}_+/D_{\text{eff}})_s \,, \tag{4.23}$$

where the index (s) designates the sheath edge and α is a numerical factor of order of magnitude unity.

For a three-carrier-component system the situation is more complicated, but in a gas with sufficiently high pressure $\lambda \ll R$ it is possible to approximate the boundary condition by the simple Schottky condition

$$n_i \simeq 0 \,. \tag{4.24}$$

The second requirement can be stated in the form

$$\Big(\sum_i e_i \Gamma_{ri}\Big)_s = 0 \,. \tag{4.25}$$

As we see it, the problem of solving the four-component system presents itself as an eigenvalue problem of three simultaneous nonlinear differential equations. The eigenvalues may be taken to be the field E and the axial densities of the positive and negative ions ($n_{\pm 0}$). The axial density of the electrons n_{e0} enters as an experimental parameter representing, *e.g.*, the discharge current.

We can not present here the details of the mathematical solution of this problem, which are given elsewhere (*). We sketch only the results.

For a fictitious gas with quantities similar to oxygen, we evaluated as well the eigenvalues E, $y_0 = n_{-0}/n_{e0}$, $Z_0 = n_{+0}/n_{e0}$, as the radial electron density distribution $n(r)$ for various values of the experimental parameters.

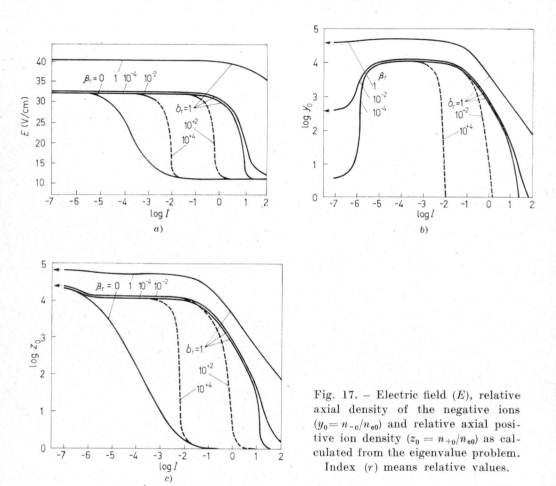

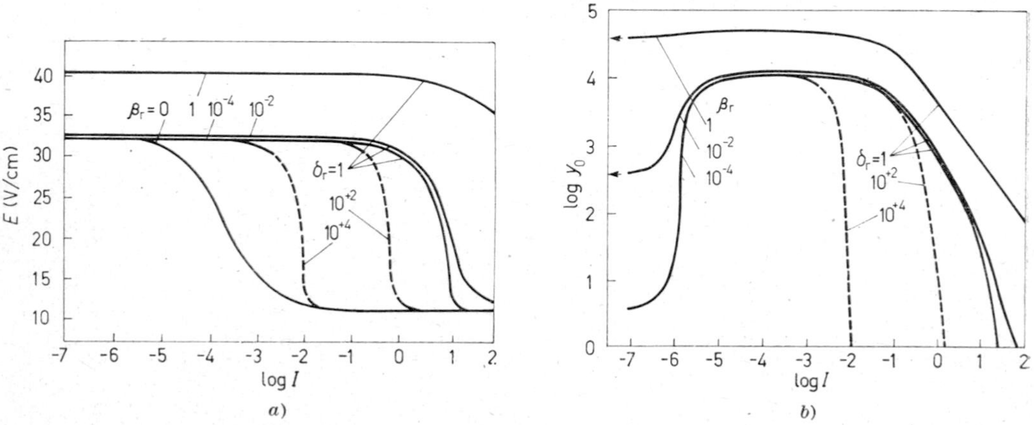

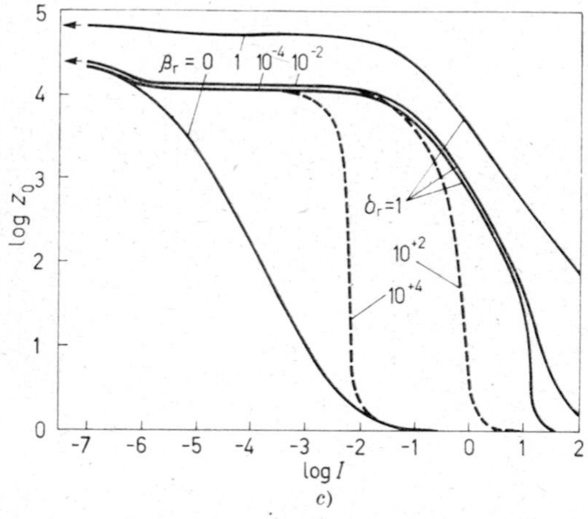

Fig. 17. – Electric field (E), relative axial density of the negative ions ($y_0 = n_{-0}/n_{e0}$) and relative axial positive ion density ($z_0 = n_{+0}/n_{e0}$) as calculated from the eigenvalue problem. Index (r) means relative values.

The results of the eigenvalues for various values of the coefficients are presented in the Fig. 17. They show the following interesting facts:

as is well known from Schottky's simple theory we have for sufficiently large currents quasi-neutrality and a field E independent of the current. However with decreasing current we reach a region where the electric field increases

(*) The asterisk refers here and in the following to the literature at the end.

rapidly due to the fact that charge densities are too small to build up the ambipolar field. Consequently the electrons leave the discharge at a higher rate approaching the coefficient D_e. To have equal ion and electron currents flowing to the wall the ion density must surpass the electron density. The concept of quasi-neutrality fails and we are in the so-called « subnormal region ».

The presence of a negative ion component changes this situation. The subnormal region is now broken up into two parts. Starting from very low currents we see the negative ion component increase. At the same time the positive ion component decreases. The same is true also for the field E but it is too small to be recognisable in the Fig. 17*a*. These changes are due to the fact that an ambipolar field develops. Consequently the lifetime of an electron in the discharge before it reaches the wall increases. That means it has a better chance to attach and form a negative ion. However, with the increase in the negative ion density the effective diffusion coefficient of the negative particles decreases and consequently the build-up of the ambipolar field is interrupted. Therefore also the increase in the negative ion density ceases. From then on all parameters change proportionally to the variation of the current. The situation is only interrupted when we reach the range where negative-positive ion recombination comes into play. Then with the decrease in the negative ion density, the increase in the ambipolar field is taken up again and this causes the well-known decrease to the normal Schottky conditions. This « two-step subnormal region » represents the essential influence of the negative ions on the characteristic.

There are two parameters Rp and pI for the evaluation of the radial distributions $n_e(r)$.

We found however, that after normalization of the ordinate and abscissa the double multiplicity of curves could be described by only one parameter for which we used the half-width h/R. The results are shown in Fig. 18. We see that the negative ion component causes a constriction compared to the Bessel distribution ($h/R = 0.63$) of the two-component Schottky theory.

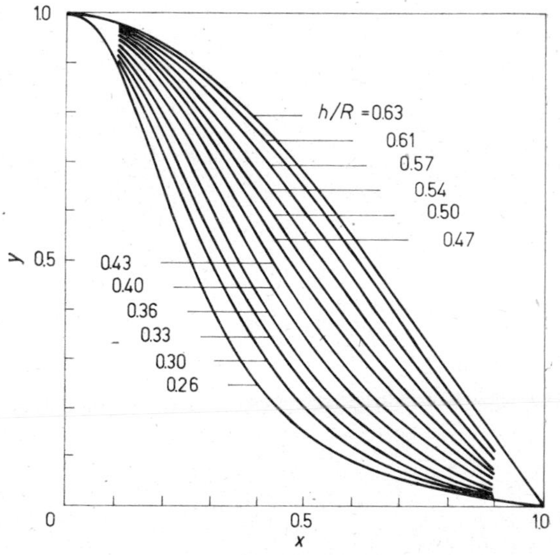

Fig. 18. – Calculated relative electron density distribution $y = n_e/n_{e0}$ as a function of the radial coordinate $x = r/R$. h/R is the relative half-width.

In Fig. 19 we have given the contraction parameter h/R as a function of the experimental data (Rp, Ip) so that we can see how the constriction due to negative ions changes with the experimental situation. Obviously there is a contraction increasing with the product Ip. We stress the fact that our

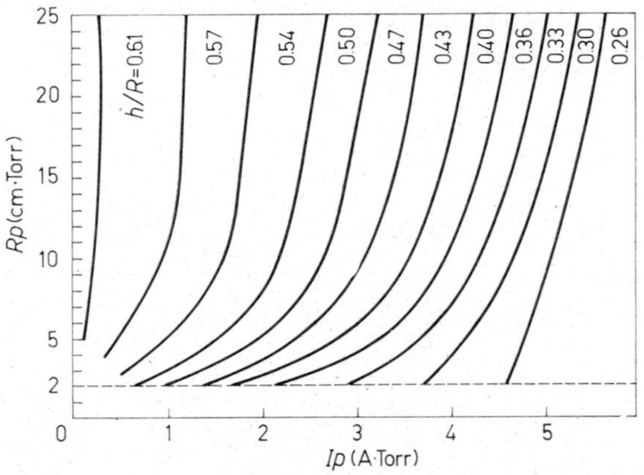

Fig. 19. – Calculated contraction parameter h/R as a function of the experimental parameters Rp and Ip.

constriction process calculated here has nothing to do with the temperature changes which may also cause a constriction but only for values of Ip larger than those given in Fig. 19.

4·2. *Quasi-neutral three-component system in a magnetic field.* – For a stationary state of a three-component system with quasi-neutral conditions $(n_+ \simeq n_e)$ the electron and ion momentum balances may be written in the form

$$\text{(4.26)} \qquad \begin{cases} \boldsymbol{B}\times\boldsymbol{\Gamma}_+ + \boldsymbol{\Gamma}_+/\mu_+ + \eta en(\boldsymbol{\Gamma}_+ - \boldsymbol{\Gamma}_e) = n\boldsymbol{X} - \nabla(\overleftrightarrow{P}_+/e)\,, \\ -\boldsymbol{B}\times\boldsymbol{\Gamma}_e + \boldsymbol{\Gamma}_e/\mu_e + \eta en(\boldsymbol{\Gamma}_e - \boldsymbol{\Gamma}_+) = -n\boldsymbol{X} - \nabla(\boldsymbol{P}_e/e)\,. \end{cases}$$

For an axially homogeneous discharge only the radial components are of interest and the congruence condition

$$\text{(4.27)} \qquad \boldsymbol{\Gamma}_{r+} = \boldsymbol{\Gamma}_{re} = \Gamma_r$$

holds for an isolated wall. It allows the elimination of the field X_r from the radial component of (4.26) which produces, after elementary transformations, the radial current

$$\text{(4.28)} \qquad \Gamma_r = \frac{1}{1 + B^2\mu_+\mu_e/(1+\sigma_+ + \sigma_e)}\cdot\frac{\partial}{\partial r}\, n\left\{\frac{D_+\mu_e + D_e\mu_+}{\mu_+ + \mu_e}\right\},$$

where the symbols D, μ are defined in eq. (4.17) and the abbreviation

$$\sigma_{\substack{+\\ e}} = e\eta n\mu_{\substack{+\\ e}} \tag{4.29}$$

has been used. By the way, if we also want to account for Simon or Allis diffusion, we can simply replace the condition of congruence (4.27) by

$$\Gamma_{re} = \varepsilon\Gamma_{r+}\,, \tag{4.30}$$

where ε is a parameter varying between zero and unity. If we introduce eq. (4.28) into the particle conservation law assuming that we have only particle production by electron collisions and no recombination within the discharge, then we find

$$\frac{D_+\mu_e + D_e\mu_+}{\mu_e + \mu_+}\cdot\frac{1}{r}\frac{d}{dr}\,r\left\{\frac{1}{1 + B^2\mu_+\mu_e/(1+\sigma_+ + \sigma_e)}\,\frac{dn}{dr}\right\} = \alpha n\,. \tag{4.31}$$

If electron-ion interaction is negligible ($\sigma_{\substack{+\\ e}} \ll 1$) then the differential eq. (4.31) is again of the Bessel type. As in the simple Schottky theory we have therefore a radial electron density distribution given by a zero-order Bessel function. Only the eigenvalue is changed by the magnetic field. We have now

$$\alpha = \left(\frac{2.4}{R}\right)^2 \frac{D_{am}}{1 + B^2\mu_+\mu_e}\,. \tag{4.32}$$

If the parameter σ is not negligible then we can not find a simple solution of the eq. (4.31). However, it is still possible to determine the radial potential distribution by integrating the equation of the radial field X_r. For the case of ambipolar diffusion we find

$$\frac{eu}{kT_-} = -\frac{1 + u_+u_e[(\mu_+/\mu_e) - (T_+/T_e)]}{1 + u_+u_e}\cdot\ln x + \frac{u_+u_e[1 + (T_+/T_e)]}{1 + u_+u_e}\cdot \ln\left\{\frac{1 + (\sigma_{0e} + \sigma_{0+})(1/x)(1 + u_+u_e)}{1 + (\sigma_{0e} + \sigma_{0+})(1 + u_+u_e)}\right\}, \tag{4.33}$$

where we used the abbreviations

$$X = n(0)/n(r)\,,\qquad u_{\substack{+\\ e}} = \mu_{\substack{+\\ e}}B\,,\qquad \sigma_0 = e\eta\mu n(0)\,. \tag{4.34}$$

The corresponding distributions are shown in Fig. 20. It is interesting to note that the potential distribution shows a minimum for a certain range of magnetic field values between $B_0 < B < B_1$ with B_0 and B_1 defined by

$$(4.35)\qquad \begin{cases} B_0^2 \cdot \mu_+ \mu_e (T_+/T_e) = 1\,, \\ B_1^2 \cdot \mu_+ \cdot \mu_e \cdot (T_+/T_e) = 1 + \sigma_{0e} + \sigma_{0+}\,. \end{cases}$$

The point where the potential minimum occurs is called the « point of ambipolar field reversal ». The phenomenon of ambipolar field reversal is interesting for the nonstationary modes of the positive column.

The problem of a three-component cylindrical column in a magnetic field is also of interest with reference to the experiments on the « Cs discharge » and the « hollow cathode discharge ». Here we have a situation where particles are introduced in a plasma column from a core inside of the column. They then diffuse — in the model — without additional particle production or volume recombination to the wall where they recombine. We have described this situation by the model of the « fully ionized plasma column with external particle production ».

In this case eqs. (4.26) simplify further since we have no neutral particle component and consequently may use $\mu_+, \mu_- \to \infty$. Under these circumstances the radial particle current is simply

$$(4.36)\qquad \Gamma_r = -\frac{\eta n}{B^2}\frac{\mathrm{d}}{\mathrm{d}r}[nk(T_+ + T_e)]\,.$$

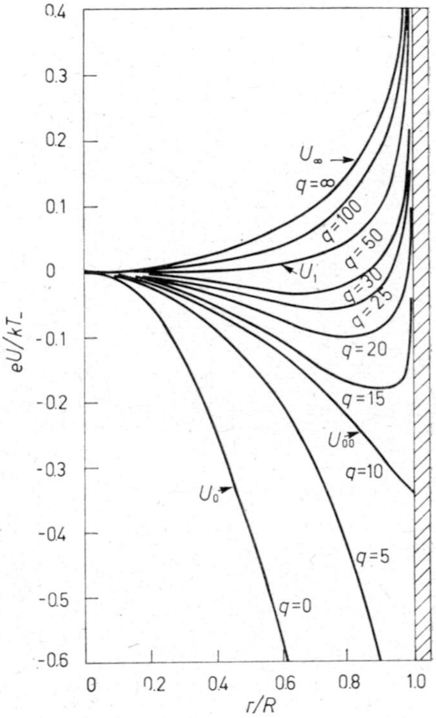

Fig. 20. – Ambipolar radial potential distribution as a function of the relative distance r/R for $\sigma_{0+} + \sigma_{0e} = 5$ and $T_e/T_+ = 10$. $q = \mu_+ \mu_e B^2$ describes the influence of the magnetic field. $U_{00} < U < U_1$ is the region of field reversal.

Since there is no net particle production we have the current continuity equation

$$(4.37)\qquad n\frac{\mathrm{d}}{\mathrm{d}r}[n(T_+ + T_e)] = -\frac{\Gamma_0 r_0 B^2}{\eta_0 k}\left(\frac{T_e}{T_0}\right)^{\frac{3}{2}} \cdot \frac{1}{r}\,,$$

where the index (0) designates quantities at the edge (r_0) of the effective particle source.

But now an additional difficulty arises, since in the case of the fully ionized column with external particle production we can no longer introduce the assumption of constant temperature. Here the temperature of the particles is not governed by the energy gain in a homogeneous electrical field and an energy loss due to collisions with uniformly distributed balance. We omit the details of this calculation (‡) which, starting from eq. (4.14), produces two simultaneous equations for the quantities $z = n(x)/n(x_0)$ and $y = T_+(x)/T_0$ with $x = r/R$ and $x_0 = r_0/R$. T_0, R, Γ_0, r_0 and B are experimental parameters, $n(x_0)$ is determined as the eigenvalue of the problem.

Two sets of typical density and temperature distributions are shown in Fig. 21*a*) and *b*). The electron temperature follows the relation

$$2T_0 = T_+ + T_e . \tag{4.38}$$

Obviously the density distribution is no longer of the Bessel-type. Rather the slope of the density first increases and later towards the wall decreases again. Also there is a finite density at the wall. The temperature T_+—and with that T_e—changes remarkably across the discharge. The ion temperature decreases, the electron temperature increases.

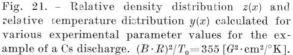

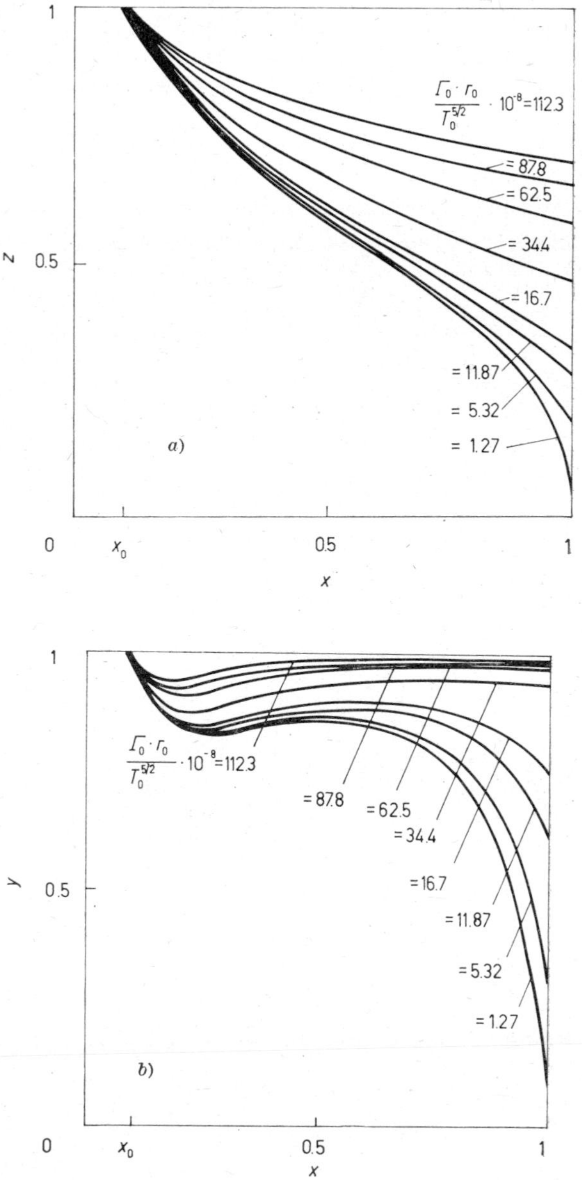

Fig. 21. – Relative density distribution $z(x)$ and relative temperature distribution $y(x)$ calculated for various experimental parameter values for the example of a Cs discharge. $(B \cdot R)^2/T_0 = 355$ [$G^2 \cdot cm^2/°K$].

Both distributions depend on the parameters $(BR)^2/T_0$ and $\Gamma_0 r_0/T^{\frac{5}{2}}$. They are open to a qualitative physical discussion and explanation.

4·3. *Instabilities of a weakly ionized quasi-neutral plasma column in a magnetic field.* – To study the instabilities of the collision-dominated column one considers in a perturbation theory the growth of small perturbations superimposed on the stationary mode. Since we consider a three-component plasma in a longitudinal magnetic field the stationary zero-order solution is already given in the preceding chapter if we use $\sigma = 0$ due to the assumption of weak ionization. If the index (0) designates the zero order approximation, then we use

$$n = n_0 + n', \qquad V = V_0 + V', \tag{4.39}$$

where the dash (′) designated the perturbations which are assumed small in comparison to the stationary mode.

The solution of the nonstationary first-order problem is quite involved. KADOMTSEV and NEDOSPASOV applied therefore the same assumptions as in the zero order: *a*) neglection of inertia, *b*) quasi-neutrality, *c*) scalar pressure approximation, *d*) electron and ion temperature constant and unperturbed across the discharge, *e*) Schottky boundary conditions. For further simplification one requires that ion diffusion and the influence of the magnetic field on the ion mobility be negligible. Both assumptions are justified in the range of experimental conditions in which we are interested. We call the instability characterized by these assumptions the K-N instability.

For the perturbations we have then the following two differential equations

$$\left\{\begin{aligned} &\frac{\partial n'_+}{\partial t} - \boldsymbol{\nabla}(n_0\mu_+\boldsymbol{\nabla} V') - \boldsymbol{\nabla}(n'_+\mu_+\boldsymbol{\nabla} V_0) = \alpha n', \\ &\frac{\partial n'_e}{\partial t} - \frac{\partial^2}{\partial z^2}(D_e n'_e) + \mu_e\frac{\partial}{\partial z}\left(n'_e\frac{\partial V_0}{\partial z}\right) + \mu_e\frac{\partial}{\partial z}\left(n_0\frac{\partial V'}{\partial z}\right) - \frac{\mu_e}{(\omega\tau)}\boldsymbol{b}\cdot(\boldsymbol{\nabla}_\perp n_0\times\boldsymbol{\nabla}_\perp V') - \\ &-\frac{\mu_e}{\omega\tau}\boldsymbol{b}(\boldsymbol{\nabla}_\perp n'_e\times\boldsymbol{\nabla}_\perp V_0) - \frac{1}{(\omega\tau)^2}\Delta_\perp(D_e n') + \\ &\qquad + \frac{\mu_e}{(\omega\tau)^2}\boldsymbol{\nabla}_\perp(n_0\boldsymbol{\nabla}_\perp V') + \frac{\mu_e}{(\omega\tau)^2}\boldsymbol{\nabla}_\perp(n'_e\boldsymbol{\nabla}_\perp V_0) = \alpha n'_e, \end{aligned}\right. \tag{4.40}$$

where τ is the collision time and $\omega = eB/m$ the cyclotron frequency of the electrons $\boldsymbol{b} = \boldsymbol{B}/B$. Applying a development of the type

$$\begin{pmatrix} n'(r,\varphi,z,t) \\ V'(r,\varphi,z,t)\end{pmatrix} = \sum \begin{pmatrix} n'_{mk}(r) \\ V'_{mk}(r)\end{pmatrix}\exp[i(\bar\omega t + m\varphi + kz)] \tag{4.41}$$

we get two equations for the radial distributions $V'_{mk}(r)$ and $n'_{mk}(r)$ of the potential and density in the first order.

With the boundary conditions

$$(4.42)\qquad n'(0)=0\,,\qquad n'(R)=0\,,\qquad V'(0)=0\,,\qquad \Gamma'_+(R)=\Gamma'_e(R)\,,$$

this is again an eigenvalue problem which defines a whole set of functions and the eigenvalues $\bar{\omega}$. We should remember that in general both eigenfunctions and the eigenvalue itself may be complex quantities.

If we had solved this problem then we could judge by the imaginary part of the eigenvalues whether a given mode grows or decays, that means whether the mode is unstable or stable. This question is of interest for the interpretation of certain experimental data on the occurence of critical magnetic fields in the positive column.

Unfortunately the exact solution of the eqs. (4.40), (4.42) encounters formidable difficulties and therefore KADOMTSEV and NEDOSPASOV assume trial solutions of the form

$$(4.43)\qquad n'=\bar{n}J_1(\beta_1 r/R)\,,\qquad V'=\bar{V}\cdot J_1(r\beta_1/R)\,,$$

where $\bar{n}$ is a real quantity and β_1 the first zero of J_1. They determine the eigenvalue $\bar{\omega}$ by averaging with the weight function $J_1(\beta_1 r/R)$. The results seem to give good agreement with the experimental data.

However, the averaging process excludes the influence of those regions on the eigenvalue where the averaging function is small. That means in particular that the influence of the wall losses is treated incorrectly. Also the use of a trial function is only of value if one can show that the results do not depend critically on the radial distribution.

One can give more general information about the problem if one simply integrates the two eqs. (4.40) after introduction of the equations (4.41) over the radial co-ordinate. We get two complex equations for the two complex quantities

$$(4.44)\qquad N'=\int_0^R n'(r)\,r\,\mathrm{d}r\,,\qquad \mathscr{V}'=\int_0^R V' r\,\mathrm{d}r\,.$$

It is possible to eliminate the function $\mathscr{V}'$ from these equations and express the eigenvalue only in terms of the following parameters which characterize the radial distribution functions of the eigensolutions

$$(4.45)\qquad \begin{pmatrix}\xi_n\\ \xi_v\end{pmatrix}=\frac{m^2}{k^2}\frac{\mu_+}{\mu_e}\int_0^R\frac{1}{r}\begin{pmatrix}n'\\ V'\end{pmatrix}\mathrm{d}r\bigg/\begin{pmatrix}N'\\ \mathscr{V}'\end{pmatrix},\qquad \begin{pmatrix}\bar{\xi}_n\\ \bar{\xi}_v\end{pmatrix}=\frac{m^2}{k^2}\frac{\mu_+}{\mu_e}\int_0^R\frac{1}{r}\cdot\frac{1}{n_0}\frac{\mathrm{d}n_0}{\mathrm{d}r}\begin{pmatrix}n'\\ V'\end{pmatrix}\mathrm{d}r\bigg/\begin{pmatrix}N'\\ \mathscr{V}'\end{pmatrix}.$$

With the help of this procedure one is able to determine whether the instabilities are strongly dependent on the type of radial distribution and whether the averaging process is of importance.

The distribution functions we used were given by

$$f(x) \propto n'(x) \propto n_0 V'(x) \propto x^{\nu} \cdot J_1(\beta_1 x) . \tag{4.46}$$

(The choice $V' \propto J_1 \cdot (\beta_1 \cdot r/B)$ in (4.43) contradicts the boundary conditions (4.42), since V' is not zero at the wall). The distributions (4.46) are shown in Fig. 22 for various values of the parameter ν.

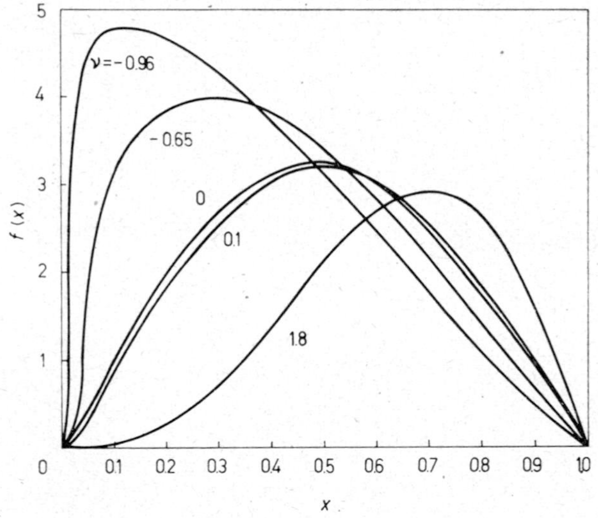

Fig. 22. – Radial density distributions used for trial solutions

$$f(x) = n'/\bar{n} = 2 \cdot x^{\nu} \cdot J_1(\beta_1 x) / \int_0^1 x^{\nu+1} \cdot J_1(\beta_1 x)\, dx .$$

The results for the instabilities in helium are demonstrated in Fig. 23 for the example $\nu = 0$. y, $\bar{\alpha}$ and ξ_n are related to the experimental parameters according to

$$\lambda = \frac{2\pi}{k} = 2\pi R \sqrt{\xi_n}, \qquad B = 1325 \cdot p \cdot \sqrt{y}, \qquad \bar{\alpha} = p \cdot R , \tag{4.47}$$

where p is measured in Torr, B in gauss and λ, R in cm. The discharge is unstable for wavelengths λ and corresponding k values in the range between the intersection points of a horizontal line through $y(B)$ with the corresponding $\bar{\alpha}$-curve.

The value of the magnetic field where the horizontal line is tangent to the $\bar{\alpha}$-curve defines the critical magnetic field B_c for this pressure and radius. The corresponding values shown in Fig. 24 are different from those with the results of KADOMTSEV and NEDOSPASOV for the reasons outlined above.

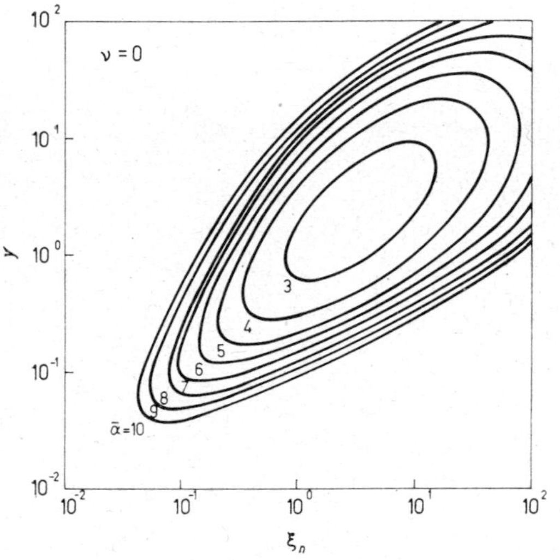

Fig. 23. – The $\bar{\alpha}$-curves define the range of k-values for which the column is unstable. For the meaning of the symbols and the interpretation of the figure see text.

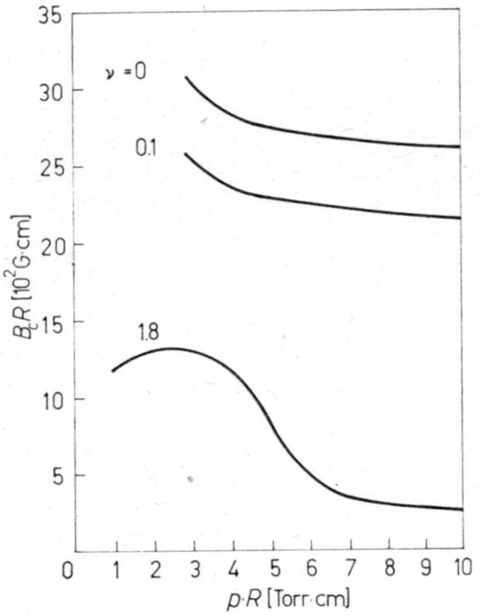

Fig. 24. – Critical magnetic field for the instability onset as a function of the product pR for three different trial solutions.

We calculated the instabilities also for the other distributions shown in Fig. 22. $\nu = 0.96$ and $\nu = 0.65$ were stable in the whole pressure range considered. For $\nu = 0.1$ and $\nu = 1.8$ we find qualitatively similar results as for $\nu = 0$ but remarkable quantitative differences. This can be seen from Fig. 24 where we have plotted the critical field *vs.* pressure for three values of ν. Particularly noteworthy is the difference between the results for $\nu = 0$ and $\nu = 0.1$. These radial distributions are quite similar but the critical fields show strong differences.

Remembering the restriction and uncertainty of the assumptions (4.46), we believe these results show that the trial solution method can only predict the order of magnitude of the effect and that for quantitative comparison at least a good approximate solution of the eigenvalue problem is required. This is, however, a very difficult problem which we can not treat in this lecture.

BIBLIOGRAPHY

G. Ecker: *Proc. Phys. Soc.*, **17**, 485 (1954); *Theorie der nicht-kontrahierten positiven Säule unter dem Einfluß der negativen Ionen*, in *Zeits. f. Naturfor.*, in print; *Theorie der Kontraktion der positiven Säule*, in *Zeits. f. Naturfor.*, in print; *Theory of the fully ionized plasma column with external particle production*, UCRL Rep. 9988; *Phys. of Fluids*, **4**, 127 (1961).

G. B. Kadomtsev and A. V. Nedospasov: *Journ. Nucl. Energy*, C **1**, 230 (1960).

B. A. Paulikas and R. V. Pyle: *Phys. of Fluids.*, **5**, 348 (1962).

Topics in Microinstabilities.

M. N. Rosenbluth

Johns Jay Hopkins Laboratory for Pure and Applied Science,
General Atomic Division of General Dynamics Corporation - San Diego, Cal.
University of California - San Diego-La Jolla, Cal.

1 - Introduction.

A microinstability is one which is not derivable from the standard magneto-hydrodynamic equations—which are obeyed for example by liquid metals—but which depends on the detailed microscopic equations for the plasma. Only a rapid survey is attempted here.

Almost all the theoretical work to date has been based on the collisionless Boltzmann equation which, as we shall see, already permits of a vast number of instabilities. One has the feeling that collisions, being a dissipative process, would serve to stabilize the system; but it is worth remembering that introducing the viscosity into the Navier-Stokes equation does make the Couette flow unstable. We shall hear more about resistive effects from Furth.

In fact, it is quite obvious that instability and impurity radiative losses are the two most serious threats, in principle, to the fusion reactor. We have heard previously that, to lowest order in an expansion which considers the Larmor radius to be small and the Larmor frequency large, a slightly modified version of the magneto-hydrodynamic equations is valid. Moreover, while they are difficult to construct, there seems to be no doubt that such systems as the hard-core pinch, helical stellarator, and shorted-out mirror machine are in fact hydromagnetically stable. Thus, we are led to a consideration of microinstabilities.

In the first place, it should be stressed that these are not likely to be as violent as the hydromagnetic instabilities due to their high frequency or short wave length. Thus if an instability had frequency $\omega > \omega_c$, then by the time the amplitude of motion had built up to a Larmor radius the velocity would be greater than the original thermal velocity of the particles and there would be no energy left to drive the instability. Similarly for a short wave-length disturbance when the amplitude exceeds the wave length, the linear theory becomes invalid. Hence, such instabilities do not remain for long in the exponentially growing phase. We would not expect then that these micro-instabilities would lead to a gross disassembly of the plasma such as occurs for example in the hydromagnetic pinch instabilities. Rather, we would expect to see such phenomena as turbulence, enhanced resistivity and enhanced diffusion across the magnetic field. This will be discussed further by STURROCK.

At this point, I would like to interject a remark on magnetohydrodynamic turbulence in a low-β plasma. Hydrodynamic turbulence is a characteristically three-dimensional phenomenon. However, a three-dimensional disturbance must distort the magnetic field lines and in a low-β plasma there is insufficient energy for this to happen since the field energy greatly exceeds the particle energy. This contrasts with the situation for instability which is often a two-dimensional phenomenon. In fact, we will often notice that microinstabilities are purely electrostatic in nature so as not to perturb the field.

It is well known that the only true thermodynamic equilibrium of a plasma in a static magnetic field is one in which the density and temperature are constant over all space. One mechanism which may bring this about is collisional heating and diffusion. These processes are ordinarily slow enough to be neglected. However we know on general grounds that any collective motions which do exist, like microinstabilities, will at least move in the direction of thermodynamic equilibrium. One may pose then the question whether the purely collective motions of the system are in themselves sufficient to produce thermodynamic equilibrium. We will now demonstrate a slightly nontrivial counter-example.

Consider a uniform infinite plasma, which may contain a static uniform magnetic field, and in which the distribution function f_0 depends only on the magnitude of particle velocity and satisfies the condition $(\partial f_0/\partial v^2) < 0$. Now, from Liouville's theorem, we know that the phase-space $d^3x\, d^3v$ occupied by a fluid element is a constant of the motion, as is the distribution function f. Hence let us consider a generalized entropy

$$S = \int G(f)\, d^3\boldsymbol{x}\, d^3\boldsymbol{v}\ , \tag{1.1}$$

where G is any functional of f like $f \ln f$. Let us compare $S(t_1)$ and $S(t_2)$ where t_1 and t_2 are two different times in the course of the motion. Consider an element of phase-space $d^3x_2\, d^3v_2$ at time t_2. At time t_1 the fluid occupying this element was at some other position in phase-space $d^3x_1\, d^3v_1 = d^3v_2\, d^3x_2^2$. Moreover, f is the same at the two times, as is $G(f)$, and the contribution to the integral S. Hence it follows that

$$\frac{\partial S}{\partial t} = 0 , \tag{1.2}$$

as may also be proved by direct substitution from the collisionless Boltzmann equation. So the constancy of S provides a constraint on the possible class of motions.

Let us now minimize, with respect to f, the total kinetic energy of the system

$$E = \int \frac{mv^2}{2} f \, \mathrm{d}^3\boldsymbol{x} \, \mathrm{d}^3\boldsymbol{v} ,$$

subject to the constancy of the total number of particles and of S by the method of Lagrange multipliers. We find

$$\frac{mv^2}{2} + \alpha + \beta G'(f_0) = 0 . \tag{1.3}$$

Let us now insert into (3) the distribution function $f_0(v^2)$ with which we are concerned. Then

$$\frac{\partial G}{\partial v^2} = \frac{-(mv^2/2) - \alpha}{\beta} \frac{\partial f_0}{\partial v^2} .$$

We may then solve for $G(v^2)$; and if v^2 is a single-valued function of f, as it will be for a monotonically decreasing f_0, we may find

$$G(f_0) = G\big(v^2(f_0)\big) .$$

Hence for every such f_0, we can define an entropy S which is a constant of the motion and which has the property that the lowest state of the energy of the system consistent with this constraint is that defined by the initial distribution f_0. In other words, any departure from the initial state increases the kinetic energy of the system. Any electric and magnetic fields which may develop must also increase the energy, since the original constant magnetic field had the lowest possible energy consistent with the constraint of flux conservation.

Since energy must be conserved in any motion, it follows that the system cannot move away from its initial state; *i.e.*, it is stable against all possible disturbances. While the stability of such an infinite system is not particularly interesting, there are two significant points here: 1) collective modes by themselves will not produce the Maxwell-Boltzmann distribution; 2) in a uniform medium there are no instabilities which transfer energy between ions and electrons at different temperatures. One also obtains nonlinear limits by this method.

In view of the complexity of the microinstability problem, the bulk of the work to date has concerned itself with the situation in an infinite homogeneous medium in which more complex forms of f than the above are considered. The equilibrium equation

$$\boldsymbol{v}\cdot\nabla f_0+\frac{e}{m}\,\boldsymbol{v}\times\boldsymbol{B}_0\cdot\nabla_v f_0=0\,,$$

allows the solution $f_0(v^2, v_\parallel)$.

In other words the distribution function may depend not only on energy but on velocity parallel to the field line. The simplest such distribution is one in which the electrons have no energy transverse to the field and consist of two equal streams at velocity $+V$ and $-V$. We will make the usual assumption of unmovable heavy ions to provide neutrality.

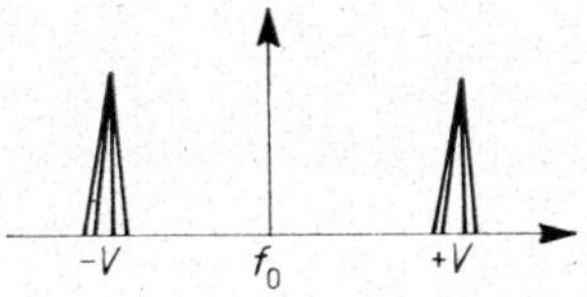

Fig. 1.

If the streams have no thermal spread then the moment equations of hydrodynamics remain exact in the linearized theory where no particle crossing occurs. This is so because all particles at a given point have had exactly the same history so that there can be no spread in particle velocity, *i.e.*, pressure. Taking the initial velocities of the streams in the x direction, we assume a perturbation of the form $\exp[i(kx+\omega t)]$. The hydrodynamic equations are

$$(1.4)\qquad \begin{cases} \dfrac{\partial n}{\partial t}+\dfrac{\partial}{\partial x}\,nv=0\,,\\[2mm] \dfrac{\partial v}{\partial t}+v\dfrac{\partial v}{\partial x}=\dfrac{e}{m}\,E\,. \end{cases}$$

Linearizing and labelling those particles initially moving in the direction of positive x by the subscript $+$ we have

$$i[\omega\pm kV]n_\pm+ikn_0v_\pm=0\,,$$

$$i[\omega\pm kV]v_\pm=\frac{e}{m}\,E\,.$$

Here $n_\pm$ and $v_\pm$ are the perturbed densities and velocities respectively, n_0 and V the initial density and velocity. Solving, we find

$$n_\pm = ikn_0 \frac{e}{m} \frac{E}{[\omega \pm kV]^2} . \tag{1.5}$$

The perturbed densities are now to be substituted in the Poisson equation

$$\nabla \cdot \boldsymbol{E} = ikE = 4\pi\varrho = 4\pi e[n_+ + n_-] .$$

This gives immediately the dispersion relation

$$1 = \frac{\omega_p^2}{[\omega + kV]^2} + \frac{\omega_p^2}{[\omega - kV]^2} , \tag{1.6}$$

where $\omega_p^2 = 4\pi n_0 e^2/m$. Equation (6) is easily generalized to the case of different initial density and masses in the stream by using the appropriate ω_p^2.

Equation (6) now yields a quadratic equation for ω^2:

$$[\omega^2 - k^2V^2]^2 = 2\omega_p^2[\omega^2 + k^2V^2]$$

$$\omega^4 - 2[k^2V^2 + \omega_p^2]\omega^2 + k^4V^4 - 2\omega_p^2 k^2V^2 = 0$$

which we may solve to obtain

$$\omega^2 = k^2V^2 + \omega_p^2 \pm \sqrt{(k^2V^2 + \omega_p^2)^2 + 2\omega_p^2 k^2V^2 - k^4V^4} . \tag{1.7}$$

Hence ω^2 is negative and we have instability if

$$kV < \sqrt{2}\,\omega_p .$$

For small kV there are two branches, the stable one with

$$-\omega = \pm\sqrt{2}\,\omega_p , \tag{1.8a}$$

and the unstable one,

$$\omega = \pm ikV . \tag{1.8b}$$

It is interesting to note that the instability rate given by (8b) is independent of the plasma properties, ω_p^2. The reason for this is that the cause of the instability is that by bringing each other to rest two coupled streams can conserve momentum and liberate energy for the instability. The nature of the coupling between them is more or less irrelevant. Thus hydrostatic coupling gives the same dispersion relation as (8b) for the Helmholtz instability of superposed fluids.

Let us now briefly discuss what happens if the two streams instead of being δ-functions possess some thermal spread as shown in the diagram.

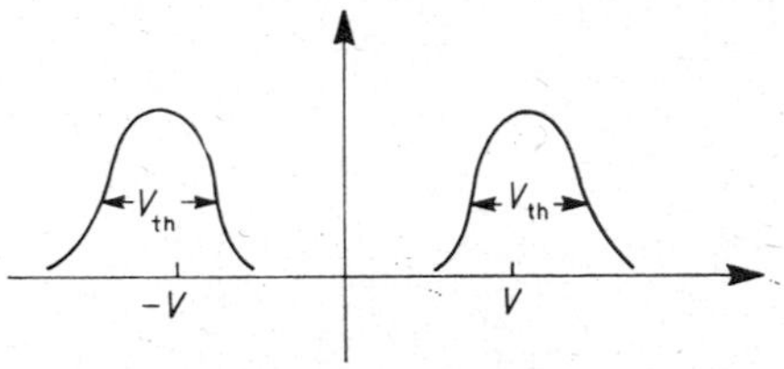

Fig. 2.

Then, we would expect that the instability would be smeared out if, during the time required to e-fold, the different particles within a distribution spread out over a wave length. This would destroy the coherent phases needed for the oscillation. Hence for stability

$$V_{th}\frac{k}{\omega} > 1 \quad \text{or using (8b)} \quad V_{th} > V\,. \tag{1.9}$$

In the detailed study of the Boltzmann equation precisely the condition (9) arises from the Landau damping of the oscillation. This two-stream instability is met in many connections in plasma theory. Thus a stream of high-energy electrons may be produced by accelerating electric fields. Moreover the very existence of an electric current parallel to the magnetic field implies a relative drift of electrons and ions. It turns out that if the electron and ion temperatures are about equal the condition for stability is

$$V_{\text{drift}} < V_{\text{th electrons}}\,.$$

This is satisfied in almost any conceivable case.

On the other hand for $T_e > 10\,T_i$ it turns out that for stability

$$V_{\text{drift}} < V_{\text{th ions}}\,,$$

a rather stringent condition. This is discussed further in the next lecture.

Let us now turn to a very brief survey of instabilities brought about in the case $T_\perp \neq T_\parallel$, *i.e.*, the mean orbital energy in motion perpendicular to the field lines differs from that in motion parallel to the field. In the case $T_\parallel > T_\perp$ an instability occurs in the Alfvén wave due to the fact that when the field line is perturbed, the centrifugal force acting on the large parallel energy of the particles tends to distort the field further. The criterion for instability is

$$p_{\parallel} - p_{\perp} > \frac{B^2}{4\pi}. \tag{1.10}$$

On the other hand if $T_{\perp} \gg T_{\parallel}$ then the magnetic mirrors produced by the change in field strength of the perturbation will heavily concentrate the particles where the field is weakest so that the pressure there becomes high, pushing the field lines apart and making the field still weaker. For this instability

$$\frac{p_{\perp}^2}{p_{\parallel}} - p_{\perp} > \frac{B^2}{8\pi}. \tag{1.11}$$

In fact, it turns out that there is a residual instability of the Alfvén wave for any anisotropy; but, the growth is quite small going as

$$\omega \approx \omega_{h_i} \exp\left[-\frac{1}{\beta}\left(\frac{T_{\perp}}{T_{\parallel}} - 1\right)^2\right]. \tag{1.12}$$

Here β is the ratio of particle to magnetic pressure.

There is still another instability of electrostatic nature which occurs near the cyclotron frequency for anisotropic distributions. In some ways this instability may be more dangerous as it is independent of β over a wide range of values. However a sufficient criterion for stability seems to be $T_{\perp}/T_{\parallel} < 2$. The above catalogue of instabilities is not inclusive even for microinstabilities of an infinite homogeneous plasma.

As high frequencies and short wavelengths are often characteristic of these microinstabilities, it is clearly a main task of the theory to undertake the investigation of non linear studies in order to assess what the effect of these instabilities on the plasma, *e.g.*, enhanced diffusion, may be.

2 - Study of the two-stream instability.

We shall consider a situation in which the electrons and ions are described by displaced Maxwellian distributions:

$$\begin{cases} f_{0i} = n\left(\dfrac{m_i}{2\pi k T_i}\right)^{\frac{3}{2}} \exp\left[-\dfrac{mv^2}{2kT_i}\right], \\ f_{0e} = n\left(\dfrac{m_e}{2\pi k T_e}\right)^{\frac{3}{2}} \exp\left[-\dfrac{m(\boldsymbol{v}-\boldsymbol{u})^2}{2kT_e}\right]. \end{cases} \tag{2.1}$$

The object is to find the dispersion relation for plasma oscillations and, in particular, to find the critical velocity for the onset of instability.

Using the method of characteristics or Green's function techniques, we solve the linearized Vlasov equation,

$$\frac{\partial f_1}{\partial f} + \boldsymbol{v}_n \cdot \boldsymbol{\nabla}_r f_1 = -\frac{e}{m} \boldsymbol{\mathscr{E}} \cdot \boldsymbol{\nabla}_v f_0 \,,$$

and obtain

$$f_1 = -\frac{e}{m} \int_{-\infty}^{t} \boldsymbol{\mathscr{E}} \cdot \boldsymbol{\nabla}_v f_0 \, \mathrm{d}t' \,, \tag{2.2}$$

where the integration is over the unperturbed particle orbits. Writing

$$\boldsymbol{\mathscr{E}} = -\boldsymbol{\nabla} \Phi$$

and

$$\Phi = \varphi \exp[pt] \exp[i\boldsymbol{k} \cdot \boldsymbol{r}] \,,$$

we have

$$f_1 = -f_0 \frac{e}{kT} \int_{-\infty}^{t} (\boldsymbol{v} \cdot \boldsymbol{\nabla}) \Phi \, \mathrm{d}t' = -\frac{e}{kT} f_0 \left[\Phi - p \int_{-\infty}^{t} \Phi(t') \, \mathrm{d}t' \right]. \tag{2.3}$$

For the case of no magnetic field or $\boldsymbol{k} \| \boldsymbol{B}_0$,

$$\Phi(t') = \Phi(t) \exp[p(t' - t)] \exp[ik(z' - z)] = \Phi(t) \exp[p\tau] \exp[ikv_z \tau] \,,$$

where

$$\tau = t' - t$$

and the z axis has been chosen along $\boldsymbol{k}$; hence,

$$f_1 = -\frac{e}{kT} f_0 \left[\Phi - p\Phi \int_{-\infty}^{0} \exp[p\tau] \exp[ikv_z \tau] \, \mathrm{d}\tau \right],$$

and

$$\varrho = e \int f_1 \, \mathrm{d}^3 v = -\frac{ne^2}{kT} \Phi \left[1 - p \int_{0}^{\infty} \exp[-p\tau] \exp[-k^2 \tau^2 v_{\mathrm{th}}^2] \, \mathrm{d}\tau \right],$$

where

$$v_{\mathrm{th}}^2 = \frac{kT}{2m} \,.$$

We have solved for each species in its own rest frame.

Defining the characteristic function

$$W(\lambda) = i\lambda \int_0^\infty \exp\,[-y^2]\exp[-i\lambda y]\,\mathrm{d}y - 1\,, \tag{2.4}$$

and using Poisson's equation

$$\nabla^2 \Phi = -4\pi\varrho\,,$$

one obtains the dispersion relation

$$F(p) = k^2 - \frac{1}{\lambda_{\mathrm{Di}}^2}\, W\left(\frac{-ip}{|k|\,v_{\mathrm{thi}}}\right) - \frac{1}{\lambda_{\mathrm{De}}^2}\, W\left(\frac{p+iky}{i\,|k|\,V_{\mathrm{the}}}\right) = 0\,. \tag{2.5}$$

It can be shown that the asymptotic forms of W are

$$\begin{cases} W_r(Z) = -1 \quad |Z| \ll 1\,, \\ W_r(Z) = \dfrac{1}{2Z^2} \quad |Z| \gg 1\,, \\ W_i(Z) = i\sqrt{\pi}\, Z \exp\,[-Z^2]\,, \end{cases} \tag{2.6}$$

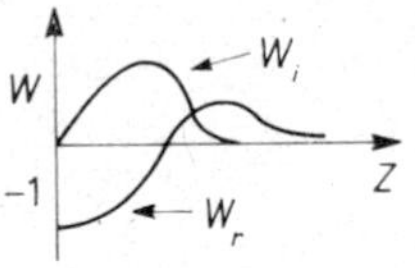

Fig. 3.

A rough plot of W is given in Fig. 3.

The zero-temperature theory results from replacing W by its asymptotic value for large z. This would be appropriate for the case where the drift velocity u exceeds the thermal velocities and results in the equation

$$\frac{\omega_{\mathrm{pi}}^2}{\omega^2} + \frac{\omega_{\mathrm{pe}}^2}{(\omega - ku)^2} = 1\,, \tag{2.7}$$

with the well known instability for $ku < \omega_{\mathrm{pe}}$ having most rapid growth for $ku \sim \omega_{\mathrm{pe}}$ with growth rate $(\omega_{\mathrm{pe}})^{\frac{1}{3}}(\omega_{\mathrm{pi}})^{\frac{2}{3}}$. A more interesting case comes for $V_{\mathrm{the}} > u > V_{\mathrm{thi}}$. In this case we may expect

$$\frac{\omega}{|k|\,v_{\mathrm{thi}}} > 1 \qquad \text{and} \qquad \frac{\omega + ku}{|k|\,v_{\mathrm{the}}} < 1\,.$$

Making the appropriate expansions of the W functions in eq. (5), small z for electrons, large z for ions gives the result (for $k^2\lambda_{\mathrm{De}}^2 \ll 1$)

$$\omega^2 = 2k^2 v_{\mathrm{thi}}^2 \frac{T_{\mathrm{e}}}{T_{\mathrm{i}}} \cdot \left\{1 - 2\,\frac{\sqrt{\pi}}{2}\,\frac{(\omega - ku)}{|k|\,v_{\mathrm{th\,e}}} - i\,\frac{\sqrt{\pi}}{2}\,\frac{\omega}{|k|\,v_{\mathrm{th\,i}}} \exp\left[-\left(\frac{\omega}{2\,|k|\,v_{\mathrm{th\,i}}}\right)^2\right]\frac{T_{\mathrm{e}}}{T_{\mathrm{i}}} + \dots\right\}. \tag{2.8}$$

Neglecting for the moment the growth or damping as represented by the imaginary terms, we see that eq. (8) represents ion sound waves driven by electron pressure with ion inertia. The imaginary terms proportional to ω represent the usual Landau damping. The possibility of instability of course arises from the term proportional to the drift velocity. For $T_e \approx T_i$ the ion term dominates, since $V_{th\,i} \ll V_{th\,e}$, giving a heavily damped wave.

We may now write down the critical condition $\left(\text{using } \omega \sim kV_{th\,i}\sqrt{T_e/T_i}\right)$. For instability,

$$\frac{u}{\sqrt{2}} > v_{th\,i}\sqrt{\frac{T_e}{T_i}} + \left(\frac{T_e}{T_i}\right)^{\frac{3}{2}} \exp\left[-\frac{T_e}{2T_i}\right] v_{th\,e}\,. \tag{2.9}$$

This shows that for $T_e \sim T_i$ extremely large drifts are necessary to produce instability—of the order of electron thermal velocities. On the other hand for quite cold ions $T_e/T_i > 10$, as may occur in the early stages of ohmic heating, the critical velocity approaches the ion thermal velocity. In fact, if we had considered the case $k\lambda_{De} \sim 1$ instead of $k\lambda_{De} < 1$, the term $\sqrt{T_e/T_i}$ in the first term of eq. (9) would have disappeared.

We may now take a brief look at the nonlinear limit of this instability from the point of view of the generalized entropy discussed earlier. We recall that, consistent with the conservation of momentum, the lowest possible internal energy state for the plasma constituents is that of the Maxwell distribution. The ions are already in this lowest energy state; the electrons differ only slightly from it as $f \sim \exp[-(v-u)^2]$ instead of $\exp[-v^2]$. Hence, the maximum free energy available for the instability is $\frac{1}{2}nmu^2$. This limit allows one to make a rough estimate of the crossed-field diffusion as discussed by Sturrock which turns out to be very small for $u \sim v_{th\,i}$. However, it should be noted that the situation which usually obtains in practice is that in addition to the initially displaced Maxwellian there is also present an applied electric field which may be continually pumping energy into the instability. Thus, the relevant nonlinear theory has not yet been studied.

It is now of interest to consider the same instability problem for the case of an infinite homogeneous plasma in a constant magnetic field. Strictly speaking one should consider here the full set of Maxwell's equations instead of just Poisson's equation so that the dispersion relative is a 3×3 determinant. However, in the limit of small β, the only possible instabilities are those which do not perturb the magnetic field as any perturbation of the field would represent an increase of energy of order $1/\beta$ compared to any decrease which might result from lowering the plasma energy. Mathematically this comes about when the full determinant is written down by a separation of the determinant into a form $D \simeq D_L + \beta D_T$ where D_L is the dispersion relation obtained from the Poisson equation.

Thus our task is reduced to the evaluation of eq. (3) for the orbits in the constant magnetic field. We take B along the z axis and k in the (y, z) plane. The orbit integral appearing in (3) is thus

$$\int_{-\infty}^{0} \exp[p\tau] \exp[ik_{\parallel} Z] \exp[ik_{\perp} y] \, d\tau \, ,$$

where

$$\begin{cases} z = v_z \tau \\ y = \dfrac{v_{\perp}}{\Omega} [\sin(\Omega\tau + \varphi) - \sin\varphi] , \\ x = -\dfrac{v_{\perp}}{\Omega} [\cos(\Omega\tau + \varphi) - \cos\varphi] . \end{cases} \tag{2.10}$$

Here Ω is the cyclotron frequency and φ a phase angle in velocity space determining the phase of the particle at $t = 0$. In order to utilize Poisson's equation, we must now perform the time integral indicated above as well as the integral over all velocities. Thus we have to do the integrals

$$\int v_{\perp} \, dv_{\perp} \int d\varphi \int dv_z \int d\tau \, \exp[p\tau] \exp[ik_z v_z \tau] \exp[ik_{\perp} y] \exp\left[-\frac{mv^2}{2kT}\right] .$$

It is convenient to use the identity

$$\exp[i\alpha \sin u] = \sum_{n=-\infty}^{\infty} \exp[inu] J_n(\alpha) \, ,$$

where J is the Bessel function of the first kind. Thus

$$\exp[ik_{\perp} y] = \sum_{n,m} J_n\left(\frac{k_{\perp} v_{\perp}}{\Omega}\right) J_m\left(\frac{k_{\perp} v_{\perp}}{\Omega}\right) \exp[in(\Omega\tau + \varphi)] \exp[-im\varphi] . \tag{2.11}$$

The integral over φ is now easily performed to give δ_{nm}. The V_z integral is done as before and the integral over $V_{\perp}$ is done with help of the integral

$$\int_0^{\infty} \exp[-ax^2] J_n^2(\beta x) x \, dx = \frac{1}{2\alpha} \exp\left[\left(-\frac{\beta^2}{2\alpha}\right) I_n\left(\frac{\beta^2}{2\alpha}\right)\right] . \tag{2.12}$$

where I_n is the Bessel function of the second kind.

Finally, the time integral can be expressed in terms of the W functions to give the result for each species

$$(2.13)\qquad 4\pi\varrho = \varphi\left\{\frac{1}{\lambda_D^2}\exp\left[\frac{-k_\perp^2 v_{th}^2}{\Omega^2}\right]\sum_n I_n\left(\frac{k_\perp^2 v_{th}^2}{\Omega^2}\right)\cdot\right.$$
$$\left.\cdot\left[W\left(\frac{-\omega+k_z u+n\Omega}{k_z v_{ht}}\right)-\frac{n\Omega}{n\Omega+k_z u-\omega}\left[1+W\left(\frac{n\Omega+k_\parallel u-\omega}{k_z v_{th}}\right)\right]\right]\right\}.$$

Insertion of (13) into the Poisson's equation finally allowes us to examine the 2-stream instability in the presence of magnetic field. If we had also kept different $T_\parallel$ and $T_\perp$ a similar form for the Harris instability could easily have been found while the full 3×3 determinant would be necessary to study such effects as the mirror and Alfvén instabilities.

3. – We discuss next another version of the two-stream instability appropriate to the magnetic field.

In a recent experiment, D'ANGELO and coworkers have observed oscillations near the ion cyclotron frequency in a tepid Cs plasma when the electron drift velocity parallel to the magnetic field exceeds about three times the ion thermal velocity. Under the conditions of this experiment the ratio of kinetic to magnetic pressure is $\beta \approx 10^{-7}$ and $T_e \cong T_i$. The usual theory of the two-stream instability, which considers only waves propagating parallel to the magnetic field, would predict stability until the electron drift velocity becomes comparable to the electron thermal velocity. Only in situations where $T_e \gtrsim T_i$ does the critical velocity approach the ion thermal velocity. In the following, we shall show that for a collisionless plasma instability occurs at a much lower velocity for electrostatic waves near the ion cyclotron frequency propagating at large angle to the field.

We consider a homogeneous infinite plasma in which the ions and electrons each have a Maxwellian distribution at a characteristic temperature with the center of the Maxwellians displaced by a drift velocity, u. We will work in the frame where the ions are at rest. If $u=0$, the plasma is evidently stable. We would therefore expect that instability could only occur if $k_\parallel u > \nu$ where ν is the wave frequency and $k_\parallel$ the wave number parallel to the field. This is the condition that the peak of electron distribution be moving slightly faster than the wave—the usual condition for being able to put energy into the wave.

In this case, if we put $\nu \sim \Omega_i$ the ion cyclotron frequency, and $u \sim v_{th\,i}$—the case we wish to discuss—we obtain the condition $k_\parallel R_{L_i} > 1$. The significance of this is that for large k and low β only pure electrostatic waves are possible. Thus, if we write down the dispersion matrix

$$\nabla\times\nabla\times\boldsymbol{E}+\ddot{\boldsymbol{E}} = -4\pi\boldsymbol{j}$$

in a co-ordinate system in which one of the axes is parallel to k, determining $\boldsymbol{j}$ from the Boltzmann equation; we find the dispersion matrix has the following structure:

$$\begin{pmatrix} k^2 - \nu^2 + (\alpha_{11}) & (\alpha_{12}) & (\alpha_{13}) \\ (\alpha_{21}) & k^2 - \nu^2 + (\alpha_{22}) & (\alpha_{23}) \\ (\alpha_{31}) & (\alpha_{32}) & -\nu^2 + (\alpha_{33}) \end{pmatrix} = 0\,, \tag{3.1}$$

where the quantities (α_{ij}) arise from the plasma currents and are all of order $\nu^2/(k\lambda_D)^2$. We note that for $k^2 \gg \nu^2$; $\nu^2/(k\lambda_D)^2$, the only possible root is $\nu^2 = (\alpha_{33})$—the pure electrostatic mode in which $E \| \boldsymbol{k}$. If we put

$$\nu \sim \Omega_i\,, \qquad k \sim \frac{\Omega_i}{v_{\text{th i}}}\,, \quad \text{we find} \quad (k\lambda_D)^2 \approx \frac{B^2}{4\pi n m_i c^2} \ll 1\,,$$

and

$$\frac{k^2(k^2\lambda_D^2)}{\nu^2} = \frac{B^2}{4\pi n m_i v_{\text{th i}}^2} = \frac{1}{\beta} \gg 1\,.$$

Thus we need only consider pure electrostatic modes.

For this case, the dispersion relation has been given by many authors, *e.g.*, BERNSTEIN [1]

$$\sum_{j=i,e} \sum_n \frac{1}{T_j} \Gamma_n(k_\perp^2 R_{L_j}^2) \left\{ W\left(\frac{\nu + k_\| u_j + n\Omega_j}{|k_\||\, v_{\text{th}\, j}}\right) - \frac{n\Omega_j}{\nu + k_\| u_j + n\Omega_j} \cdot \left[1 + W\left(\frac{\nu + k_\| u_j + n\Omega_j}{|k_\||\, v_{\text{th}\, j}}\right)\right]\right\} = 0\,. \tag{3.2}$$

Here T is the temperature, u_j the drift velocity of the j species, $\Gamma_n(x) = \exp[-x] I_n(x)$, I_n the usual Bessel function of imaginary argument, and

$$W(x) = -1 + \frac{x}{\sqrt{\pi}} \int_{-\infty}^{\infty} \frac{\exp[-y^2]}{x + \tau}\, d\tau\,. \tag{3.3}$$

The contour of the integral may be taken along the real axis for x in the lower half-plane (growing waves). Limiting forms are given by

$$\begin{cases} W \approx -1 + i\sqrt{\pi}\, x & |x| \ll 1\,, \\ W \approx \dfrac{1}{2x^2} & |x| \gg 1\,. \end{cases} \tag{3.4}$$

For all real x, $IP(W) = i\sqrt{\pi}\, x \exp[-x^2]$. The limiting form given for large x is not valid for highly damped waves which are of no concern here.

Since we are concerned with wave lengths comparable to the ion-gyroradius, $k_\perp R_{Le} \ll 1$; $\Gamma_{0,e} = 1$; $\Gamma_{n\neq 0,e} = 0$. Moreover we see that since $u/v_{\text{th e}} \ll 1$ the argument of the electron W function is very small. Since we are concerned with a wave nearly at resonance with the ion-gyrofrequency, we may neglect all the ion terms except $n = -1$. This gives us

$$(3.5)\qquad 0 = \left(-1 + i\sqrt{\pi}\,\frac{(\nu + k_\parallel u)}{|k_\parallel|\, v_{\text{th e}}}\right) + \\ + \frac{T_e}{T_i}\Gamma_1\left[W\left(\frac{\nu - \Omega_i}{|k_\parallel|\, v_{\text{th i}}}\right) + \frac{\Omega_i}{\nu - \Omega_i}\left\{1 + W\left(\frac{\nu - \Omega_i}{|k_\parallel|\, v_{\text{th i}}}\right)\right\}\right].$$

We may obtain an approximate solution by noting that a condition for solution is a large argument for the ion W function as otherwise the large imaginary part (ion cyclotron damping) will give a damped solution. If

$$(3.6)\qquad \left|\frac{\nu - \Omega_i}{k_\parallel v_{\text{th i}}}\right| \gg 1\,,$$

we have simply

$$(3.7)\qquad \begin{cases} \dfrac{\Omega_i}{\nu - \Omega_i}\dfrac{T_e}{T_i}\Gamma_1 = \left[1 - i\sqrt{\pi}\left(\dfrac{\Omega_i - k_\parallel u}{k_\parallel v_{\text{th e}}}\right)\right], \\ \text{or} \\ \nu - \Omega_i = \Omega_i \dfrac{T_e}{T_i}\Gamma_1\left\{1 + i\sqrt{\pi}\left(\dfrac{\Omega_i}{k_\parallel v_{\text{th e}}} - \dfrac{u}{v_{\text{th e}}}\right)\right\}. \end{cases}$$

Here we have chosen $k_\parallel$ negative as the direction of propagation for instability and used $\nu \sim \Omega_i$.

We note that Γ_1 has a very flat maximum at $k_\perp^2 R_L^2 \approx 1.5$ attaining there a value .22. We conclude therefore that the maximum growth rate is given by

$$(3.8)\qquad -IP(\nu) \approx \cdot 4\,\frac{T_e}{T_i}\,\Omega_i\,\frac{u}{v_{\text{th e}}}\,,$$

occurring for $k_\perp^2 R_L^2 \sim 1$ and $k_\parallel > \Omega_i/u$.

Moreover, from eq. (6), we must have

$$\frac{\nu - \Omega_i}{k_\parallel v_{\text{th i}}} \approx \cdot 2\,\frac{T_e}{T_i}\,\frac{\Omega_i}{k_\parallel v_{\text{th i}}} > 1\,,$$

and also $\Omega/k_\parallel V_{\text{th e}} < u/V_{\text{th e}}$ so that a rough criterion for instability is given as

$$\frac{u}{v_{\text{th i}}} > 5 .$$

To refine the stability criterion, we return to eq. (5) and look for a critical value of u which will lead to real frequency ν. The imaginary part of eq. (5) then becomes

$$\frac{\nu}{k_\parallel v_{\text{th e}}} - \frac{u}{v_{\text{th e}}} + \frac{T_e}{T_i}\Gamma_1 \frac{\nu}{k_\parallel v_{\text{th i}}} \exp\left[-\left(\frac{\nu - \Omega_i}{k_\parallel v_{\text{th i}}}\right)^2\right] = 0 , \tag{3.9a}$$

and the real part

$$\frac{T_e}{T_i}\Gamma_1 \frac{\Omega_i}{\nu - \Omega_i} = 1 . \tag{3.9b}$$

We have neglected the small real part of W here as its argument is large. Substituting (9b) into (9a), we have

$$\frac{u}{v_{\text{th e}}} = \left(1 + \frac{T_i}{T_e}\Gamma_1\right)\left[\frac{\nu - \Omega_i}{k_\parallel v_{\text{th i}}}\frac{v_{\text{th i}}}{v_{\text{th e}}}\frac{T_i}{\Gamma_1 T_e} + \frac{\nu - \Omega_i}{k_\parallel v_{\text{th i}}} \exp\left[-\left(\frac{\nu - \Omega_i}{k_\parallel v_{\text{th i}}}\right)^2\right]\right] .$$

The minimum comes for

$$\left(\frac{\nu - \Omega}{k_\parallel v_{\text{th i}}}\right)^2 \approx -\ln \frac{v_{\text{th i}}}{v_{\text{th e}}}\frac{T_i}{\Gamma_1 T_e} \approx \frac{1}{2}\ln\frac{m_i}{m_e} ,$$

or

$$\frac{\Omega}{k_\parallel v_{\text{th i}}} \approx \sqrt{\frac{1}{2}\ln\frac{m_i}{m_e}\frac{T_i}{T_e \Gamma_1}} ,$$

and the critical drift is then

$$\frac{u}{v_{\text{th i}}} \approx \left(\frac{T_i}{\Gamma_1 T_e} + 1\right)\sqrt{\frac{1}{2}\ln\frac{m_i}{m_e}} \approx 10\,\frac{T_i}{T_e} + 2 . \tag{3.10}$$

This formula is not reliable for $T_e > T_i$ as then higher n values must be considered in eq. (9b). None the less, one can see by inspection that as T_e/T_i increases the critical drift decreases.

It would appear then that this instability near the ion-cyclotron frequency reduces the amount of current which can be drawn parallel to the field by about an order of magnitude in the case of equal temperature compared to previous theories which consider only $k_\perp = 0$. None the less, it appears far from a satisfactory explanation of d'Angelo's results for the reason that this theory is based on the collisionless equations. In d'Angelo's experiments $\Omega_i \tau_{coll} \sim 10$, where τ_{coll} is the ion-ion collision time. In view of the low growth rates we calculate, it is difficult to see why collisional damping would not dominate or, in fact, how any coherent wave near the ion cyclotron frequency can exist.

4. – It is of considerable interest to discuss the stability of the simplest sort of magnetic confinement with respect to microinstabilities. In particular, we study a low β situation which implies electrostatic instabilities and consider simple localized instabilities. We also choose the simplest geometry: magnetic field in the z direction, and plasma parameters and field strength varying with χ. Thus, for the equilibrium field, we can take, locally,

$$B = B_0(1 + \varepsilon\chi) \tag{4.1}$$

and construct arbitrary equilibrium distribution functions from the constants of the motion V^2 and $x + V_y/\Omega$. A particularly simple choice is

$$f = n_0 \left(\frac{\alpha}{\pi}\right)^{\frac{3}{2}} \exp[-\alpha v^2] \left\{1 + (\gamma + \alpha\delta v^2)\left(x + \frac{v_v}{\Omega}\right)\right\}. \tag{4.2}$$

Both ions and electrons must of course have such a distribution function. We keep both the terms γ and δ in order to allow for the possibility of both density and temperature gradients. The importance of the temperature gradients in leading to instability was first pointed out by TSERKOVNIKOV.

From (2), we have

$$\frac{n'}{n} = \gamma + \frac{3}{2}\delta ,$$

$$\frac{T'}{T} = \delta , \tag{4.3}$$

and from charge neutrality, we must have

$$\left.\frac{n'}{n}\right)_i = \left.\frac{n'}{n}\right)_e .$$

We will also assume for simplicity that the temperatures are equal.

In deciding which perturbation to study, we recall from our earlier discussions that this equilibrium is certainly stable hydromagnetically; and from Low's theorem in the limit of low β and n'/n it is necessary to study new types of modes in order to find the possibility of instability. In particular, we note that in the presence of an inhomogeneity there exists a new phenomenon, the well-known guiding center drift across the field given by $V_0 \sim \lambda V_\perp^2 y$ where $\lambda \approx + \varepsilon/\partial\Omega$. This leads to a new possibility for resonant interaction which can detach particles from field lines and we will hence concentrate our attention on electrostatic perturbations $\varphi \sim \exp[i(ky+\omega t)]$ where $\omega \sim kv_0$, and where the variation of φ with x will be assumed negligible. We consider only modes with $(k \cdot B) = 0$.

As usual, we solve the Boltzman equation by the method of characteristics:

$$f_1 = \frac{e}{m}\int \nabla\varphi \cdot \frac{\partial f_0}{\partial \boldsymbol{v}}\, \mathrm{d}t\,, \tag{4.4}$$

with

$$\frac{\partial f_0}{\partial \boldsymbol{v}} \approx \left\{-2\alpha^* \boldsymbol{v} + \frac{(\gamma + \delta\alpha v')}{\Omega}\,\hat{y}\right\} f_0\,, \qquad \alpha^* = \alpha\left(1 - \delta\left(x + \frac{v_y}{\Omega}\right)\right).$$

In evaluating the time integral, we remember that f_0 is a constant of the motion. After performing the usual integration by parts

$$\frac{m}{e}\, f_1 \approx \left\{-2\alpha^*\varphi + i\left(2\alpha^*\omega + \frac{k\gamma + k\,\delta\alpha v^2}{\Omega}\right)\int \varphi\, \mathrm{d}t\right\} f_0\,. \tag{4.5}$$

As before, we must do the time integration, and then integrate over all velocities to find the charge density. Fortunately in this case we need only use the condition of quasi-neutrality instead of the full Poisson equation since the Debye length is very small.

In evaluating the orbit integral, we recall that, to first order in smallness of the Larmor radius, the only effect on the orbit is the inhomogeneous field drift we have mentioned above. As this is proportional to $v_\perp^2$, we are now unable to perform the integration over $v_\perp$ as before. On the other hand, since we have chosen $k \cdot B = 0$, the Landau damping is absent and the time integral may be performed trivially, as well as the v_z and φ integrations. This reduces the expression to a single integral over $v_\perp$. We introduce a dimensionless notation $\alpha v_\perp^2 = x$. As before, we get terms of the form $\sum_n J_n^2(kR_L)/(\omega - n\Omega)$. Since $\omega \ll \Omega$ we, keep only the term $n = 0$; and, as $kR_L \ll 1$, we put $J_0^2 = 1$ to obtain

$$0 = -b^2 + \sum \frac{1}{\lambda_D^2}\left\{-1 + \int_0^\infty \exp[-x]\,\mathrm{d}x \left[\frac{\omega + \left(\left((\gamma + \frac{1}{2}\delta) + \delta x\right)/2\alpha\Omega\right)k}{\omega + (k\varepsilon/2\alpha\Omega)x}\right]\right\}.$$

Here the summation refers to ions and electrons. For the special case of equal temperatures, the Debye lengths, λ_D, are equal for the two species and $\alpha\Omega)_i \equiv \alpha\Omega)_e$: A slightly more convenient form, for this case, is

$$F(\omega) \equiv -k^2\lambda_D^2 + \sum_{+-}\frac{1}{\varepsilon}\int_0^\infty \frac{(\gamma+\frac{1}{2}\delta)+(\delta-\varepsilon)x}{x\pm 2\alpha\Omega\omega/k\varepsilon}\exp[-x]\,dx = 0\,. \tag{4.6}$$

The $\pm$ indicates that when summing over ions and electrons the proper, respective sign must be chosen.

First, we study the case $\delta = 0$: no temperature gradient. In this case, we see from the equilibrium pressure balance, $\varepsilon \approx -\beta\gamma$. Thus, the numerator, $\gamma - \varepsilon\chi$, is always negative if we adopt the sign convention $\varepsilon > 0$. In drawing the Nyquist diagram, we regard $\omega = \omega - i\Delta$ while performing the contour plot of $F(\omega)$. We note that for $\omega > 0$ only the electrons, *i.e.* the minus sign in eq. (6), contribute an imaginary part and for $\omega < 0$ only the ions contribute.

We conclude further that for $\omega > 0$, $\operatorname{Im} F > 0$ and for $\omega < 0$, $\operatorname{Im} F < 0$. Hence $\operatorname{Im} F = 0$ only at $\omega = 0$ and the question of stability *i.e.* whether the origin is encircled, may be decided by looking at $\operatorname{Re} F$ at $\omega = 0$ and $\omega = \infty$. At $\omega = \infty$, $\operatorname{Re} F = -(k\lambda_D)^2$, *i.e.*, $\operatorname{Re} F < 0$. At $\omega = 0$, $\operatorname{Re} F = (\gamma/\varepsilon)\ln|\omega|$, *i.e.*, $\operatorname{Re} F < 0$. Hence the Nyquist plot of $F(\omega)$ looks as in Fig. 4. We conclude that the origin is not encircled and that we have a stable situation for the case $\delta = 0$.

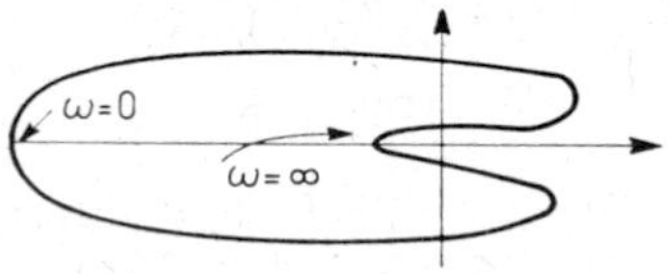

Fig. 4.

For the case $\delta \neq 0$, the Nyquist argument is more involved. First, we show that when $\gamma + \frac{1}{2}\delta > 0$, there exists a simple growing mode. If we put $2\alpha\Omega\omega/k\varepsilon = i\chi$, eq. (6) becomes

$$F(\chi) = \int_0^\infty \frac{(\gamma+\frac{1}{2}\delta)+(\delta-\varepsilon)x}{x^2+\chi^2}\,x\exp[-x]\,dx = 0\,, \tag{4.7}$$

where we have neglected $k^2\lambda_D^2$ as being very small. As $\chi \to 0$,

$$F(\chi) \to \left(\gamma + \frac{1}{2}\delta\right)\ln|\chi| > 0;\ \text{as}\ \chi\to\infty,\ F(\chi)\to\left(\gamma+\frac{5}{2}\delta\right) - 2\varepsilon \approx -\varepsilon\left(2+\frac{1}{\beta}\right),$$

i.e., $F(\chi) < 0$. Hence, there must be at least one purely growing root $\omega = -i\chi$. Thus, $(\gamma + \frac{1}{2}\delta) > 0$ corresponds to an unstable equilibrium; or, if $(P'/P - 2(T'/T))\,(B'/B) > 0$, the plasma is unstable.

From the complete Nyquist analysis, to be given in the Appendix, we also conclude that the case $(\gamma + \frac{1}{2}\delta) < 0$, $(\delta - \varepsilon) > 0$ corresponds to instability with complex frequency. We may summarize the results by requiring for stability

$$\frac{1}{2} > \frac{\partial \ln T}{\partial \ln P} > -\frac{1}{2}\beta\,. \tag{4.8}$$

In all cases, the growth rate is of order $\omega \sim kv_{\text{D}}$. The physical mechanism is obscure except that we may note that for frequencies of this order particles are not tied to field lines and interchange may occur. In particular, the lower limit of (8) corresponds to the case $(T'/T)(B'/B) > 0$ so that an interchange at constant pressure conserving the magnetic moment would lower the energy.

APPENDIX TO SECTION 4 (by N. KRALL).

The Nyquist method allows us to determine directly whether or not eq. (4.6) has any roots, ω, corresponding to instability, *i.e.*, with $\operatorname{Im}\omega < 0$. We observe the change in the argument of $F(\omega)$ as ω goes from $-\infty$ to $+\infty$ along the real axis. We define the integrals in eq. (4.6) by letting ω have a negative imaginary part, $\lim_{\Delta\to 0}(\omega_R - i\Delta)$. Then, we have for the imaginary part of $F(\omega)$

$$\operatorname{Im} F(\omega) = -\frac{\pi}{\varepsilon}\left[\left(\gamma + \frac{1}{2}\,\delta\right) + (\delta - \varepsilon)\,\frac{2\alpha\Omega\omega}{k\varepsilon}\right]\exp\left[-\left(\frac{2\alpha\Omega\omega}{k\varepsilon}\right)\right] \qquad \omega > 0\,,$$

and

$$\operatorname{Im} F(\omega) = \frac{\pi}{\varepsilon}\left[\left(\gamma + \frac{1}{2}\,\delta\right) + (\delta - \varepsilon)\,\frac{2\alpha\Omega\,|\omega|}{k\varepsilon}\right]\exp\left[-\left|\frac{2\alpha\Omega\omega}{k\varepsilon}\right|\right] \qquad \omega < 0\,. \tag{4.9}$$

We take the magnetic field gradient ε to be positive. Note that $\operatorname{Im} F(\omega)$ changes sign as ω goes through $\omega = 0$. We continue our mapping of $F(\omega)$ for real ω by noting that

$$\operatorname{Re} F(\omega) = -\,k^2\lambda_{\text{D}}^2\,, \qquad \omega \to \pm\infty$$

and

$$\operatorname{Re} F(\omega) = (\gamma + \tfrac{1}{2}\,\delta)\,g(\omega)\,, \qquad g(\omega) \to +\infty \quad \text{as} \quad \omega \to 0^{\pm}\,. \tag{4.10}$$

In order to determine the change in the argument in $F(\omega)$ as ω goes from $-\infty$ to $+\infty$, it is unnecessary to actually calculate $\operatorname{Re} F(\omega)$ for all values of ω. The only significant values of $\operatorname{Re} F$ are at values of ω at which $\operatorname{Im} F$ changes sign. In fact, only the sign of $\operatorname{Re} F$ where $\operatorname{Im} F$ changes sign is necessary. This is true because, since $F(\omega = +\infty) = F(\omega = -\infty)$, a change in the argument of F implies that the parametric plot $F(\omega)$ encircles the origin $(F = 0)$. For example, if $\operatorname{Im} F(\omega) > 0$ for $\omega < 0$, and $\operatorname{Im} F < 0$ for $\omega > 0$

then the two possible Nyquist diagrams are given by Fig. 5 and Fig. 6. Figure 5 corresponds to the equation $F(\omega) = 0$ having no roots corresponding to instability; Fig. 6 means there exists one root, ω_0, corresponding to instability, since in Fig. 6 $F(+\infty) = F(-\infty)\exp[2\pi i]$, the argument of F having changed by 2π. The sign of $\operatorname{Re} F(\omega = 0)$ thus determines the stability.

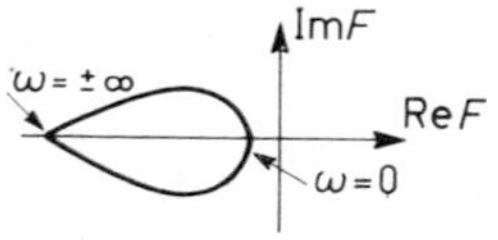

Fig. 5.

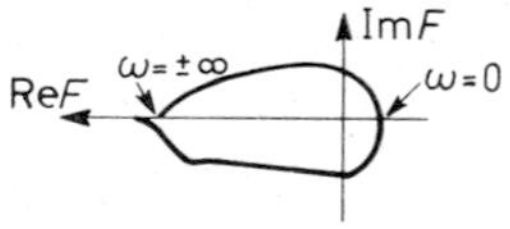

Fig. 6.

We observe that for the case $(\gamma + \frac{1}{2}\delta) > 0$, $\delta > \varepsilon$, it is indeed true that $\operatorname{Im} F < 0$ for $\omega > 0$ and $\operatorname{Im} F > 0$ for $\omega < 0$. However, since $p'/p < 0$ and

$$\gamma + \tfrac{1}{2}\delta = \left(\gamma + \frac{5}{2}\delta\right) - 2\delta = \frac{p'}{p} - 2\delta\,,$$

we cannot have simultaneously $(\gamma + \frac{1}{2}\delta) > 0$ and $\delta > 0$.

We now systematically investigate all other possible ranges of γ and δ. For $\gamma + \frac{1}{2}\delta > 0$, $\delta < 0$, $\operatorname{Im} F(\omega)$ changes sign for values of ω other than $\omega = 0$.

In particular, when $\gamma + \frac{1}{2}\delta > 0$, $\delta - \varepsilon < 0$, we have

$$(4.11)\qquad \begin{cases} \operatorname{Im} F(\omega) > 0\,, & \omega > \dfrac{(\gamma + \frac{1}{2}\delta)}{2\alpha\Omega}\,\dfrac{k\varepsilon}{|\delta - \varepsilon|} \equiv \omega_1 > 0\,, \\ \operatorname{Im} F(\omega) < 0\,, & 0 < \omega < \omega_1\,, \\ \operatorname{Im} F(\omega) > 0\,, & 0 > \omega > -\omega_1\,, \\ \operatorname{Im} F(\omega) < 0\,, & \omega < -\omega_1\,. \end{cases}$$

We, therefore, need to know the signs of $\operatorname{Re} F$ at $\omega = 0$, $\pm\infty$, $\pm\omega_1$. From eq. (10), we conclude that $\operatorname{Re} F(\pm\infty) < 0$, $\operatorname{Re} F(0) > 0$, and we need only calculate the sign of $\operatorname{Re} F(\pm\omega_1)$. From eq. (6), we find directly that

$$\operatorname{Re} F(\pm\omega_1) = -k^2\lambda_D^2 + \frac{2}{\varepsilon}(\delta - \varepsilon) + \frac{2}{\varepsilon}\int_0^\infty \frac{(\gamma + \frac{1}{2}\delta)\exp[-x]\,\mathrm{d}x}{x + 2\alpha\Omega\omega_1/k\varepsilon}\,,$$

where

$$(4.12)\qquad \omega_1 = \frac{k\varepsilon}{2\alpha\Omega}\,\frac{(\gamma + \frac{1}{2}\delta)}{|\delta - \varepsilon|}\,.$$

By omitting the x in the denominator of

$$\int_0^\infty \frac{1}{x + x}\exp[-x]\,\mathrm{d}x\,,$$

we obtain the inequality

$$\operatorname{Re} F(\pm\,\omega_1) < -\,k^2\lambda_D^2 + \frac{2}{\varepsilon}\,(\delta-\varepsilon) + \frac{2}{\varepsilon}\,|\,\delta-\varepsilon\,| = -\,k^2\lambda_D^2 < 0\,,$$

where we have observed that we are solving the case $\delta-\varepsilon<0$. We have, therefore, established that the Nyquist diagram has the form indicated in Fig. 7.

Since the argument of F changes by 2π in this diagram as ω goes from $+\infty$ to $-\infty$, we conclude that the equation $F(\omega)=0$ has one root corresponding to instability when $\varepsilon'+\delta/2>0$, $\delta-\varepsilon<0$, which, expressed in terms of temperature and pressure gradients becomes

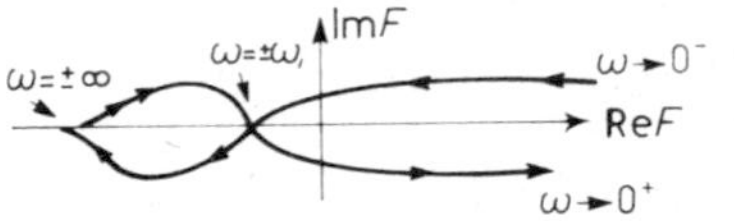

Fig. 7.

$$\frac{P'}{P}-2\,\frac{T'}{T}>0\,,\qquad \frac{T'}{T}-\frac{B'}{B}<0\,. \tag{4.13}$$

We next investigate the case $(\gamma+\frac{1}{2}\,\delta)<0$, $\delta-\varepsilon<0$. Here we see at once from eq. (9) that the $\operatorname{Im} F(\omega)$ changes sign only at $\omega=0$. Since, from eq. (10), $\operatorname{Re} F(0)<0$ in this case, we conclude that the Nyquist diagram has the form of Fig. 5, and there are no unstable roots for this range of γ and δ.

Finally, we consider $(\gamma+\frac{1}{2}\,\delta)<0$, $\delta-\varepsilon>0$. In this case, we again write from eq. (9),

$$\left\{\begin{array}{lll}\operatorname{Im} F(\omega)<0\,, & \omega>\dfrac{k\varepsilon}{2\alpha\Omega}\,\dfrac{|\gamma+\frac{1}{2}\,\delta|}{\delta-\varepsilon}\equiv\omega_1\,, \\ \operatorname{Im} F(\omega)>0\,, & 0<\omega<\omega_1 \\ \operatorname{Im} F(\omega)<0\,, & 0>\omega>-\,\omega_1\,, \\ \operatorname{Im} F(\omega)>0\,, & \omega<-\,\omega_1\,. \end{array}\right. \tag{4.14}$$

We again have $\operatorname{Re} F(\pm\,\omega_1)$ given by eq. (12). However, since in this case $\gamma+\frac{1}{2}\,\delta<0$ the inequality

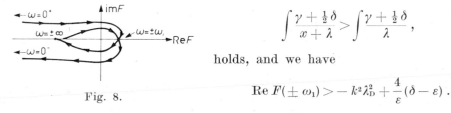

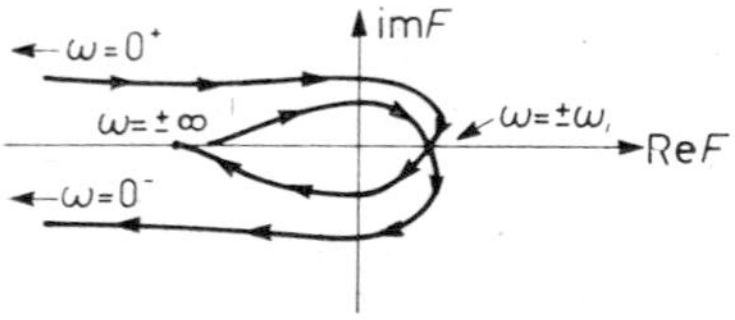

Fig. 8.

$$\int\frac{\gamma+\frac{1}{2}\,\delta}{x+\lambda}>\int\frac{\gamma+\frac{1}{2}\,\delta}{\lambda}\,,$$

holds, and we have

$$\operatorname{Re} F(\pm\,\omega_1)>-\,k^2\lambda_D^2+\frac{4}{\varepsilon}(\delta-\varepsilon)\,.$$

The Nyquist diagram then takes the form of Fig. 8 when

$$\frac{4}{\varepsilon}\,(\delta-\varepsilon)>k^2\lambda_D^2\,,$$

certainly true for low β.

The plasma is therefore unstable in this case, which translated into T'/T and P'/P is

$$\frac{P'}{P}-\frac{2T'}{T}<0\,, \quad \frac{T'}{T}-\frac{B'}{B}>\frac{k^2\lambda_D^2}{4}\frac{B'}{B}\,. \tag{4.15}$$

There are two unstable roots in this case since $F(+\infty)=F(-\infty)\exp[4\pi i]$. We can combine eq. (13) and eq. (15) into one stability condition if we set $k^2\lambda_D^2(B'/B)\sim 0$: equivalent to taking $\beta=P/(B^2/8\pi)\ll 1$.

Writing

$$\frac{T'}{T}\frac{P}{P'}=\frac{\partial \ln T}{\partial \ln P} \quad \text{and} \quad \frac{B'}{B}=-\frac{\beta}{2}\frac{P'}{P}\,,$$

we have

$$\frac{1-2(\partial \ln T/\partial \ln P)}{\partial \ln T/\partial \ln P+\beta/2}>0 \qquad \text{for stability}\,, \tag{4.16a}$$

or, rewriting,

$$\frac{1}{2}>\frac{\partial \ln T}{\partial \ln P}>-\frac{1}{2}\beta \qquad \text{for stability}\,. \tag{4.16b}$$

REFERENCE

[1] I. B. BERNSTEIN: *Phys. Rev.*, **109**, 10 (1958).

Instabilities due to Finite Resistivity or Finite Current-Carrier Mass.

H. P. Furth

Lawrence Radiation Laboratory - Livermore, Cal.

This lecture series will be composed of two parts. Part A consists mainly of extracts from the paper « Finite-resistivity instabilities of a plane sheet pinch », written in collaboration with M. N. Rosenbluth and J. Killeen. The complete paper is available by request. Part B discusses some of the material on collisionless sheet-pinch instabilities and related instabilities in references [2] and [3]. An excellent generalization of some of this material was given in the seminar by D. Pfirsch (see also ref. [4]).

The purpose of presenting the material of Parts A and B together is to bring out certain similarities between hydromagnetic instabilities due to finite resistivity and instabilities of the collisionless sheet-pinch type, which may be said to be due to the finite mass of the current carriers. In both cases the instability depends on $\boldsymbol{E}+\boldsymbol{v}\times\boldsymbol{B}\neq 0$. In the first case, the $\eta\boldsymbol{j}$ term in Ohm's law is the important one; in the second case, the $(mc^2/ne^2)\,\mathrm{d}\boldsymbol{j}/\mathrm{d}t$ term.

Part A: Instabilities due to Finite Resistivity.

1. – Introduction.

A principal result of pinch and stellarator research has been the observed instability of configurations that the hydromagnetic theory would predict to be stable in the limit of perfect electrical conductivity. In order to establish the cause of this observed instability, the extension of the hydromagnetic analysis to the case of finite conductivity becomes of considerable interest.

In the present analysis, general equations are derived for the plane resistive current layer in the incompressible hydromagnetic approximation. A dispersion relation is obtained for high but finite conductivity that describes purely growing modes. An interchange mode driven by a gravitational field perpendicular to the plane layer is discussed briefly. A more thorough treatment of this mode is given in the complete version of our paper.

The analysis for the plane current layer is particularly significant in the high-conductivity limit, since the problem then separates into the analysis of two regions: 1) a narrow central region, where finite conductivity permits relative motions of field and fluid, and where geometric curvature may be neglected; 2) an outer region, where field and fluid are coupled, as in the infinite-conductivity case, and where generalizations to nonplanar geometry can be introduced as desired.

In Section **2** of this paper we derive the basis equations. In Section **3** we demonstrate some general properties of the unstable modes. In Section **4** we find the solutions in the outer, infinite-conductivity region, and obtain boundary conditions for the finite-conductivity region. In Section **5**, we find solutions for the finite-conductivity region, which, combined with the boundary conditions, give us an eigenvalue relation that determines the instability growth rates. Particular attention is called to Section **6**, where the instabilities are derived in heuristic terms, and their properties are summarized.

2. – Assumptions and basic equations.

We treat an infinite plane current layer specified by

$$\boldsymbol{B}_0 = \hat{x} B_{x0}(y) + \hat{z} B_{z0}(y) \,. \tag{1}$$

The following assumptions are made.

1) The hydromagnetic approximation is assumed to be valid, and the ion pressure and inertia terms are neglected in Ohm's law,

$$\frac{\partial \boldsymbol{B}}{\partial t} = \nabla \times (\boldsymbol{v} \times \boldsymbol{B}) - \nabla \times \left[\frac{\eta}{4\pi} \nabla \times \boldsymbol{B}\right] . \tag{2}$$

As the analysis will show, these assumptions are violated in the treatment of a plasma of sufficiently high conductivity, since the «resistive» modes then develop increasingly sharp discontinuities, and we must expect «finite-Larmor-radius» effects. Plasma stability in the limit of high but finite conductivity (the limit of maximum practical interest) thus depends critically on nonhydromagnetic effects. An isotropic resistivity is assumed in eq. (2), and the mass of the electrons is neglected.

2) The fluid is assumed to be incompressible,

$$\nabla \cdot \boldsymbol{v} = 0 \,. \tag{3}$$

Generalization of the present equations to include the compressible case, however, is straightforward, and does not basically alter the present modes.

3) Viscosity is neglected, so that the equation of motion may be written as

$$\nabla\times\left(\varrho\frac{\mathrm{d}\boldsymbol{v}}{\mathrm{d}t}\right)=\nabla\times\left[\frac{1}{4\pi}(\nabla\times\boldsymbol{B})\times\boldsymbol{B}+\boldsymbol{g}\varrho\right], \tag{4}$$

where ϱ is the mass density and g the acceleration due to gravity. (Allowance for finite viscosity increases the effective mass density and thus decreases the growth rates somewhat).

4) Perturbations in plasma resistivity are assumed to result only from convection,

$$\frac{\partial\eta}{\partial t}+\boldsymbol{v}\cdot\nabla\eta=0\,. \tag{5}$$

The neglect of thermal conductivity along magnetic field lines, however, cannot be justified for a high-temperature plasma. The associated stabilizing effect generally suppresses the « rippling » mode above something like 30 eV. The neglect of ohmic heating in eq. (5) is unimportant in the high-conductivity, short-wave-length limit.

5) Perturbations in $\boldsymbol{g}\varrho$ are assumed to result only from convection

$$\frac{\partial(\boldsymbol{g}\varrho)}{\partial t}+\boldsymbol{v}\cdot\nabla(\boldsymbol{g}\varrho)=0\,. \tag{6}$$

6) The zero-order distribution will be assumed to have $\boldsymbol{v}_0=0$. Strictly speaking, this condition implies

$$\nabla\times(\eta_0\,\nabla\times\boldsymbol{B}_0)=0\,, \tag{7}$$

which will be referred to as the standard case. For modes with sufficiently large growth rates, however, the approximation of null zero-order drift velocity may be valid even if eq. (7) is not strictly satisfied.

Denoting perturbed quantities by the subscript 1,

$$f_1(\boldsymbol{r},t)=f_1(y)\exp[i(k_x x+k_z z)+\omega t]$$

we obtain to first order the set of equations

$$\omega\boldsymbol{B}_1=\nabla\times(\boldsymbol{v}_1\times\boldsymbol{B}_0)-\frac{1}{4\pi}\nabla\times[\eta_0\nabla\times\boldsymbol{B}_1+\eta_1\nabla\times\boldsymbol{B}_0]\,, \tag{8}$$

$$\omega\nabla\times\varrho_0\boldsymbol{v}_1=\nabla\times\left[\frac{1}{4\pi}\{(B_0\cdot\nabla)\boldsymbol{B}_1+(\boldsymbol{B}_1\cdot\nabla)\boldsymbol{B}_0\}+(\boldsymbol{g}\varrho)_1\right], \tag{9}$$

(10) $$\nabla \cdot \boldsymbol{v}_1 = \nabla \cdot \boldsymbol{B}_1 = 0 ,$$

(11) $$\omega \eta_1 + (\boldsymbol{v}_1 \cdot \nabla)\eta_0 = 0 ,$$

(12) $$\omega(\boldsymbol{g}\varrho)_1 + (v_1 \cdot \nabla)(\boldsymbol{g}\varrho)_0 = 0 .$$

From this set of equations, we may separate two that involve only B_{yl} and v_{yl}. In dimensionless form, we have

(13) $$\frac{\psi''}{\alpha^2} = \psi\left(1 + \frac{p}{\tilde{\eta}\alpha^2}\right) + \frac{W}{\alpha^2}\left(\frac{F}{\tilde{\eta}} + \frac{\tilde{\eta}' F'}{\tilde{\eta} p}\right),$$

(14) $$\frac{(\tilde{\varrho} W')'}{\alpha^2} = W\left[\tilde{\varrho} - \frac{S^2 G}{p^2} + \frac{F S^2}{p}\left(\frac{F}{\tilde{\eta}} + \frac{\tilde{\eta}' F'}{\tilde{\eta} p}\right)\right] + \psi S^2\left(\frac{F}{\tilde{\eta}} - \frac{F''}{p}\right),$$

where

$$\begin{aligned}
\psi &= B_{yl}/B ,\\
W &= -\, i v_{yl} k \tau_R ,\\
F &= (k_x B_{x0} + k_z B_{z0})/kB ,\\
k &= (k_x^2 + k_z^2)^{\frac{1}{2}} ,\\
\alpha &= ka ,\\
\tau_R &= 4\pi a^2/\langle\eta\rangle ,\\
\tau_H &= a(4\pi\langle\varrho\rangle)^{\frac{1}{2}}/B ,\\
S &= \tau_R/\tau_H ,\\
p &= \omega\tau_R ,\\
\tilde{\eta} &= \eta_0/\langle\eta\rangle ,\\
\tilde{\varrho} &= \varrho_0/\langle\varrho\rangle ,\\
G &= -\, \tau_H^2 (g/\varrho_0)\frac{\partial \varrho_0}{\partial y} .
\end{aligned}$$

The primes denote differentiation with respect to a dimensionless variable $\mu = y/a$, where a is a measure of the thickness of the current layer. The quantities B, $\langle\eta\rangle$, and $\langle\varrho\rangle$ are measures of the field strength, resistivity, and mass density respectively.

For thermal plasmas, we have approximately $S \sim 0.1 a\, T^2 \beta_0^{-\frac{1}{2}}$, with T in eV. The parameter S exceeds a hundred for most present-day hot-plasma experiments and must become much larger yet in experiments of thermonuclear interest. Accordingly, our primary interest will be in the case $S \to \infty$. Note that the growth rate p is expressed in units of the resistive diffusion time.

Only one component of the $\boldsymbol{B}_0$-field, namely $\boldsymbol{k}(\boldsymbol{k}\cdot\boldsymbol{B})$ appears in eqs. (13) and (14). For any given $\boldsymbol{B}_0$-field having finite shear, we may choose $\boldsymbol{k}$ so that $\boldsymbol{k}\cdot\boldsymbol{B}\sim F$ passes through a null. The typical μ-dependence of F and η that will be considered here is illustrated in Fig. 1*a*.

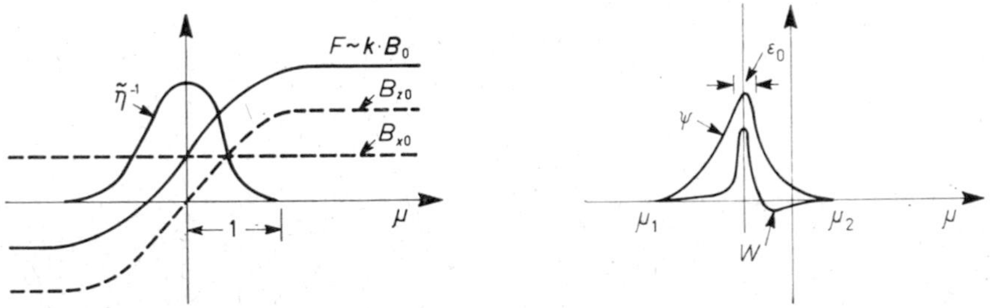

Fig. 1. – *a*) Equilibrium configuration of sheet pinch; *b*) form of the perturbation.

The zero-order equilibrium condition of eq. (7) may be written as

$$\frac{\tilde{\eta}'}{\tilde{\eta}}F' = -F''. \tag{15}$$

The usual boundary conditions are that both ψ and W should vanish at infinity or at conducting boundaries, located at $\mu=\mu_1,\ \mu_2$.

3. – General remarks.

Nonexistence of overstable modes. – Equations (13) and (14) can be solved in the limit $S\to\infty$, to give an oscillatory mode [$\mathrm{Re}\,(p)=0$]. Expanding about this solution in powers of S^{-1}, one finds that when $\mathrm{Im}\,(p)\neq 0$ either $\mathrm{Re}\,(p)=0(1)$ (so that the growth rate is insignificant); or else the zero-order current layer must have sharp resistivity gradients, which become increasingly so as $S\to\infty$. In a more general analysis, including plasma compressibility, etc., we would expect the same result. This follows since the equations can always be expanded in powers of S^{-1}, as long as the zero-order conductivity is large everywhere and has finite gradients. Thus the modes of greatest practical interest are new modes that do not exist at all for $S=\infty$. The situation is similar to that for hydrodynamic sher-flow stability at high Reynolds number.

In the incompressible case, we can show that no overstable modes exist at all, provided that nonequilibrium zero-order configurations are excluded by requiring eq. (7) (or (15)) to be satisfied. This condition is appropriate for configurations with sharp resistivity gradients, since the zero-order drift velocity could not otherwise be neglected. For convenience, we will use a definition of the quantity $\langle\eta\rangle$ such that $\tilde{\eta}F' = 1$. Equations (13), (14) can then be rewritten in the form

$$\frac{p^2}{\alpha^2 S^2 F}\left[(\tilde{\varrho}W')' + \alpha^2 W\left(\frac{S^2 G}{p^2} - \tilde{\varrho}\right)\right] = (p\psi + WF)\left(pF' - \frac{F''}{F}\right), \tag{16}$$

$$= p\psi'' - p\psi\left(\alpha^2 + \frac{F''}{F}\right). \tag{17}$$

Equations (16) and (17) yield the condition

$$\int_{\mu_1}^{\mu_2} d\mu \left[\frac{p^2}{|p|^2\alpha^2 S^2}\left\{\tilde{\varrho}|W'|^2 + \alpha^2|W|^2\left(\tilde{\varrho} - \frac{S^2 G}{p^2}\right)\right\} + \right. \tag{18}$$

$$\left. + \frac{pF' - F''/F}{|pF' - F''/F|^2}|\psi'' - \psi(\alpha^2 + F''/F)|^2 + |\psi'|^2 + |\psi|^2(\alpha^2 + F''/F)\right] = 0 .$$

Taking the imaginary part of eq. (18), we find that if $\mathrm{Im}(p) \neq 0$, then $\mathrm{Re}(p) \leqslant 0$.

Characteristics of unstable modes. – We will devote primary attention to those unstable modes for which $S \to \infty$ and $p \sim S^\nu$ where $0 < \nu < 1$. The lower limit on ν corresponds to a growth rate that is of the same order as the rate of resistive diffusion, and is therefore insignificant. The upper limit on ν is reached only by modes that exist also in the standard infinite-conductivity treatment.

Since the growth rates of the modes to be considered are slow compared with the hydromagnetic rates, the flow is subsonic; *i.e.*, the incompressibility approximation is satisfactory. On the other hand, since the growth-rates are fast compared with resistive diffusion rates, the effect of ohmic heating is negligible.

For unstable modes, all quantities in eq. (18) are real. In the limit $S \to \infty$, eq. (18) can be satisfied in three distinct ways, each corresponding to a negative contribution from one of the three terms: 1) if $G > 0$, there can be gravitationally driven modes; 2) if ψ is peaked near the point $F = 0$, and if we can have $F''/F > 0$ at this point (*i.e.*, if $\eta' \neq 0$ there), then there are modes corresponding to the « rippling » instability; 3) since F''/F is predominantly negative, for sufficiently small α^2 there are modes corresponding to the « tearing » instability.

The behavior of the solutions over most of the range in μ can be established on a general basis. As $S \to \infty$, we must have

$$p\psi \approx -FW \tag{19}$$

everywhere except in a small interval near $F=0$. This condition follows from the consideration that either W or ψ would diverge strongly at large μ if the right-hand term in eq. (16) were either negative or positive except in a small interval. Equation (19) is, of course, the condition that the fluid remains « frozen » to magnetic field lines.

Using eq. (19), we then see from eqs. (16) and (17) that the condition

$$\psi'' - \psi\left(\alpha^2 + \frac{F''}{F} - \frac{G}{F^2}\right) = 0\,, \tag{20}$$

must be satisfied everywhere except in a small interval. The general procedure in the $S\to\infty$, $0<\nu<1$ limit is therefore as follows. We obtain solutions to eq. (20) that vanish at $\mu=\mu_1$, μ_2, the external boundaries. These solutions cannot, in general, be joined without a discontinuity in ψ',

$$\Delta' = \frac{\psi_2'}{\psi_2} - \frac{\psi_1'}{\psi_1}\,, \tag{21}$$

where the subscripts refer to values on either side of the point of juncture. The typical behavior of ψ is illustrated in Fig. 1*b*. The discontinuity in ψ'/ψ corresponds to large local values of ψ''. From eqs. (16) and (17) we see that such values can be obtained only near the point $F=0$. Equation (13) implies that large local values of W are also obtained near the same point and only there. The second stage of the general solution therefore consists in solving for ψ and W in a small region R_0 about the point $F=0$, with the boundary conditions that ψ'/ψ matches the solutions of eq. (20), and that W is well behaved outside the region R_0.

In more formal terms, we may say that eqs. (19) and (20) provide an asymptotic solution of eqs. (16) and (17), which breaks down near $F=0$. We note that if $F\neq 0$ everywhere, then eq. (20) applies throughout, and there is no solution unless $G/(F')^2=0$ [1], in which case the layer is unstable even in the $S=\infty$ limit.

The argument of this section has, for reasons of convenience, made use of eqs. (16) and (17), which refer specifically to the standard case (*i.e.*, eq. (15) holds). The conclusions can, however, be extended to more general choice of F, if desired.

4. – Solutions in the outer region.

We assume that eq. (20) holds everywhere outside a small region R_0 with a width of order ε_0 around the point μ_0 at which $F=0$. Equation (20) is to be solved subject to the boundary condition $\psi=0$ at the points μ_1, μ_2, which we will take for convenience at $\mp\infty$. We will calculate the quantity Δ' of eq. (21) for the case $\psi_1 = \psi_2$ which is of principal interest in Section **5**. Equation (20) yields the expression

$$\Delta' = -2\alpha - \frac{1}{\psi_1}\int_{-\infty}^{\infty} d\mu \exp[-\alpha|\mu|]\,\psi\,\frac{F''}{F} + 0\left\{\frac{G}{(F')^2\varepsilon_0}\right\}. \tag{22}$$

Note that when $F'' \neq 0$ at μ_0, there is a singularity in the integrand on the right side of eq. (22). Difficulties arising from the corresponding logarithmic singularity in ψ' are avoided here, since we consider only Δ' instead of the individual values of ψ_1' and ψ_2'. In this and the following sections we will restrict ourselves to the case where $G/(F')^2$ is sufficiently small so that the G term in eq. (22) can be neglected.

For the case $\alpha^2 \gg 1$, one obtains from eq. (22)

$$\Delta' = -2\alpha + 0\left(\frac{1}{\alpha}\right). \tag{23}$$

For the case $\alpha^2 \ll 1$ we expand

$$\psi = \exp[-\alpha|\mu - \mu_1|]\{\psi_{(0)} + \alpha\psi_{(1)} + \alpha^2\psi_{(2)} + \ldots\} \tag{24}$$

and find

$$-(\psi_{(n)}F' - \psi'_{(n)}F)' = \frac{2|\mu - \mu_0|}{\mu - \mu_0}\,F\psi'_{(n-1)}. \tag{25}$$

The well-behaved solution is characterized by

$$\psi_{(0)} = |F| \tag{26}$$

and near the point μ_0 by

$$\begin{cases} \psi_{(1)} = \dfrac{F^2_{-\infty}}{F'} & \mu < \mu_0, \\ \quad\;\; = \dfrac{F^2_{\infty}}{F'} & \mu > \mu_0. \end{cases} \tag{27}$$

In calculating Δ', the derivative of ψ_1 may be neglected, though it has a logarithmic singularity at μ_0, since the contributions it makes near μ_0 cancel out and the rest is of order α. Thus we obtain

$$\Delta' = \frac{1}{\alpha}(F')^2\left(\frac{1}{F^2_{-\infty}} + \frac{1}{F^2_{\infty}}\right). \tag{28}$$

In the case of symmetric F''/F, it will be of interest to obtain Δ' for arbitrary α. For this purpose it is convenient to choose a specific model. When

$$F = \operatorname{tgh} \mu \, , \tag{29}$$

then eq. (20) may be solved explicitly in terms of associated Legendre functions, and we have

$$\Delta' = 2\left(\frac{1}{\alpha} - \alpha\right) . \tag{30}$$

Note that Δ' goes monotonically from ∞ to $-\infty$ as α goes from 0 to ∞. There is a null of Δ' at the point $\alpha = \alpha_c = 1$.

5. – Solutions in the region of discontinuity.

Basic equations. – In the small region R_0 about the point μ_0 we may take the quantities F', F'', $\tilde{\eta}$, $\tilde{\eta}'$, G and $\tilde{\varrho}$ in eqs. (13) and (14) to be constant. We may approximate F as $F'(\mu - \mu_0)$ and neglect the term $\tilde{\varrho}'W'$ relative to $\tilde{\varrho}W''$.

Defining a new independent variable

$$\theta = \frac{1}{\varepsilon}\left(\mu - \mu_0 + \frac{\tilde{\eta}'}{2p}\right) , \tag{31}$$

equations (13) and (14) may be written as

$$\frac{d^2\psi}{d\theta^2} - \varepsilon^2\alpha^2\psi = \varepsilon\Omega[4\psi + U(\theta + \delta_1)] \, , \tag{32}$$

$$\frac{d^2 U}{d\theta^2} + U\left(\Lambda - \frac{\theta^2}{4}\right) = \psi(\theta - \delta) \, , \tag{33}$$

where

$$\varepsilon = \left\{ \frac{p\tilde{\eta}\tilde{\varrho}}{4\alpha^2 S^2 (F')^2} \right\} , \tag{34}$$

$$U = W \frac{4\varepsilon F'}{p} , \tag{35}$$

$$\Omega = \frac{p\varepsilon}{4\tilde{\eta}} , \tag{36}$$

$$\delta = \frac{1}{4\Omega}\left(\frac{F''}{F'} + \frac{\tilde{\eta}'}{2\tilde{\eta}}\right) , \tag{37}$$

(38) $$\delta_1 = \frac{1}{8\Omega}\left(\frac{\tilde{\eta}'}{\tilde{\eta}}\right),$$

(39) $$\Lambda = \frac{(\tilde{\eta}')^2}{16\varepsilon^2 p^2} + \frac{S^2\alpha^2\varepsilon^2 G}{p^2\tilde{\varrho}} - \alpha^2\varepsilon^2 .$$

Note that in the standard case (cf. eq. (15)) we have $\delta = -\delta_1$.

Let us expand U in terms of normalized Hermite functions

(40) $$U = \sum_{n=0}^{\infty} a_n u_n ,$$

where

(41) $$\frac{d^2 u_n}{d\theta^2} + \left(n + \frac{1}{2} - \frac{\theta^2}{4}\right) u_n = 0 ,$$

and

(42) $$u_n = \frac{(-1)^n}{(2\pi)^{\frac{1}{4}}(n!)^{\frac{1}{2}}} \exp\left[\frac{\theta^2}{4}\right] \frac{d^n}{d\theta^n} \exp\left[-\frac{\theta^2}{2}\right] .$$

Then eq. (33) can be written in the form

(43) $$a_n = \frac{1}{\Lambda - (n + \frac{1}{2})} \int_{-\infty}^{\infty} d\theta_1 u_n \psi(\theta_1 - \delta) .$$

Equations (32) and (33) are valid over the range $(\mu - \mu_0)^2 \ll 1$. We will apply these equations over a region R_0 of width ε_0, outside of which eqs. (19) and (20) are to be valid. From eqs. (32) and (33) it follows than that we must require $\varepsilon_0 > \varepsilon$.

The most important class of unstable modes corresponds to the approximation $\psi = \text{const}$ in R_0. For this case we obtain the « tearing » and « rippling » modes over a range of α consistent with $\varepsilon_0 |\psi/'\psi| < 1$ in R_0, or roughly

(44) $$\varepsilon_0 |\Delta'| < 1 .$$

Using the requirement $\varepsilon_0 > \varepsilon$ and the results of eqs. (23) and (28), we may rewrite eq. (44) as

(44a) $$\varepsilon (F')^2 \left(\frac{1}{F^2_{-\infty}} + \frac{1}{F^2_{\infty}}\right) < \alpha < \frac{1}{2\varepsilon} .$$

For $\psi = \text{const}$ in R_0 we also obtain the low-G gravitational interchange mode. This mode is discussed analytically in the complete version of this paper, and is derived in a heuristic manner in Section **4**.

Solutions with constant ψ. – If $\psi=\text{const}$ and eqs. (44) and (45) are satisfied, we obtain from eqs. (32) and (43)

$$\Delta' = \Omega \sum_{n=0}^{\infty} 4 \left[\int_{-\infty}^{\infty} d\theta_1 u_n\right]^2 + \frac{1}{\Lambda - (n+\frac{1}{2})} \left[\int_{-\infty}^{\infty} d\theta_1 u_n(\theta_1+\delta_1)\right]\left[\int_{-\infty}^{\infty} d\theta_2 u_n(\theta_2-\delta)\right]. \tag{45}$$

Using the integrals

$$\left\{\begin{aligned} \int_{-\infty}^{\infty} d\theta_1 u_n &= 2^{\frac{3}{4}} \left[\frac{\Gamma(n/2+\frac{1}{2})}{\Gamma(n/2+1)}\right]^{\frac{1}{2}}, && n \text{ even} \\ &= 0, && n \text{ odd} \\ \int_{-\infty}^{\infty} d\theta_1 \theta_1 u_n &= 0, && n \text{ even} \\ &= 2^{\frac{9}{4}} \left[\frac{\Gamma(n/2+1)}{\Gamma(n/2+\frac{1}{2})}\right]^{\frac{1}{2}}, && n \text{ odd.} \end{aligned}\right. \tag{46}$$

we obtain the eigenvalue equation

$$\Delta' = 2^{\frac{7}{2}} \Omega \sum_{m=0}^{\infty} \frac{\Gamma(m+\frac{1}{2})}{\Gamma(m+1)} \left\{\frac{\Lambda - \frac{1}{2}}{\Lambda - (2m+\frac{3}{2})} - \frac{\delta\delta_1/4}{\Lambda - (2m+\frac{1}{2})}\right\}, \tag{47}$$

where Δ' is determined by the « outside » solutions (cf. Section **4**). We have replaced even n by $2m$, odd n by $2m+1$.

The sums can be evaluated as hypergeometric series of argument 1 to give

$$\Delta' = 2^{\frac{7}{2}} \pi \Omega \left[\frac{\Gamma(\frac{3}{4} - \Lambda/2)}{\Gamma(\frac{1}{4} - \nabla/2)} + \frac{\delta\delta_1}{8} \frac{\Gamma(\frac{1}{4} - \Lambda/2)}{\Gamma(\frac{3}{4} - \Lambda/2)}\right]. \tag{48}$$

The following general remarks can be made about the solutions of eq. (48).

1) If $\delta\delta_1 < 0$ (the most common case), then Δ'/Ω goes from ∞ to $-\infty$ as Λ goes from $\frac{1}{2}$ to $\frac{3}{2}$, from $\frac{3}{2}$ to $\frac{5}{2}$, etc. The quantity Ω, related to Λ by eqs. (36) and (39), is finite for finite Λ. Hence for any given Δ', as obtained from eq. (21), there is an infinite sequence of eigenvalues $\Lambda \sim 1, 2, 3, \ldots$. There is also an eigenvalue below $\frac{1}{2}$, which moves to 0 as $\Delta' \to \infty$, while Ω becomes large.

2) When $|\Lambda| \ll \frac{1}{2}$, eq. (47) reduces to

$$\Delta' = \Omega(12 + 13\,\delta\delta_1), \tag{49}$$

In that case Ω is to be determined by the value of Δ' in eq. (21), and the condition on Λ is to be verified by means of eqs. (36) and (39). Evidently this case arises only for positive Δ'.

We can now identify a number of basic modes.

The «rippling» mode. – The «rippling» mode is characterized by the finiteness of Λ and the predominance of the $(\eta')^2$ term on the right side of eq. (39). In that case

(50) $$p = \left\{ \frac{(\tilde{\eta}')^2 \alpha S |F'|}{8\Lambda \tilde{\eta}^{\frac{1}{2}} \tilde{\varrho}^{\frac{1}{2}}} \right\}^{\frac{2}{5}},$$

(51) $$\Omega = \frac{|\tilde{\eta}'|}{16\tilde{\eta}\Lambda^{\frac{1}{2}}},$$

(52) $$\delta_1 = 2\Lambda^{\frac{1}{2}},$$

(53) $$\varepsilon = \frac{|\tilde{\eta}'|}{4p\Lambda^{\frac{1}{2}}}.$$

In the standard case we have $\delta\delta_1 = -4\Lambda$, and the remarks made above in paragraph 1) then apply to the Λ-spectrum. (For other reasonable choices of $\delta\delta_1 < 0$ the Λ-spectrum is modified only slightly.)

In the limit $\alpha \gg 1$, which according to eq. (23) corresponds to large negative Δ', we find that the eigenvalues Λ lie slightly below the points $\frac{1}{2}, \frac{3}{2}, \frac{5}{2}, \dots$. For the fastest growing mode, which corresponds to a solution U that is basically symmetric near μ_0, we have $\Lambda \approx \frac{1}{2}$. As we move towards the other limit, $\alpha \ll 1$ (*i.e.*, large positive Δ') the eigenvalues that were slightly below $\frac{3}{2}, \frac{5}{2}, \dots$ move to points slightly above $\frac{1}{2}, \frac{3}{2}, \dots$. The fastest growing mode of this series again occurs for $\Lambda \approx \frac{1}{2}$, and corresponds to a basically symmetric U in the neighborhood of μ_0. From eq. (50) we see that the growth rates of these modes become small as $\alpha \to 0$. The eigenvalue lying below $\frac{1}{2}$ moves toward 0 as $\alpha \to 0$. This mode goes over into the «tearing» mode (see below). These results hold for a range of α defined by eq. (44*a*).

The «tearing» mode. – The «tearing» mode is characterized by the condition $|\Lambda| \ll \frac{1}{2}$. Equation (49) shows that this mode is limited to positive Δ', *i.e.*, to $\alpha \leqslant \alpha_c \sim 1$ (cf. eq. (30)). The growth rate is obtained from eqs. (34) and (36)

(54) $$p = 4\{\alpha S \Omega^2 \tilde{\eta}^{\frac{3}{2}} \tilde{\varrho}^{-\frac{1}{2}} |F'|\}^{\frac{2}{5}}$$

and the condition on Λ can be expressed by means of eq. (39)

(55) $$\Lambda = \frac{1}{\Omega^2}\left[\left(\frac{\tilde{\eta}'}{16\tilde{\eta}}\right)^2 + \frac{pG}{64\tilde{\eta}(F')^2}\right] \ll 1 .$$

For $\alpha \ll 1$, we have from eqs. (28) and (49)

$$\Omega = \frac{1}{12\alpha}(F')^2\left(\frac{1}{F_{-\infty}^2} + \frac{1}{F_{\infty}^2}\right), \tag{56}$$

so that

$$p = (F')^2\left\{\frac{2S\tilde{\eta}^{\frac{3}{2}}}{9\alpha\tilde{\varrho}^{\frac{1}{2}}}\right\}^{\frac{2}{5}}\left\{\frac{1}{F_{-\infty}^2} + \frac{1}{F_{\infty}^2}\right\}^{\frac{4}{5}}. \tag{57}$$

The fastest growing mode is generally obtained for the « symmetric case » where $F''=0$ at $F=0$. A lower limit to α is set by eq. (44a).

$$\alpha > \left(\frac{1}{F_{-\infty}^2} + \frac{1}{F_{\infty}^2}\right)\left\{\frac{\tilde{\eta}\tilde{\varrho}^{\frac{1}{2}}|F'|^9}{3.3S}\right\}^{\frac{1}{4}}. \tag{58}$$

The maximal growth rate p_m thus goes as $S^{\frac{1}{2}}$.

If the current layer is perfectly symmetric, so that the nulls of $\tilde{\eta}'$ and G occur at the same point $\mu=0$, eq. (55) is always satisfied for a mode where $\mu_0=0$. More generally, we see that the $(\eta')^2$ term in eq. (55) is always negligible for $\alpha \ll 1$. The effect of the gravitational term when $G \neq 0$ at μ_0, however can be appreciable.

For modes of the « tearing » type, the solution U is basically antisymmetric, since the $\tilde{\eta}'$ terms can usually be neglected by symmetry or because $\alpha \ll 1$.

6. – Summary and elucidation of principal results.

In the high-S limit, a current layer with finite gradients has three basic unstable modes and no overstable modes. The approximate properties of the

TABLE I.

Mode	Range of instability	Growth rate p	Region of disc. ε	Relevant equations	Valid range of equations
« Rippling »	$\tilde{\eta}' \neq 0$	$\alpha^{\frac{2}{5}}S^{\frac{2}{5}}$	$\alpha^{-\frac{2}{5}}S^{-\frac{2}{5}}$	(50)-(53)	$S^{-\frac{2}{7}} < \alpha < S^{\frac{2}{3}}$ $\lvert G\rvert < \alpha^{-\frac{2}{5}}S^{-\frac{2}{5}}$
« Tearing »	$\alpha < 1$	$\alpha^{-\frac{2}{5}}S^{\frac{2}{5}}$	$\alpha^{-\frac{3}{5}}S^{-\frac{3}{5}}$	(54)-(58)	$S^{-\frac{1}{4}} < \alpha$ $\lvert G\rvert < \alpha^{-\frac{8}{5}}S^{-\frac{2}{5}}$
Gravitational interchange	$G > 0$	$\alpha^{\frac{2}{3}}S^{\frac{2}{3}}G^{\frac{2}{3}}$	$\alpha^{-\frac{1}{3}}S^{-\frac{1}{3}}G^{\frac{1}{6}}$	See complete paper	$S^{-\frac{1}{4}}G^{\frac{1}{8}} < \alpha < S^{\frac{1}{2}}G\ \frac{1}{4}$ $G < 1,\ \alpha^2$ $G > \alpha^{-\frac{2}{5}}S^{-\frac{2}{5}},\ \alpha^{-\frac{8}{5}}S^{-\frac{2}{5}}$

unstable modes in their characteristic parametric ranges are summarized in Table I. Here it has been assumed that the dimensionless quantities F', F'', etc., are all of order unity. References are given to the more exact equations of the previous sections. We shall now discuss and rederive the modes of Table I in heuristic terms.

The existence of the three « resistive » instabilities depends on the local relaxation of the constraint that fluid must remain attached to magnetic field. For a zero-order field that is not a vacuum field, possibilities of lowering potential energy are always present; the introduction of finite conductivity makes some energetically possible modes topologically accessible. In the case of the infinite-conductivity modes, lines of force that are initially distinct must remain so during the perturbation. For the three « resistive » modes, lines of force that are initially distinct link up during the perturbation. These modes have no counterpart in the infinite-conductivity limit and disappear altogether, their characteristic times becoming infinite.

The growth rates of the « resistive » modes are sufficiently small on the hydromagnetic time scale so that the fluid motion is subsonic, *i.e.*, incompressible. This feature is of critical importance in simplifying the analysis of the plane current layer: it permits us to consider the magnetic field and velocity component within the ky-plane independently of the components in the direction $\hat{n}$ normal to the ky-plane. The reasons for this decoupling effect are readily seen.

The co-ordinate along $\hat{n}$ is ignorable; therefore, the field lines of the component $\boldsymbol{B}_n$ are not distorted during the perturbation. The only manner in which the magnitude of $\boldsymbol{B}_n$ could affect the motion in the ky-plane is by way of the magnetic pressure $B_n^2/8\pi$. The gradients of this pressure, however, merely tend to induce plasma compression or expansion. An incompressible fluid automatically provides compensating hydrostatic pressure gradients, so that there is no net effect on the dynamics. As for Ohm's law, there the resistive diffusion term does not couple the field components if the resistivity is isotropic, and the convective term couples the magnetic field and velocity components within the ky-plane to each other. Finally, the two equations specifying $\boldsymbol{B}$ and $\boldsymbol{v}$ to be solenoidal hold as well for the vector components in the ky-plane taken alone. Thus we have four equations for two unknown two-component vectors, and we may restrict ourselves in what follows to the analysis of the two-dimensional problem. Typical field and velocity components in the ky-plane are illustrated in Fig. 2.

To understand the basic character of the unstable modes, let us consider the mechanism whereby the fluid resists detachment from flux lines. Starting with Ohm's law

$$\eta \boldsymbol{j} = \boldsymbol{E} + \boldsymbol{v} \times \boldsymbol{B} . \tag{59}$$

Let us suppose that the fluid is moving but the flux lines are not, *i.e.*, $\boldsymbol{E} \equiv 0$. Then we find $\boldsymbol{j} = (\boldsymbol{v} \times \boldsymbol{B})/\eta$, with a resultant motor force

$$\boldsymbol{F}_s = \boldsymbol{j} \times \boldsymbol{B} = \frac{\boldsymbol{B}(\boldsymbol{v} \cdot \boldsymbol{B}) - \boldsymbol{v}B^2}{\eta}, \tag{60}$$

that opposes the fluid motion. In the limit $\eta \to 0$, this force, of course, prevents any separate fluid motion from taking place. We note, that the restraining force becomes arbitrarily weak near the point where $\boldsymbol{B}$ vanishes, and this is the key to the situation. Since the quantity $\boldsymbol{B}$ in the present discussion

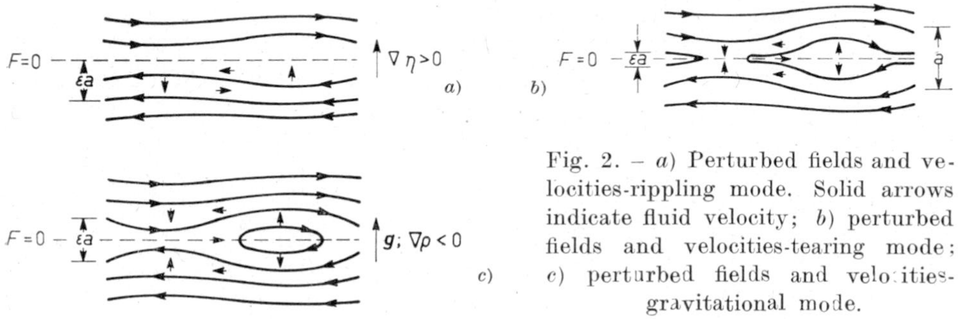

Fig. 2. – *a*) Perturbed fields and velocities-rippling mode. Solid arrows indicate fluid velocity; *b*) perturbed fields and velocities-tearing mode; *c*) perturbed fields and velocities-gravitational mode.

refers only to the magnetic-field components in the ky-plane, we can generally select $\boldsymbol{k}$ so that $\boldsymbol{B}$ has a null at any desired value of y. We may expect that detached fluid motion can take place within a region of order εa about such a null point. For each unstable mode, we will find a driving force $\boldsymbol{F}_d$ that dominates the restraining force $\boldsymbol{F}_s$ within the inner region, and that is itself dominated by $\boldsymbol{F}_s$ outside this region.

We can relate the « skin depth » εa to the growth rate of the instability. Since $\boldsymbol{F}_d$ is comparable in magnitude to $\boldsymbol{F}_s$, the rate at which work is done on the fluid is given by

$$P \sim -\boldsymbol{v} \cdot \boldsymbol{F}_s \sim v_y^2 \frac{(B')^2(\varepsilon a)^2}{\eta}, \tag{61}$$

where we have used $B \sim B'\varepsilon a$. The driving force gives rise to motion both in the $\hat{y}$ and $\boldsymbol{k}$ directions, since $\nabla \cdot \boldsymbol{v} = 0$. In general the instability wavelength will be much larger than εa, and therefore the fluid kinetic energy in the $\boldsymbol{k}$ direction is dominant. Equating the rate of change of this energy to the driving power, we have

$$\omega \varrho \frac{v_y^2}{k^2(\varepsilon a)^2} = \frac{v_y^2(B')^2(\varepsilon a)^2}{\eta}.$$

The skin depth is then given by

$$\varepsilon a \sim \left\{\frac{\omega\varrho\eta}{k^2(B')^2}\right\}^{\frac{1}{4}}, \tag{62}$$

which agrees with eq. (34) when expressed in the appropriate dimensionless variables. To arrive at the instability growth rates, we must next determine εa by comparison of $\boldsymbol{F}_d$ with $\boldsymbol{F}_s$. For this purpose we turn to consideration of specific modes.

In the «rippling» mode of Fig. 2*a*, the circulatory motion of the fluid creates a ridge of lower-resistivity fluid into which the local current is channeled. In other words, when a resistivity gradient exists, Ohm's law in its linearized form has an extra term

$$\eta_0 \boldsymbol{j}_1 = -\eta_1 \boldsymbol{j}_0 + \boldsymbol{v}\times\boldsymbol{B}, \tag{63}$$

where η_1 is given by the convective law

$$\eta_1 = -\frac{\boldsymbol{v}\cdot\nabla\eta_0}{\omega}, \tag{64}$$

and where $\boldsymbol{E}=0$ has again been used, as is appropriate within the small region of decoupled flow. The η_1-term in eq. (63) gives rise to a motor force

$$\boldsymbol{F}_{dr} = \boldsymbol{j}_1\times\boldsymbol{B} = \frac{(\boldsymbol{v}\cdot\nabla\eta_0)}{\omega\eta_0}\,\boldsymbol{j}_0\times\boldsymbol{B} \tag{65}$$

that changes sign as $\boldsymbol{B}$ passes from one side of the null point to the other. Hence, $\boldsymbol{F}_{dr}$ is a stabilizing force on the side of higher resistivity and a destabilizing force on the side of lower resistivity. An unstable mode is obtained if the region of decoupled flow lies on the lower-resistivity side and has a width εa such that the driving power

$$\boldsymbol{v}\cdot\boldsymbol{F}_{dr} \sim v_y^2\,\frac{\eta_0'(B')^2\varepsilon a}{4\pi\eta_0\omega} \tag{66}$$

just dominates $\boldsymbol{v}\cdot\boldsymbol{F}_s$ inside the region. Comparison of eqs. (61) and (66) yields

$$\varepsilon a \sim \frac{\eta_0'}{4\pi\omega}. \tag{67}$$

From eqs. (61) and (67) we can then obtain a growth rate that agrees with eq. (50) and Table I. We note incidentally that the fluid flow and the perturbation current density are strongly peaked in the decoupled region, while the magnetic-field perturbation falls off over a region in y that is of order k^{-1}. In this outer region the fluid and field are well coupled, and a fluid motion of small magnitude accompanies the field perturbation.

We turn next to the gravitational interchange mode, which is quite similar in character to the «rippling» mode. In the presence of a mass-density gradient, and a y-directed gravitational field, the fluid motion gives rise to a force

(68) $$\boldsymbol{F}_{\text{dg}} = \varrho_1 \boldsymbol{g} = \frac{-v_y \varrho_0'}{\omega} \boldsymbol{g} ,$$

which is destabilizing if $\boldsymbol{g}$ points toward decreasing density. Comparison of $\boldsymbol{v}\cdot\boldsymbol{F}_{\text{dg}}$ with eq. (61) gives

(69) $$\varepsilon a \sim \left\{ \frac{\varrho_0' g \eta}{(B')^2 \omega} \right\}^{\frac{1}{2}} .$$

From eqs. (62) and (69) we then obtain a growth rate that agrees with Table I. The mode that decouples the fluid and field most effectively in this case is the counter-circulatory mode shown in Fig. 2c. In the infinite-conductivity case, such a fluid motion would lead to local compressions of $\boldsymbol{B}$, and so could not proceed unless $\boldsymbol{g}$ is large. This phenomenon is known as shear-stabilization. In the mode of Fig. 2c, the opposing flux components brought together at the null in $\boldsymbol{B}$ can cancel out, and so the mode can grow for arbitrarily small $\boldsymbol{g}$. As $\boldsymbol{g}$ increases, the region of substantial motion becomes wider, until conditions for infinite-conductivity instability are reached. We note that if $\boldsymbol{g}$ points in the stabilizing direction, the possibility exists of using $\boldsymbol{F}_{\text{dg}}$ to overcome the driving force $\boldsymbol{F}_{\text{dr}}$, thus stabilizing the «rippling» mode.

The «tearing» mode of Fig. 2b differs from the other two modes in that it is typically a long-wave rather than a short-wave mode relative to the dimension of the current layer. The driving force is due to the structure of the magnetic field outside the region of decoupled flow. (The nature of this force is readily perceived by applying the «rubber-band» argument to the diagram of Fig. 2b, but as we are not dealing with a localized perturbation the argument is not so simple.) Even in the inner region, the flow is not perfectly decoupled, and a term

(70) $$\boldsymbol{E} \sim \frac{\omega B_y}{k} \hat{n} ,$$

must be taken into account in Ohm's law. This term corresponds to the generation of the perturbation flux that links the field regions on either side of $\boldsymbol{B}=0$. We have then

$$\eta_0 \boldsymbol{j}_1 = \boldsymbol{E}_1 + \boldsymbol{v}\times\boldsymbol{B} \tag{71}$$

and we must select εa so that the first term on the right in eq. (71) dominates the second in the region of the partly decoupled flow. Using $\nabla\cdot\boldsymbol{B}=0$, we have

$$\boldsymbol{j}_1 = \frac{B''_y}{4\pi k}\,\hat{n}\,. \tag{72}$$

For wavelengths that are much greater than the current-layer thickness a, we find

$$B''_y \sim \frac{B'_y}{\varepsilon a} \approx \frac{B_y}{\varepsilon k a}\,, \tag{73}$$

(cf. eq. (28)). If we now choose ε so that $\eta_0 \boldsymbol{j}_1 \sim \boldsymbol{E}_1$, we find

$$\varepsilon a \sim \frac{\eta_0}{4\pi k \omega}\,. \tag{74}$$

The growth rate obtained from eqs. (61) and (74) approximates the result of eq. (57) and Table I. This analysis is applicable only for $ka < 1$, since otherwise eq. (73) breaks down, B''_y/B_y becoming negative.

Part B: Instabilities due to Finite Current-Carrier Mass.

In Part A, we were forced to conclude that the finite-resistivity hydromagnetic stability analysis inevitably breaks down in the limit of very high-conductivity, since the width of the critical « region of discontinuity » becomes smaller than the particle gyro-radii. A rigorous analysis, based on the Boltzmann equation, has not been carried out, and appears rather difficult. If we go to the Vlasov equation, the analysis is much simpler, but we then lose the finite resistivity. Fortunately, the effect of resistivity is very similar to that of current-carrier inertia, so that we may expect even in the collisionless case to obtain instabilities closely analogous to those of Part A.

A simple example [2] will be treated here. A plane sheet-pinch or E-layer has the self-consistent zero-order equilibrium

(75) $$B_{z0} = 2v(2\pi n_0 m)^{\frac{1}{2}} \operatorname{tgh}(hy)\,,$$

(76) $$f_0 = \frac{n_0}{2(2\pi)^{\frac{3}{2}} v^3} \frac{\exp\left[-(1/2v^2)\{(v_x - v_{sx})^2 + v_z^2 + v_y^2\}\right]}{\cosh^2(hy)}\,,$$

(77) $$\begin{cases} h^2 = \dfrac{2\pi n_0 r_c v_{sx}^2}{v^2}\,, \\[2ex] r_c = \dfrac{e^2}{mc^2}\,, \end{cases}$$

where f_0 is the electron distribution function, with n_0 the electron density at $y = 0$. The ions are assumed to be cold and to have the same density distribution as the electrons. We consider a perturbation

(78) $$B_{y1} = B\,\psi(y) \exp[\omega t + ikz]\,,$$

(79) $$B_{z1} = i\,\frac{B}{k}\,\frac{d\psi}{dy} \exp[\omega t + ikz]\,,$$

(80) $$E_x = \frac{i\omega}{ck} B\psi \exp[\omega t + ikz]\,.$$

From the first-order Vlasov and Maxwell equations, we find that for $\omega = 0$ we must have

(81) $$\frac{4\pi n_0 r_c}{hk}\left(\frac{v_{sx}}{v}\right)^2 = \mu\,,$$

where the eigenvalue μ is obtained by solving the equation

(82) $$\frac{1}{hk}\,\frac{d^2\psi}{dy^2} = \psi\left\{\frac{k}{h} - \frac{\mu}{\cosh^2(hy)}\right\}.$$

Now using the zero-order pressure-balance condition (eq. (77)), we find that eq. (81) reduces to $\mu = 2h/k$. Equation (81) is then seen to be equivalent to the first-order pressure-balance condition of eq. (20), where $hy = \mu$ and $F = \operatorname{tgh}\mu$. Thus we see that the condition for $\omega = 0$ is the same in the exact collisionless treatment as in the hydromagnetic treatment. Solving eq. (82) for $\psi = 0$ at $y = \pm\infty$, we find $\mu = 1 + (k/h)$, so that we have $k = h$ for $\omega = 0$, just as in Part A.

It remains to ask whether the point $\omega = 0$ is a dividing point between stable and unstable regimes. That this is so, can be made plausible by a simple heuristic argument. For $\psi = 0$ at $y = \pm\infty$, we may write eq. (81) in the form

$$\frac{1}{2} A n_0 r_c \left(\frac{v_{sx}}{v}\right)^2 = 1 , \tag{83}$$

where $A = 8\pi/k(h+k)$. Let us compare eq. (83) with the Bennett pinch condition

$$I^2 > 2 A_s n m v^2 ,$$

or

$$\frac{1}{2} A_s n r_c \left(\frac{v_{sx}}{v}\right)^2 > 1 ,$$

where I is the current of a cylindrical particle stream, A_s its area, and v_s the directed particle velocity. We propose to interpret the condition that the left-hand side of eq. (83) be greater than the right-hand side as a condition that the zero-order sheet-pinch be capable of subdivision into first-order cylindrical pinches. The area A_s of each first-order pinch is in this case to be identified with A, which is roughly the product of the sheet-pinch thickness $2/h$ and the instability wave-length $2\pi/k$. To say that $k = h$ represents marginal stability, while $k < h$ represents instability, is therefore equivalent to the plausible assertion that an element of width $2/h$ cut from a sheet pinch will by itself marginally satisfy the Bennett condition for a cylindrical pinch and thus be static, while an element of greater width will overfulfill the Bennett condition, and thus be subject to unstable contraction. The proposition that $\omega = 0$ is a marginal point between regions of stability and instability is demonstrated on a more mathematical level in ref. [3].

We turn next to the question of growth rates. The appropriate Vlasov-equation calculation has been carried out only in the limit $v_{sx}^2/v^2 \to \infty$, which we see from eq. (77) to be the same as $n_0 \to 0$. In ref. [4] the problem is the same as that defined here by eqs. (75)–(80), but the procedure is to average over y and v_y before solving the first-order equations. This approximation, which is readily justifiable for $v_{sx}^2 \gg v^2$, yields the result [4]

$$\omega^2 = \frac{4\pi k n_0 e^2 v_{sx}^2}{M c^2 h} , \tag{84}$$

where M is the ion mass.

From the point of view of the hydromagnetic finite-resistivity analysis, we approach the same problem by defining a generalized resistivity

$$\eta^* = \eta + \omega/n r_c \tag{85}$$

or

$$\tilde{\eta}^* = \tilde{\eta} + p/N \,, \tag{86}$$

where

$$N = 4\pi n r_c a^2 \,.$$

The extra term corresponds to the dj/dt term in Ohm's law. We can use all the results of Section A, simply replacing $\tilde{\eta}$ by $\tilde{\eta}^*$. For the case of very small real resistivity, where only the inertial term matters in eq. (86), we may as well set $S = 1$, and we obtain for the « tearing » mode

$$p = \frac{2\,|F'|^5}{9\alpha\tilde{\varrho}^{\frac{1}{2}} N^{\frac{3}{2}}} \left\{\frac{1}{F^2_{-\infty}} + \frac{1}{F^2_{\infty}}\right\}^2 , \tag{87}$$

in the limit of large N, (which now corresponds to the limit of high conductivity). The case treated in ref. [4] corresponds to the opposite limit, $N \to 0$. To obtain a comparison, we may make use of the low-conductivity result, derived in Appendix B of ref. [1], that

$$p^2 = \alpha S^2 \tag{88}$$

or

$$\omega^2 = \frac{kB^2}{4\pi a \varrho} \,. \tag{88}$$

Using eqs. (75) and (77), this reduces identically to eq. (84). This agreement between the hydromagnetic and Vlasov-equation approaches in the simple cases considered here, suggests that the hydromagnetic results of Section A together with eq. (86) may provide guide-lines for the full Vlasov-equation stability analysis, and that conversely the Vlasov-equation analysis will shed light on the role of finite-gyro-radius effects in the hydromagnetic finite-resistivity problem.

REFERENCES

[1] H. P. Furth, J. Killeen and M. N. Rosenbluth: *Phys. Fluids* **6**, 459 (1963).
[2] H. P. Furth: *Suppl. Nucl. Fusion*, 169 (1962) Part I.
[3] V. K. Neil: *Phys. Fluids*, **5**, 14 (1962).
[4] D. Pfirsch: *Mikroinstabilitäten vom Spiegeltyp in inhomogenen Plasmen* (Max-Planck Inst. 1962).

Nonlinear Theory of Electrostatic Waves in Plasmas (*).

P. A. STURROCK

Microwave Laboratory, W. W. Hansen Laboratories of Physics,
Stanford University - Stanford, Cal.

1. – Introduction.

It is conventional to discuss nonlinear theory after linear theory not because this is the logical order but because it is the expedient one. Nonlinear theory of waves in plasmas, or of any other nontrivial phenomenon, does not consist of a new sweeping treatment of the subject which contains linear theory as a simple special case; rather, it constitutes a number of cautious excursions from the familiar and comparatively safe territory of linear theory into the jungle of nonlinearity. This article will contain a brief account of some of the approaches to the nonlinear theory of waves in plasmas which have been developed in the last few years, and a more extended account of the «fairly-small-amplitude» approximation in which the nonlinear terms of the wave equations are regarded as small by comparison with linear terms, and their effects treated by a perturbation procedure. In order to make the subject more manageable, we shall restrict our attention to plasma oscillations in uniform plasmas free from magnetic fields. This branch of the wave theory of plasmas is in fact that which has received most attention, but it is not entirely fortunate for expository purposes, since plasma oscillations are in an important respect atypical of waves in plasmas and indeed of waves in general. A review of contributions to the nonlinear theory of plasma waves in general has been given in ref. [1].

The dispersion relation for electrostatic waves in a cold homogeneous plasma, free from magnetic field, is

$$\omega^2 = \omega_p^2 , \tag{1.1}$$

(*) This study was supported by the Air Force Office of Scientific Research under grant no. AF-AFOSR-62-343.

where ω_p is the familiar plasma frequency given by

$$\omega_p^2 = \frac{4\pi n e^2}{m} . \tag{1.2}$$

Equation (1.1) shows that, in linear theory, any initial disturbance of a plasma oscillates with frequency ω_p, so that there can be no propagation of a disturbance, a fact which can otherwise be seen by noting that the group velocity is zero. We shall in the next section point out another respect in which plasma oscillations are atypical. If the motion is purely one-dimensional, oscillations remain (in a certain sense) strictly periodic, with frequency ω_p, for finite amplitudes up to a certain critical amplitude.

In Section **2** we shall discuss certain simple but instructive models for the excitation of a cold homogeneous plasma, following the work of DAWSON [2]. We show that one-dimensional excitation of a plasma leads to steady-state oscillations provided the amplitude is not sufficient to lead to « cross-over » of the electrons. On the other hand, cylindrically and spherically symmetric modes of excitation necessarily lead to cross-over within a finite time. Hence the study of one-dimensional disturbances above a critical amplitude, and the study of more general types of disturbance, lead necessarily to the problem of the influence of cross-over on the development of oscillation patterns. The effect of cross-over has been studied by numerical techniques by Dawson, for the case of one-dimensional disturbances exceeding the critical amplitude. As might be expected, it is found that cross-over leads to randomization of the energy, whether the energy is partitioned among particles or among waves.

In Section **3**, we present material due to BERNSTEIN, GREENE and KRUSKAL [3], which, in a sense, contradicts the assertion that cross-over leads to degradation of wave patterns. It is shown that even in a multistream plasma (a stream of nonzero temperature), a judicious choice of the initial state of the disturbance of a plasma leads to a wave of finite amplitude which can persist indefinitely without change of form. However, the electron energy spectrum must be tailored carefully to fit the electric-potential profile, and we may conclude that an initial state differing slightly from that required by the Bernstein-Greene-Kruskal model would lead to a disturbance which becomes degraded in time.

Section **4** treats what might be termed a « fairly-small-amplitude » theory of plasma oscillations [4]. One assumes that amplitudes are small; not so small that linear theory is perfectly acceptable, but sufficiently small that the departure from linearity may be treated by a perturbation technique. In this context, nonlinear effects may be treated by considering the waves which arise in linear theory to interact with each other. This gives one a feeling for some of the effects of nonlinearity, and it also enables one to make certain simple calculations. We shall show that one effect is to modify the dispersion relation (1.1) into a form which resembles the dispersion relation for a warm plasma [5]

$$\omega^2 = \omega_p^2 + \alpha \langle v^2 \rangle k^2 \,. \tag{1.3}$$

We shall also show that another effect of this wave interaction is to damp an initial wave, if there is a random excitation of « background » waves in the plasma.

One of the consequences of nonlinearity of the equations governing wave propagation in a plasma is that the division of disturbances into various types of waves (plasma oscillations, electromagnetic waves, etc.) is not clear-cut. In the fairly-small-amplitude approximation, we may still represent what goes on in a plasma instantaneously in terms of these waves, but we must allow for coupling between them so that the presence of one wave affects the dispersion relation of another, and there is the possibility of exchange of energy between waves of different types [6, 7]. This type of interplay will be discussed briefly in Section **5**, for the special case of the coupling between plasma oscillations and electromagnetic waves in a cold homogeneous plasma.

In Section **6**, we discuss the energy exchange between waves in a general way. Certain relations of this type [8] may be closely related to the wave-interaction formalism underlying the fairly-small-amplitude approximation. In the case that the excitation of the entire system occurs at combinations of two (or more) « fundamental » frequencies, one may set up relations between power exchange between various input and output connections under quite general conditions [9, 10], and without limitation on the amplitudes. These are known as the « Manley-Rowe » relations [11] which are useful in discussing possible microwave applications of plasma systems.

One of the most important aspects of nonlinearity in wave equations is the significance of nonlinearity in limiting the growth of instabilities predicted by linear theory. This is a very important aspect of nonlinear theory, which we shall not go into in detail in this article. A technique for handing this problem has recently been developed by DRUMMOND and PINES [12] and it is anticipated that the further development and applications of their method will be an important ingredient of the future studies of nonlinear effects in plasmas.

2. – Limitations of the single-stream model [2].

We first consider the case of plane oscillations, assuming that electrons vibrate back and forth in the x-direction and so ignoring the y and z co-ordinates. If x_0 is the position of an electron in an assumed quiescent unperturbed state, we write

$$x = x_0 + \xi(x_0, t) \tag{2.1}$$

for its perturbed state. When the plane of electrons initially at $x = x_0$ is displaced to $x = x_0 + \xi(x_0)$, it passes over an amount of positive charge equal to $en_0\xi$ per unit area, where n_0 is the equilibrium (uniform) number density of electrons. Here and throughout this article, we neglect ion motion. If there is no cross-over, that is if the order of electrons in the x-direction is not changed, the amount of charge to either side of the electron sheet due to the electron gas is unchanged, so that the electric field now experienced by the electron sheet is due to the excess positive charge, and is therefore given by

$$E = 4\pi e n_0 \xi . \tag{2.2}$$

The equation of motion of the electron sheet is

$$m \frac{\mathrm{d}^2\xi}{\mathrm{d}t^2} = - eE , \tag{2.3}$$

which becomes

$$\frac{\mathrm{d}^2\xi}{\mathrm{d}t^2} + \omega_p^2 \xi = 0 . \tag{2.4}$$

We therefore see that, if there is no cross-over, each sheet oscillates sinusoidally with the frequency ω_p. Note, however, that the charge density and the electric field at any one point are not related in a linear way to the displacement of the charge sheets, so that the « Eulerian » quantities will not exhibit simple harmonic time variation. We may see this very simply by noting that the local charge density is related to the unperturbed charge density and electron displacement by the Lagrange expansion [13]

$$n(x_0) = n_0 \left[1 - \frac{\partial \xi}{\partial x_0} + \frac{1}{2} \frac{\partial^2 \xi^2}{\partial x_0^2} - \dots \right] . \tag{2.5}$$

It is easy to see from (2.1) that if the inequality

$$\partial\xi/\partial x_0 > -1 \tag{2.6}$$

is satisfied, the order in the electrons is preserved, that is, there is no cross-over. It may now be shown that if the initial amplitude is sufficiently small, this condition (2.6) can be satisfied for all time. To see this, we note that since ξ is simple harmonic with frequency ω_p, so is $\partial\xi/\partial x_0$, so that

$$W = (\partial^2\xi/\partial t\,\partial x_0)^2 + \omega_p^2(\partial\xi/\partial x_0)^2\,, \tag{2.7}$$

is constant. Hence if $W(x_0)$ is everywhere less than ω_p^2 initially, it is always everywhere less than ω_p^2, and hence (2.6) is always satisfied everywhere.

We may carry out a similar analysis for cylindrical oscillations. If the electrons initially on a cylinder of radius r_0 are displaced to a cylinder of radius $r_0 + \xi$, evaluation of the electric field at the displaced position by means of Gauss's theorem leads to the equation of motion

$$m\,\frac{\mathrm{d}^2\xi}{\mathrm{d}t^2} = \frac{2\pi n_0 e^2}{r_0 + \xi}\left[(r_0 + \xi)^2 - r_0^2\right], \tag{2.8}$$

provided there is no cross-over. If we write $\xi/r_0 = \lambda$, (2.8) can be written as

$$\frac{\mathrm{d}^2\lambda}{\mathrm{d}t^2} + \omega_p\left(\frac{\lambda + \frac{1}{2}\lambda^2}{1+\lambda}\right) = 0\,, \tag{2.9}$$

which shows that the periodic of the oscillations now depends upon their amplitude. Hence, unless all particles have the same value of λ initially, particles with different equilibrium radii will have different periods. We find that the period is given approximately by

$$\tau = \frac{2\pi}{\omega_p}\left(1 - \frac{1}{12}\,\lambda_{\mathrm{M}}^2 + \ldots\right), \tag{2.10}$$

where λ_{M} is the maximum value of λ.

By considering two neighboring cylindrical shells, we may now estimate the length of time it takes for these shells to cross over, due to the difference in period of oscillation of these two shells. This time is found to be given approximately by

$$T = \frac{(3\pi/\omega_p)r_0^2}{\xi_{\mathrm{M}}^2(\mathrm{d}\xi_{\mathrm{M}}/\mathrm{d}r_0)}\,. \tag{2.11}$$

This formula has the property one would expect: the smaller the amplitude, the longer the time before cross-over occurs.

3. – Steady-state large amplitude waves.

We now sketch the theory of BERNSTEIN, GREENE and KRUSKAL [3], demonstrating the existence of steady-state large amplitude waves in a thermal plasma. Assuming that such a wave exists, we may conveniently work in a co-ordinate system in which the wave is at rest; all quantities then become time-independent. The equations governing the phenomenon are then the Boltzmann equation for ions and electrons (the collisions being ignored)

$$v\,\frac{\partial f_\pm(x,v)}{\partial x} \pm \frac{e}{m_\pm}\,\frac{\partial\varphi(x)}{\partial x}\,\frac{\partial f_\pm(x,v)}{\partial v}=0\,, \tag{3.1}$$

and Poisson's equation

$$\frac{\partial^2\varphi}{\partial x^2}=4\pi e\int_{-\infty}^{\infty}\mathrm{d}v\,[f_-(x,v)-f_+(x,v)]\,. \tag{3.2}$$

If we introduce quantities E_+, E_- for the total energies of ions and electrons,

$$E_\pm=\tfrac{1}{2}\,m_\pm v^2\pm e\varphi\,, \tag{3.3}$$

the general solution of eq. (3.1) may be written as

$$f_\pm=f_\pm(E_\pm)\,. \tag{3.4}$$

It should be noted that eq. (3.4) satisfies (3.1) independently of the partition of particles with a given energy between the two possible directions of velocity.

In addition to eq. (3.1) and (3.2), there are certain other conditions which must be satisfied. Consider for this purpose a potential of the form shown in Fig. 1. Ions with an energy E_+ such that $e\varphi_{\min}\leqslant E_+\leqslant e\varphi_{\max}$ are trapped by the potential, *i.e.*, they are restricted to regions where $e\varphi<E_+$. Thus, an ion with energy $E_+=e\varphi_1$ is restricted to regions C and F of Fig. 1, but once in C cannot move into F and vice versa. That is, x_1, x_2, and x_3 are turning points at which the ion reverses its velocity. Since the distribution function must be independent of the time, ions of energy $E_+=e\varphi_1$ must be equally distributed between the two directions of the velocity, and similarly for all trapped ions. On the other hand, ions with energies $E_+>e\varphi_{\max}$ can move freely in either direction, and hence their partition between the two directions of the velocity is arbitrary. Note that the distributions of the trapped ions in regions C and F are independent since the two regions are isolated from each other.

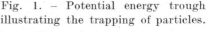

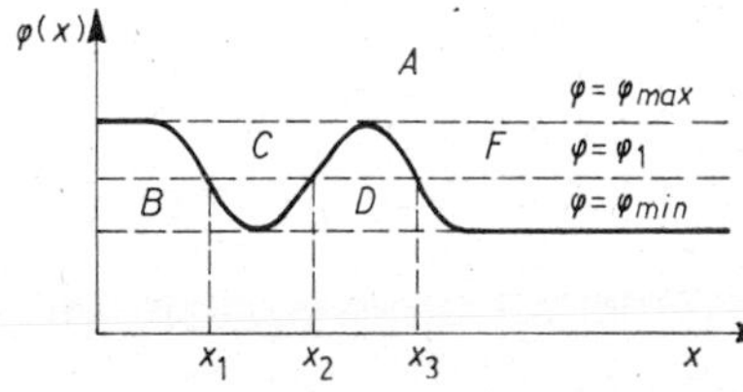

Fig. 1. – Potential energy trough illustrating the trapping of particles.

Similarly electrons of energy $E_- < - e\varphi_{\min}$ are trapped. For instance an electron of energy $E_- = - e\varphi_1$ can be only in regions B and D of Fig. 1, and the trapped electrons must be equally distributed between the two directions of velocity. Electrons of energy $E_- > - e\varphi_{\min}$ can move freely, and their partition between the two directions of velocity is arbitrary.

One may re-express the Poisson equation (3.2) in terms of the form (3.4) for the distribution functions, converting the velocity integral to an energy integral by means of (3.3). In this way, equation (3.2) takes the form

$$\frac{\mathrm{d}^2\varphi(x)}{\mathrm{d}x^2} = 4\pi e\left\{\int_{-\infty}^{\infty}\frac{\mathrm{d}E\, f_-(E)}{[2m_-(E + e\varphi(x))]^{\frac{1}{2}}} - \int_{-\infty}^{\infty}\frac{\mathrm{d}E\, f_+(E)}{[2m_+(E - e\varphi(x))]^{\frac{1}{2}}}\right\}. \tag{3.5}$$

If f_+ and f_- are regarded as given in eq. (3.5), this equation is an ordinary integro-differential equation for the potential φ. This equation may be integrated once to give

$$\left(\frac{\mathrm{d}\varphi}{\mathrm{d}x}\right)^2 + V(\varphi) = \text{const}\,, \tag{3.6}$$

where

$$V(\varphi) = 8\pi\int_{e\varphi}^{\infty}\mathrm{d}E\, f_+(E)[2(E - e\varphi)/m_+]^{\frac{1}{2}} + 8\pi\int_{-e}^{\infty}\mathrm{d}E\, f_-(E)\,[2(E + e\varphi)/m]^{\frac{1}{2}}\,. \tag{3.7}$$

One can now look upon eq. (3.6) as an integral equation for the distribution function of trapped electrons. Namely, if one writes $\mathrm{d}^2\varphi/\mathrm{d}x^2 = - N(e\varphi)$, where $N(e\varphi)$ is the net charge density, eq. (3.5) can be written in the form

$$\int_{-e\varphi}^{-e\varphi_{\min}}\mathrm{d}E\, f_-(E)[2m_-(E + e\varphi)]^{-\frac{1}{2}} = g(e\varphi)\,, \tag{3.8}$$

where

$$g(e\varphi) = -\frac{N(e\varphi)}{4\pi e} + \int_{e\varphi}^{\infty}\mathrm{d}E\, f_+(E)[2m_+(E - e\varphi)]^{-\frac{1}{2}} - \tag{3.9}$$

$$- \int_{-e\varphi_{\min}}^{\infty}\mathrm{d}E\, f_-(E)[2m_-(E + e\varphi)]^{-\frac{1}{2}}\,,$$

is the density of trapped electrons at the point x corresponding to the potential $\varphi(x)$. Thus, if one prescribes the potential $\varphi(x)$ (which determines $\mathrm{d}^2\varphi/\mathrm{d}x^2$ and hence $N(e\varphi)$), the entire ion distribution $f_+(E)$ and the distribution of *untrapped* electrons $f_-(e)$ for $E > -e\varphi_{\min}$, then $g(e\varphi)$ is a known function, and eq. (3.7) is an integral equation of the convolution type for the distribution function of the trapped electrons. It can be readily solved by the Laplace transformation, yielding

$$f_-(E) = \frac{(2m_-)^{\frac{1}{2}}}{\pi} \int_{e\varphi_{\min}}^{-E} \mathrm{d}V \frac{\mathrm{d}g(V)}{\mathrm{d}V} [-E - V]^{-\frac{1}{2}}, \qquad E < -e\varphi_{\min}, \tag{3.10}$$

a result which may be verified directly by substituting eq. (3.10) into eq. (3.8).

The choice of the arbitrary quantities in eq. (3.9) is restricted only by the weak requirements that $g(e\varphi_{\min}) = 0$, which follows from eq. (3.8), and that $f_-(E)$ as given by eq. (3.10) be nonnegative in order that $f_-(E)$ be a legitimate distribution function.

By considering various trapping regions of a wave of arbitrary form successively, filling a potential well with an appropriate distribution of trapped ions and a potential hump with an appropriate distribution of trapped electrons, one may show that it is possible to find distributions of untrapped and trapped electrons and ions to make up a self-consistent wave pattern with an arbitrary potential profile.

It will be noted that it takes a carefully tailored distribution of electron and ion energies to produce a steady-state wave of large amplitude. One may infer that departures of the distribution functions from these required forms would give rise to progressive degradation of the wave pattern.

4. – Fairly-small-amplitude theory of plasma oscillations.

The aspects of nonlinear theory which we have considered so far have involved a very specialized form for the initial excitation of the plasma. If we wish to consider a more general type of perturbation of the plasma, it is necessary to introduce some compensating simplification either in the model or in the approximation to be used. We here make the simplification that the plasma is uniform and of zero temperature, and again assume that the ions are virtually fixed, and once again assume that there is no magnetic field. We also simplify the problem by assuming that the amplitude is sufficiently small for the nonlinear terms in the equations to be regarded as a perturbation in comparison with the linear terms. The procedure will then be to work to the lowest order in a perturbation expansion which will yield nonlinear results [4]. The detailed mathematical steps are given in Appendices I through V. In this section we simply state the salient equations and discuss their implications.

It is convenient to discuss the motion of electrons by means of a displacement vector, saying that the electron which is at position x_r in the quiescent plasma is at position $x_r + \xi_r(x, t)$ at time t in the state of motion under discussion. The variables ξ are appropriate for setting up an action principle to describe the behavior of the plasma. By introducing a Green function to characterize the electrostatic force between two electrons, we may dispense with a separate description of the electric field and so obtain an action principle in the form

$$\delta S = 0 \,, \tag{4.1}$$

where S is an integral over space and time of a Lagrange density, the arguments of the Lagrange density being the displacement vectors ξ_r and their time derivatives.

We may expand the action function as follows

$$S = S^{(0)} + S^{(1)} + S^{(2)} + S^{(3)} + S^{(4)} + \dots \,, \tag{4.2}$$

where $S^{(n)}$ is of the n-th order in the dynamical variables ξ_r. One finds that the first two terms of (4.2) vanish identically, so that the lowest-order term is $S^{(2)}$ which leads to a linear equation of motion of the system. Formulas for functions appearing in (4.2) are derived in Appendix I, and the linear theory is sketched in Appendix II.

The linear equations derivable from $S^{(2)}$ are found to be

$$\frac{\mathrm{d}^2\xi_r}{\mathrm{d}t^2} + \omega_{\mathrm{P}}^2 \frac{k_r k_s}{k^2}\, \xi_r = 0 \,, \tag{4.3}$$

where we introduce the Fourier transform notation

$$\xi_r(\boldsymbol{x}, t) = \sum \xi_r(\boldsymbol{k}, t) \exp[i\boldsymbol{k}\cdot\boldsymbol{x}] \,, \tag{4.4}$$

$$\xi_r(\boldsymbol{k}, t) = V^{-1}\!\int \mathrm{d}^3\boldsymbol{x}\, \xi_r(\boldsymbol{x}, t) \exp[-\, i\boldsymbol{k}\cdot\boldsymbol{x}] \,. \tag{4.5}$$

It is here assumed for simplicity that the excitation is periodic with respect to a large cubical unit cell of volume V. If we neglect the zero-frequency « shear » waves, the solution of (4.3) may be written as

$$\xi_r(\boldsymbol{k}, t) = (2\omega_{\mathrm{P}} k^2)^{-\frac{1}{2}} k_r \{a(\boldsymbol{k}, t) \exp[i\omega_{\mathrm{P}} t] - a^*(\boldsymbol{k}, t) \exp[-\, i\omega_{\mathrm{P}} t]\} \,. \tag{4.6}$$

The choice of normalization is determined by the requirement of finding a canonically conjugate set of variables.

In linear theory, the variables $a(k, t)$ are in fact independent of time. When nonlinear terms of the equations of motion are taken into account, it is found that the resulting modification of the dynamical variables may be separated into two parts: a change of « wave-form », which one can represent by the addition of harmonics to the fundamental; and slow variation of the amplitudes with time. The slow variation of amplitude may itself be divided into two parts: one is an oscillatory part which one can regard as a shift in resonant frequency of the normal modes; the other represents an exchange of energy between sets of waves. In Appendices III, IV and V, two distinct techniques are developed for evaluating the nonlinear interaction by perturbation methods. The harmonic content can be extracted from these calculations, but will not be discussed here. The more interesting effect is the slow time variation of the amplitudes. It is found that both methods lead to the same result, that the lowest-order nonlinear interaction involves the effects of both $S^{(3)}$ and $S^{(4)}$ and is therefore a second-order perturbation calculation, which may be represented by the equation

$$\frac{\mathrm{d}a(\boldsymbol{k}_1)}{\mathrm{d}t} = i \sum_{\boldsymbol{k}_1+\boldsymbol{k}_2=\boldsymbol{k}_3+\boldsymbol{k}_4} C(\boldsymbol{k}_1, \boldsymbol{k}_2, \boldsymbol{k}_3, \boldsymbol{k}_4)\, a^*(\boldsymbol{k}_2)\, a(\boldsymbol{k}_3)\, a(\boldsymbol{k}_4)\,, \tag{4.7}$$

where

$$\begin{aligned} C(\boldsymbol{k}_1, \boldsymbol{k}_2, \boldsymbol{k}_3, \boldsymbol{k}_4) &= \\ = \tfrac{1}{4}(k_1 k_2 k_3 k_4)^{-1}[&-\boldsymbol{k}_1\cdot\boldsymbol{k}_3\boldsymbol{k}_1\cdot\boldsymbol{k}_4\boldsymbol{k}_2\cdot(\boldsymbol{k}_3+\boldsymbol{k}_4) - \boldsymbol{k}_2\cdot\boldsymbol{k}_3\boldsymbol{k}_2\cdot\boldsymbol{k}_4\boldsymbol{k}_1\cdot(\boldsymbol{k}_3+\boldsymbol{k}_4)] + \\ + \frac{1}{12}(k_1 k_2 k_3 k_4)^{-1}|\boldsymbol{k}_1+\boldsymbol{k}_2|^{-2}[&3\boldsymbol{k}_1\cdot(\boldsymbol{k}_1+\boldsymbol{k}_2)\boldsymbol{k}_2\cdot(\boldsymbol{k}_1+\boldsymbol{k}_2)\boldsymbol{k}_3\cdot(\boldsymbol{k}_1+\boldsymbol{k}_2)\boldsymbol{k}_4\cdot(\boldsymbol{k}_1+\boldsymbol{k}_2) - \\ &- \{k_1^2k_2^2 - (\boldsymbol{k}_1\cdot\boldsymbol{k}_2)^2\}\{k_3^2k_4^2 - (\boldsymbol{k}_3\cdot\boldsymbol{k}_4)^2\}]\,. \end{aligned} \tag{4.8}$$

The time dependence of the amplitude is not expressed explicitly in this equation.

It is seen from (4.7) that, to lowest order, nonlinearity may be regarded as interaction of electrostatic waves in groups of four, satisfying the selection rule

$$\boldsymbol{k}_1 + \boldsymbol{k}_2 = \boldsymbol{k}_3 + \boldsymbol{k}_4\,. \tag{4.9}$$

If any one such interaction is considered independently of the rest, it is found that energy is exchanged between such a group according to the relations

$$\frac{\mathrm{d}\varepsilon(k_1)}{\mathrm{d}t} = \frac{\mathrm{d}\varepsilon(k_2)}{\mathrm{d}t} = \frac{-\,\mathrm{d}\varepsilon(k_3)}{\mathrm{d}t} = \frac{-\,\mathrm{d}\varepsilon(k_4)}{\mathrm{d}t}\,, \tag{4.10}$$

where

(4.11) $$\varepsilon(k) = \omega_P mn\,|a(k)|^2 .$$

$\varepsilon(k)$ is the mean energy density to be associated with the wave of wave number $\boldsymbol{k}$. This is a special case of the « action-transfer relations » [8], which will be discussed further in Section **6**. The terms of (4.7) for which $\boldsymbol{k}_1 = \boldsymbol{k}_2$ and $\boldsymbol{k}_3 = \boldsymbol{k}_4$ form a special set which we term « coherent », since the phase relationship of the time-derivative to the amplitude is independent of the amplitudes involved. For this set, there is obviously a phase difference of $\pi/2$ so that there will be no change in the magnitude of the complex amplitude of any wave. In other words, this set of interactions does not give rise to energy exchange, a fact which may otherwise be seen from relations (4.10).

The contribution of the coherent terms alone to the interaction (4.7) may be written as

(4.12) $$\frac{\mathrm{d}a(k)}{\mathrm{d}t} = i \sum_{k'} C(\boldsymbol{k}, \boldsymbol{k}')\, a^*(\boldsymbol{k}')\, a(\boldsymbol{k}')\, a(\boldsymbol{k}) ,$$

where

(4.13) $$C(\boldsymbol{k}, \boldsymbol{k}') = C(\boldsymbol{k}, \boldsymbol{k}', \boldsymbol{k}, \boldsymbol{k}') .$$

We see from (4.8) that

(4.14) $$C(\boldsymbol{k}, \boldsymbol{k}') = \frac{1}{6}\,\frac{\{\boldsymbol{k}^2\boldsymbol{k}'^2 - (\boldsymbol{k}\cdot\boldsymbol{k}')^2\}^2}{\boldsymbol{k}^2\boldsymbol{k}'^2\,|\boldsymbol{k}+\boldsymbol{k}'|^2} .$$

The effect of the coherent group of terms (4.12) is to give rise to a frequency change of any wave which may be expressed by the dispersion relation

(4.15) $$\omega = \omega_P + \Delta\Omega(\boldsymbol{k}) ,$$

where

(4.16) $$\Delta\Omega(\boldsymbol{k}) = \sum_{k'} C(\boldsymbol{k}, \boldsymbol{k}')\,|a(\boldsymbol{k}')|^2 .$$

The expression (4.16) may be simplified if it is assumed that the wave under consideration, the « test » wave, is interacting with a large number of « background » waves, and that the great majority of background waves are of much shorter wave length than the test wave. If it is furthermore assumed that the energy spectrum of the background waves is isotropic, (4.16) reduces to

(4.17) $$\Delta\Omega(\boldsymbol{k}) = \frac{4}{45}\,\frac{\varepsilon_w}{\omega_P mn}\,\boldsymbol{k}^2 ,$$

where ε_w is the total energy density to be ascribed to the background waves.

Since an expression of the form (4.17) is derived in the familiar treatment of plasma oscillations in a thermal plasma [5], it is pertinent to inquire whether the effect we have just derived is identical with that which arises in the linear theory of thermal plasmas. We may see that this is not the case by noting that the dispersion relation for a wave in a thermal plasma may be obtained on a one-dimensional model, whereas, according to (4.14), the background waves parallel to the test wave make no contribution to (4.17). However, in a thermal plasma, for which the excitation of background waves is a purely thermal process, the contribution (4.17) to the dispersion relation is much less than that obtained in linear theory based on the Boltzmann equation [5]. The ratio of the two contributions is approximately the ratio of wave energy density to particle energy density which is itself of order $1/N$, where N is the number of particles in a Debye sphere.

We now consider the « incoherent » terms of (4.7), since energy-exchange is due essentially to such terms, and hence spectral decay, that is the degradation of an initial energy spectrum of waves, must be due to such terms.

In order to study the decay of energy in any particular wave due to the interaction of this wave with a large number of other waves, we should clearly suppose that the « test » wave has significantly higher energy density than any one of the « background » waves. However, if the phase relationship of all the waves is at any given time assumed random, the phase of the time derivative given by (4.7) will be random so that this expression will not indicate a preference of decay over growth. However, we may make a simple order-of-magnitude estimate of the decay rate by evaluating the second-order time derivative, first taking the first derivative of (4.7) and then expressing the first derivatives in terms of amplitudes by repeated application of (4.7). In this way we obtain the equation

$$(4.18)\qquad \frac{\mathrm{d}^2 J(\boldsymbol{k}_1)}{\mathrm{d}t^2} = 4 \sum_{k_1+k_2=k_3+k_4} \{C(\boldsymbol{k}_1, \boldsymbol{k}_2, \boldsymbol{k}_3, \boldsymbol{k}_4\}^2 \{J(\boldsymbol{k}_2)J(\boldsymbol{k}_3)J(\boldsymbol{k}_4) + J(\boldsymbol{k}_1)J(\boldsymbol{k}_3)J(\boldsymbol{k}_4) - J(\boldsymbol{k}_1)J(\boldsymbol{k}_2)J(\boldsymbol{k}_4) - J(\boldsymbol{k}_1)J(\boldsymbol{k}_2)J(\boldsymbol{k}_3)\}\,,$$

where

$$(4.19)\qquad J(k) = |a(\boldsymbol{k})|^2\,.$$

Equation (4.18) represents only the « coherent » contribution to the second derivative, *i.e.*, the terms which survive after making the random-phase approximation. This is the reason that the summation is over groups of four waves rather than larger groups.

The form of (4.18) already indicates that, if the amplitude of the test wave $\boldsymbol{k}_1$ is much larger than the amplitudes of all other background waves, the second derivative of $J(\boldsymbol{k}_1)$ is negative so that the wave begins to decay in amplitude.

If we make the approximation

$$|\boldsymbol{k}_1| \ll |\boldsymbol{k}_3|, \quad |\boldsymbol{k}_4|, \tag{4.20}$$

we find that the dominant contribution to (4.8) may be written as

$$C \approx \frac{1}{12}\frac{1}{k_1 k_3 k_4}\frac{1}{|\boldsymbol{k}_3 + \boldsymbol{k}_4|}\left[3(\boldsymbol{k}_3 + \boldsymbol{k}_4)^2(\boldsymbol{k}_3\cdot\boldsymbol{k}_4)^2 + \{k_3^2 k_4^2 - (\boldsymbol{k}_3\cdot\boldsymbol{k}_4)^2\} k_1'^2\right], \tag{4.21}$$

where $\boldsymbol{k}_1'$ is that part of $\boldsymbol{k}_1$ which is normal to the vector $\boldsymbol{k}_3 + \boldsymbol{k}_4$.

We shall not here attempt exact evaluation of the sum in (4.18), but merely estimate the magnitude of the effect. For this purpose, we note that if $\langle k_\mathrm{B}^2 \rangle$ is the mean square value of the « background » wave vectors, $\boldsymbol{k}_2$, $\boldsymbol{k}_3$, $\boldsymbol{k}_4$, and k_T^2 is the square of the magnitude of the wave-vector of the test wave $\boldsymbol{k}_1$, then

$$C \approx .01 k_\mathrm{T}^2 \langle k_\mathrm{B}^2 \rangle . \tag{4.22}$$

Hence with the help of (4.11) we find that (4.18) may be written in the form

$$\frac{\mathrm{d}^2 J(\boldsymbol{k}_\mathrm{T})}{\mathrm{d}t^2} = \frac{1}{\tau^2} J(\boldsymbol{k}_\mathrm{T}) , \tag{4.23}$$

where τ, which is clearly an estimate of the decay time, is given by

$$\tau \approx \frac{5\omega_\mathrm{P} m n}{k_\mathrm{T} \langle k_\mathrm{B} \rangle \varepsilon_w} . \tag{4.24}$$

where ε_w is the background wave energy density.

The treatment we have outlined is not applicable to a thermal plasma in which the multi-stream nature of the electron flow is taken into account and the correct dispersion relation for electrostatic waves is used. Nevertheless, it is interesting to try to apply (4.24) to the case that the background excitation is due to thermal agitation of the plasma. If we replace $\langle k_\mathrm{B} \rangle$ by λ_D^{-1}, where λ_D is the Debye length, and note that approximately N^{-1} of the total thermal energy exists in the form of wave energy, (4.24) is found to become, approximately

$$\tau \approx \frac{2N}{k_\mathrm{T} \lambda_\mathrm{D} \omega_\mathrm{P}} . \tag{4.25}$$

For sufficiently small wave-vectors, this damping would be more important than Landau damping [14].

5. – Coupling of plasma oscillations and electromagnetic waves.

In the previous section, we considered the coupling of electrostatic waves among themselves and found that the coupling occurs basically in groups of four waves. In this section, we consider very briefly the coupling which arises between plasma oscillations and electromagnetic waves due to nonlinearity of the equations of motion. The analysis follows the same lines as that adopted for longitudinal waves, and is set out in detail in Appendix VI.

The disturbance of the plasma may be characterized by the displacement vector $\xi_r(x, t)$, and the electromagnetic field may be described by the magnetic vector potential $A_r(x, t)$. In linear theory the general solution may be expanded as plane longitudinal waves (plasma oscillations) and plane transverse waves (electromagnetic waves):

$$(5.1)\qquad A_r = \sum_k l_r \left[a_L \exp[i(\boldsymbol{k}\cdot\boldsymbol{x} - \omega_P t)] + a_L^* \exp[-i(\boldsymbol{k}\cdot\boldsymbol{x} - \omega_P t)] + \right. \\ \left. + \sum_k e_r \left[a_T \exp[i(\boldsymbol{k}\cdot\boldsymbol{x} - \omega_T t)] + a_T^* \exp[-i(\boldsymbol{k}\cdot\boldsymbol{x} - \omega_T t)\right]\right],$$

and

$$\xi_r = \frac{mc\omega_P}{ie} \sum_k l_r \left[a_L \exp[i(\boldsymbol{k}\cdot\boldsymbol{x} - \omega_P t)] - a_L^* \exp[-i(\boldsymbol{k}\cdot\boldsymbol{x} - \omega_P t)] + \right. \\ \left. + \frac{ie}{mc} \sum_k \omega_T^{-1} e_r \left[a_T \exp[i(\boldsymbol{k}\cdot\boldsymbol{x} - \omega_P t)] - a_T^* \exp[-i(\boldsymbol{k}\cdot\boldsymbol{x} - \omega_P t)\right]\right].$$

l_r is the unit vector parallel to k_r, e_r is the polarization vector normal to k_r, which may be given two values for each wave vector, and ω_T satisfies the dispersion relation

$$(5.3)\qquad \omega_T^2 = \omega_P^2 + c^2 k^2 .$$

Two distinct types of coupling are possible in the lowest order of nonlinearity, that corresponding to three-wave coupling. One is the possible coupling of two electromagnetic waves to one plasma oscillation, but this is not investigated in this article. The other is the coupling of two plasma oscillations to one electromagnetic wave. This coupling takes place among such a group provided the wave vectors of the longitudinal waves are related to that for the transverse wave by

$$(5.4)\qquad \boldsymbol{k}_T = \boldsymbol{k}_{L1} + \boldsymbol{k}_{L2}$$

and that the frequency of the transverse wave is given by

$$(5.5)\qquad \omega_T = 2\omega_P .$$

It follows from (5.3) and (5.5) that $\boldsymbol{k}_T$ has magnitude $3^{\frac{1}{2}}(\omega_P/c)$.

In some problems, it would be convenient to follow the amplitudes of the plasma oscillations and electromagnetic waves as functions of the spatial co-ordinates rather than of time. This approach is appropriate, for instance, in the study of radiation of electromagnetic waves from plasma oscillations. For this reason, it is shown in Appendix VI that the derivative-expansion technique may be extended to allow for slow variations of the amplitudes with the spatial co-ordinates. In this way, one arrives at the following pair of equations for the slow variation of the amplitudes of the plasma oscillations and electromagnetic waves:

$$\frac{\partial^{\mathrm{I}} a_{\mathrm{L}1}}{\partial t} = \frac{1}{2} i \frac{e}{mc} \sum k_{\mathrm{L}1}^{-1} k_{\mathrm{L}2}^{-1} (k_{\mathrm{L}1}^2 - k_{\mathrm{L}2}^2) \boldsymbol{e} \cdot \boldsymbol{k}_{\mathrm{L}1} a_{\mathrm{T}} a_{\mathrm{L}2}^* , \tag{5.6}$$

$$\frac{\partial^{\mathrm{I}} a_{\mathrm{T}}}{\partial t} + \frac{1}{2} \frac{c^2}{\omega_{\mathrm{P}}} k_{r\mathrm{T}} \frac{\partial^{\mathrm{I}} a_{\mathrm{T}}}{\partial x_r} = \frac{1}{4} i \frac{e}{mc} \sum k_{\mathrm{L}1}^{-1} k_{\mathrm{L}2}^{-1} (k_{\mathrm{L}1}^2 - k_{\mathrm{L}2}^2) \boldsymbol{e} \cdot \boldsymbol{k}_{\mathrm{L}1} a_{\mathrm{L}1} a_{\mathrm{L}2} , \tag{5.7}$$

where the summation is over all groups of waves satisfying (5.4).

It is interesting that eq. (5.6) for the slow variation of the plasma-oscillation amplitude involves only a time derivative, whereas the corresponding eq. (5.7) for the slow variation of the electromagnetic amplitude involves both space and time derivatives. The difference is due to the fact that plasma oscillations have zero group velocity, whereas electromagnetic waves do not. In fact, it is readily verified that the operator appearing on the left-hand side of (5.7) represents the total time derivative following the group velocity of the wave. This is not unreasonable, for the growth in amplitude of a wave packet should depend upon the source terms which are distributed along the trajectory of the packet.

As one would expect, the interaction represented by (5.6) and (5.7) vanishes if the wave vectors are parallel. It is also interesting to note that the interaction vanishes if the wave vectors $\boldsymbol{k}_{\mathrm{L}_1}$ and $\boldsymbol{k}_{\mathrm{L}_2}$ are equal in magnitude.

One may also derive from (5.6) and (5.7) an interesting relationship between the variations of energy density of three interacting waves. If these energy densities are denoted $\varepsilon_{\mathrm{L}_1}$, $\varepsilon_{\mathrm{L}_2}$ and ε_{T}, we find from the above equations that

$$\frac{\partial \varepsilon_{\mathrm{L}1}/\partial t}{\omega_{\mathrm{P}}} = \frac{\partial \varepsilon_{\mathrm{L}2}/\partial t}{\omega_{\mathrm{L}}} = -\frac{\partial \varepsilon_{\mathrm{T}}/\partial t + U_{r\mathrm{T}}(\partial \varepsilon_{\mathrm{T}}/\partial x_r)}{\omega_{\mathrm{T}}} , \tag{5.8}$$

where $U_{r\mathrm{T}}$ is the group velocity of the transverse wave. This is an example of the wave-interaction relations which will be discussed further in Section **6**.

It is a simple matter to obtain from eq. (5.7) an estimate of the rate at which energy is transferred from plasma oscillations to electromagnetic waves in the case that just one set of three waves is involved. In the case that the build-up of electromagnetic energy is spatial rather than temporal, we find that the energy density of the electromagnetic waves has built up to a level comparable with the energy density of the plasma oscillations in a distance given approximately by

$$L \sim D^{-1}(c/v_{\text{ph}})^2(c/\omega_{\text{P}}) , \tag{5.9}$$

where v_{ph} is the phase velocity of the plasma oscillations and D is the « depth of modulation » of the plasma oscillations, *i.e.* the fractional change in density of the electron plasma. For parameters appropriate to a laboratory experiment, L would be sufficiently large for the energy transfer to be very small.

Although eqs. (5.6) and (5.7) are in principle applicable to the problem of the radiation of electromagnetic waves from a turbulent spectrum of plasma oscillations, considerable further development is necessary for the treatment of this problem, so that no estimates of the magnitude of the effect can be made at this time.

6. – Energy relations.

We have seen in Sections **4** and **5** that the perturbation technique which has been described as the « fairly-small-amplitude approximation » enables one to break down the nonlinear process going on in a plasma as interactions between groups of small numbers of waves. One may study each interacting group separately, and this leads to selection rules of the type (4.9) and (5.4), and it also leads to relations such as (4.10) and (5.8) governing the exchange of energy between waves. One might expect energy and momentum to be separately conserved in an interacting group, but it turns out that there is indeed a stronger set of relations which may be established [8]. If the selection rules for a given group of waves are written as

$$l\omega_\lambda + m\omega_\mu + \ldots = 0 , \tag{6.1}$$

and

$$l\boldsymbol{k}_\lambda + m\boldsymbol{k}_\mu + \ldots = 0 , \tag{6.2}$$

where $\lambda, \mu, \ldots,$ are the labels attached to the interacting waves and $l, m, \ldots,$ are small positive or negative integers, then the exchange of energy is governed by the equations

$$\frac{1}{l}\frac{\mathrm{d}J_\lambda}{\mathrm{d}t} = \frac{1}{m}\frac{\mathrm{d}J_\mu}{\mathrm{d}t} = \ldots, \tag{6.3}$$

where J_λ is the « action » (or action density) of the wave λ, etc. The above relations are strongly reminiscent of the quantum conditions governing the interaction of waves in quantum mechanics. Since energy and momentum of a wave are related to the action by [15]

$$E = J\omega \,, \tag{6.4}$$

and

$$\boldsymbol{P} = J\boldsymbol{k} \,, \tag{6.5}$$

we see that energy and momentum are separately conserved among interacting groups of waves.

The energy and momentum which appear in the preceding equations are those contributions to the total energy and momentum of a system which may be attributed to a given wave excitation. There are, indeed, reasons for preferring the terms « pseudo-energy » and « pseudo-momentum » for these quantities [16]. For instance, it is found that under certain conditions the pseudo-energy of a wave may be negative [17]. This property makes the pseudo-energy helpful in understanding certain types of instability.

A set of relations closely related to the « action-transfer relations » (6.3) may be obtained by considering multiply periodic excitation of a plasma [9, 10]. If we assume, for simplicity, that energy enters and leaves across a surface at which the equations are linear (for instance, free from plasma), then we may ascribe the total energy input S_{rs} to the frequency ω_{rs}, where

$$\omega_{rs} = r\omega_a + s\omega_b \tag{6.6}$$

where we assume for simplicity that there are just two « fundamental » frequencies ω_a, ω_b from which all other frequencies excited in the system may be composed by harmonic combination. By making use of the Hamiltonian nature of the system, we may establish the following relations

$$\sum_{r,s} \frac{rS_{rs}}{r\omega_a + s\omega_b} = 0 \,, \tag{6.7}$$

$$\sum_{r,s} \frac{sS_{rs}}{r\omega_a + s\omega_b} = 0 \,. \tag{6.8}$$

These equations are identical with the equations established by MANLEY and ROWE [11] for electrical networks, and have been very useful in the analysis of nonlinear microwave devices.

7. – Discussion.

We may now review the topics which we have treated in detail, in order to discuss the extent to which they are typical of plasma phenomena, and to list the various nonlinear effects which have so far not been treated.

As we have already remarked, the property of a cold homogeneous plasma where disturbances below a certain critical amplitude persist without change of form must be considered atypical. More generally, one would expect that any wave pattern would change slowly in time due to nonlinearity. On the other hand, the possibility of setting up steady-state travelling-wave solutions of nonlinear equations exists for almost any conservative system. In the case that the disturbance is finite in space (or time), such waves are known as « solitary » waves [18]. The existence of such solutions of the nonlinear magneto-hydrodinamic equations has been established by ADLAM and ALLEN [19].

It is interesting to compare the kernel (4.8) derived in this article with the kernel (7.1) of ref. [4]. Both describe the nonlinear behaviour of plasma oscillations as a wave-interaction process. The two formulas are not equivalent, and the difference may be traced to differences between the subsidiary conditions which appear most appropriate to the Eulerian formulation and the Lagrangian formulation of the problem. The physical implications of these differences are being studied to determine when each formulation is the more appropriate. However, it may readily be confirmed that both interaction formulas vanish identically when the spectrum is purely one-dimensional. The scattering kernel derived in the present article is in some ıespects preferable to that derived in the earlier article, in that it does not display a singularity as the directions of a pair of wave vectors coalesce. Nevertheless, it seems that the wave-interaction representation should be made more precise by considering nonlinear effects as an initial-value problem, using Laplace-transform techniques.

The discussion of the coupling between electrostatic and electromagnetic waves in plasmas may be regarded as typical of the phenomenon of mode-coupling which is a consequence of nonlinearity of wave equations. Coupling between other wave types could be discussed in a similar manner. Even in the case of coupling between electrostatic and electromagnetic waves, in the fairly-small-amplitude approximation, we could discuss in addition the phenomenon of scattering of electromagnetic waves by plasma oscillations, by considering wave interactions between two electromagnetic waves and one plasma oscillation, associated with the selection rules

$$(7.1)\qquad \begin{cases} \omega_{T1} = \omega_{T2} + \omega_P\,, \\ \boldsymbol{k}_{T1} = \boldsymbol{k}_{T2} + \boldsymbol{k}_L\,. \end{cases}$$

This would clearly give rise to a frequency shift between the « transmitted » and « reflected » waves. Interaction of this type has recently been examined in great detail in connection with the study of ionospheric densities by the « incoherent backscatter » method [20].

There are certainly many other nonlinear phenomana than those which have been discussed in this article. The limitation of instabilities by nonlinear effects is a very important process which we have not discussed. A calculation of the limitation of two-stream instability by this process has recently been made by DRUMMOND and PINES [12]. The phenomenon of anomalous diffusion due to the excitation of oscillations in a plasma may be regarded as a nonlinear wave problem but it is normally not discussed in these terms. Nonlinear wave phenomena are also important in certain models for the structure of collision-free shock waves [21], but this topic has not been discussed in this article.

This article has discussed fairly simple phenomena in a very simple model of a plasma. More detailed treatment of more realistic models will yield a wide range of phenomena, and the nonlinear theory of waves in plasmas remains a rich field for research.

* * *

The author is indebted to many colleagues at Stanford and elsewhere for helpful discussion at various stages of this work. Special thanks are due to Mr. R. BALL who is responsible for the reduction of the kernel (A-3.18) to (A-3.19).

APPENDIX I

Construction of the Lagrangian and Hamiltonian functions.

The model which we consider is that of a uniform distribution of electrons, of density n, neutralized by an equal density of ions which we assume to be stationary. We describe the motion of the electron gas by a displacement-vector $\xi_r(x, t)$, where $\boldsymbol{x}$ is used as an abbreviation for (x_1, x_2, x_3) and $r(s$, etc.) takes the values $1, 2, 3$; and $x_r + \xi_r(\boldsymbol{k}, t)$ is the position at time t of the electron which, in the quiescent state of the electron gas, is at position x_r. The quiescent state is that in which electrons are uniformly distributed and are at rest. If we describe the electrostatic interaction between electrons by the Green function $G(\boldsymbol{x})$,

$$G(\boldsymbol{x}) = \frac{1}{|\boldsymbol{x}|}, \tag{A-1.1}$$

the motion of the electron gas is given by the action principle

(A-1.2) $$\delta S = 0\,,$$

where

(A-1.3) $$S = \int \mathrm{d}^3x \frac{1}{2}\, nm \left(\frac{\partial \xi_r}{\partial t}\right)^2 - \frac{1}{2}\, e^2 n^2 \iint \mathrm{d}^3x\, \mathrm{d}^3x' \cdot \\ \cdot \big[G\big(\boldsymbol{x} + \boldsymbol{\xi}(\boldsymbol{x}, t) - \boldsymbol{x}' - \boldsymbol{\xi}(\boldsymbol{x}', t)\big) - G\big(\boldsymbol{x} + \xi(\boldsymbol{x}, t) - \boldsymbol{x}'\big) - \\ - G\big(\boldsymbol{x} - \boldsymbol{x}' - \boldsymbol{\xi}(\boldsymbol{x}', G)\big) + G(\boldsymbol{x} - \boldsymbol{x}')\big]\,.$$

To simplify calculations, we assume that the excitation is periodic with respect to a large cubical unit cell of volume V. One may show that, if the disturbance extends over a finite region of space, the interaction between disturbances in different cells may be made negligibly small by making the dimensions of the cells sufficiently large. Hence we may consider the integrations in (A-1.3) to be taken over a unit cell.

We now proceed to Fourier-analyze, writing

(A-1.4) $$\varphi(\boldsymbol{x}) = \sum \varphi(\boldsymbol{k}) \exp[i\boldsymbol{k}\cdot\boldsymbol{x}]\,,$$

and

(A-1.5) $$\varphi(\boldsymbol{k}) = V^{-1}\int \mathrm{d}^3x\, \varphi(\boldsymbol{x}) \exp[-\, i\boldsymbol{k}\cdot\boldsymbol{x}]\,.$$

The summation in (A-1.4) extends over all values of $\boldsymbol{k}$ for which $\exp[i\boldsymbol{k}\cdot\boldsymbol{x}]$ has the same value over the surface of the unit cube. We note that

(A-1.6) $$G(\boldsymbol{k}) = \frac{4\pi V^{-1}}{k^2}\,.$$

We now analyse S as follows:

(A-1.7) $$S = S^{(0)} + S^{(1)} + S^{(2)} + \dots\,,$$

where $S^{(n)}$ is of the n-th order in the dynamical variables $\xi_r(\boldsymbol{x}, t)$. Clearly, $S^{(0)}$ will not contribute to the equations of motion. Since $\xi_r \equiv 0$ is a solution of the equations of motion, we also expect that $S^{(1)}$ may be ignored, and it may be verified by calculation that $S^{(1)} = 0$. We obtain, for the next three terms, the following formulas:

(A-1.8) $$S^{(2)} = \frac{1}{2}\, nm \int \mathrm{d}^3x \left(\frac{\partial \xi_r}{\partial t}\right)^2 + \frac{1}{2}\, e^2 n^2 \iint \mathrm{d}^3x\, \mathrm{d}^3x'\, \xi_r(\boldsymbol{x})\, \xi_r(\boldsymbol{x}')\, \frac{\partial^2 G(\boldsymbol{x} - \boldsymbol{x}')}{\partial x_r\, \partial x_s}\,,$$

(A-1.9) $$S^{(3)} = \frac{1}{2}\, e^2 n^2 \iint \mathrm{d}^3x\, \mathrm{d}^3x'\, \xi_r(\boldsymbol{x})\, \xi_s(\boldsymbol{x})\, \xi_t(\boldsymbol{x}')\, \frac{\partial^3 G(\boldsymbol{x} - \boldsymbol{x}')}{\partial x_r\, \partial x_s\, \partial x_t}\,.$$

(A-1.10) $$S^{(4)} = e^2 n^2 \int \mathrm{d}^3x\, \mathrm{d}^3x' \left[\frac{1}{6}\, \xi_r(\boldsymbol{x})\, \xi_s(\boldsymbol{x})\, \xi_t(\boldsymbol{x})\, \xi_u(\boldsymbol{x}') - \right. \\ \left. - \frac{1}{8}\, \xi_r(\boldsymbol{x})\, \xi_s(\boldsymbol{x})\, \xi_t(\boldsymbol{x}')\, \xi_u(\boldsymbol{x}')\right] \cdot \frac{\partial^4 G(\boldsymbol{x} - \boldsymbol{x}')}{\partial x_r\, \partial x_s\, \partial x_t\, \partial x_u}\,,$$

where we adopt the summation convention and, for brevity, suppress the argument t of $\xi_r(\boldsymbol{x}, t)$.

We now Fourier-transform according to the scheme (A-1.4), (A-1.5), and use (A-1.6). The Lagrangian function so obtained is related to the action integral by

(A-1.11) $$S = nmV \cdot L\,.$$

The quadratic, cubic and quadratic terms of L are found to be

(A-1.12) $$L^{(2)} = \frac{1}{2} \sum_{\boldsymbol{k}+\boldsymbol{k}'=0} \left\{ \dot{\xi}_r(\boldsymbol{k}) \dot{\xi}_r(\boldsymbol{k}') - \omega_P^2 \frac{1}{\boldsymbol{k}^2} \boldsymbol{k}\cdot\xi(\boldsymbol{k}) \boldsymbol{k}\cdot\xi(\boldsymbol{k}') \right\},$$

(A-1.13) $$L^{(3)} = -\frac{1}{2} i\omega_P^2 \sum_{\boldsymbol{k}+\boldsymbol{k}'+\boldsymbol{k}''=0} \left(\frac{1}{k^2}\right) \boldsymbol{k}\cdot\boldsymbol{\xi}(\boldsymbol{k}) \boldsymbol{k}\cdot\boldsymbol{\xi}(\boldsymbol{k}') \boldsymbol{k}\cdot\boldsymbol{\xi}(\boldsymbol{k}''),$$

(A-1.14) $$L^{(4)} = \omega_P^2 \sum_{\boldsymbol{k}+\boldsymbol{k}'+\boldsymbol{k}''+\boldsymbol{k}'''=0} \left\{ \frac{1}{6}\left(\frac{1}{k^2}\right) \boldsymbol{k}\cdot\boldsymbol{\xi}(\boldsymbol{k}) \boldsymbol{k}\cdot\boldsymbol{\xi}(\boldsymbol{k}') \boldsymbol{k}\cdot\boldsymbol{\xi}(\boldsymbol{k}'') \boldsymbol{k}\cdot\boldsymbol{\xi}(\boldsymbol{k}''') - \right.$$
$$\left. - \frac{1}{8} \frac{1}{(\boldsymbol{k}+\boldsymbol{k}')^2} (\boldsymbol{k}+\boldsymbol{k}')\cdot\boldsymbol{\xi}(\boldsymbol{k}) (\boldsymbol{k}+\boldsymbol{k}')\cdot\boldsymbol{\xi}(\boldsymbol{k}') (\boldsymbol{k}+\boldsymbol{k}')\cdot\boldsymbol{\xi}(\boldsymbol{k}'') (\boldsymbol{k}+\boldsymbol{k}')\cdot\boldsymbol{\xi}(\boldsymbol{k}''') \right\}.$$

We introduce the momentum variable canonically conjugate to $\xi_r(\boldsymbol{k}, t)$ as follows:

(A.-1.15) $$\pi_r(\boldsymbol{k}, t) \equiv \frac{\partial L}{\partial \dot{\xi}_r(\boldsymbol{k}, t)} = \dot{\xi}_r(-\boldsymbol{k}, t)\,.$$

We now form the Hamiltonian function

(A-1.16) $$H = \sum_{\boldsymbol{k}} \pi_r(\boldsymbol{k}) \dot{\xi}_r(\boldsymbol{k}) - L\,,$$

and find that

(A-1.17) $$H^{(2)} = \frac{1}{2} \sum_{\boldsymbol{k}+\boldsymbol{k}'=0} \left\{ \pi_r(\boldsymbol{k}) \pi_r(\boldsymbol{k})' + \omega_P^2 \left(\frac{1}{k^2}\right) \boldsymbol{k}\cdot\boldsymbol{\xi}(\boldsymbol{k}) \boldsymbol{k}\cdot\boldsymbol{\xi}(\boldsymbol{k}') \right\},$$

and that

(A-1.18) $$H^{(3)} = -L^{(3)}\,, \qquad H^{(4)} = -L^{(4)}\,, \text{ etc.}\,.$$

Appendix II

Linear theory.

The linearized equation of motion, which is derivable from (A-1.12),

(A-2.1) $$\ddot{\xi}_r(\boldsymbol{k}) + \omega_P^2 \left(\frac{1}{k^2}\right) k_r k_s \xi_s(\boldsymbol{k}) = 0\,,$$

has solutions which fall into two categories. There are longitudinal modes for which $\boldsymbol{\xi}$ is parallel to $\boldsymbol{k}$, the frequency of which is ω_P; these are the plasma oscillations. The other modes are transverse, $\boldsymbol{\xi}$ being perpendicular to $\boldsymbol{k}$, and have zero frequency; these represent possible states of divergence-free shear motion of the electron gas. Since the « shear waves » are of zero frequency, it follows from the action-transfer relations that they can neither lose energy to nor gain energy from the plasma oscillations. These modes will affect each other only by modifying the dispersion relations. Since energy exchange among waves is a more important process, we shall neglect shear waves in what follows.

On noting that the reality of $\boldsymbol{\xi}(\boldsymbol{x}, t)$ entails the fact that

$$\xi_r(-\boldsymbol{k}, t) = \xi_r^*(\boldsymbol{k}, t)\,, \tag{A-2.2}$$

we see that the longitudinal modes are expressible as

$$\xi_r(\boldsymbol{k}, t) = k_r\{A(\boldsymbol{k})\exp[i(\omega_P t + \alpha(\boldsymbol{k}))] - A(-\boldsymbol{k})\exp[-i(\omega_P t + \alpha(-\boldsymbol{k}))]\}\,. \tag{A-2.3}$$

The corresponding formula for the momentum is

$$\pi_r(\boldsymbol{k}, t) = -i\omega_P k_r\{A(-\boldsymbol{k})\exp[i(\omega_P t + \alpha(-\boldsymbol{k}))] + A(\boldsymbol{k})\exp[-i(\omega_P t + \alpha(\boldsymbol{k}))]\}\,. \tag{A-2.4}$$

It is convenient to work with complex amplitudes as independent variables, but it is also necessary for subsequent purposes to maintain the canonical formalism. Both requirements can be met as follows: It is possible to transform to action angle variables $J(\boldsymbol{k})$, $\alpha(\boldsymbol{k})$, where $\alpha(\boldsymbol{k})$ has already been introduced and the appropriate form of $J(\boldsymbol{k})$ is found to be

$$J(\boldsymbol{k}) = 2\omega_P k^2 A^2(\boldsymbol{k})\,. \tag{A-2.5}$$

We may now form a canonical set of complex amplitudes by the formulas

$$a(\boldsymbol{k}, t) = J^{\frac{1}{2}}(\boldsymbol{k})\exp[i\alpha(\boldsymbol{k})]\,, \qquad a^\dagger(\boldsymbol{k}, t) = -iJ^{\frac{1}{2}}(\boldsymbol{k})\exp[-i\alpha(\boldsymbol{k})]\,, \tag{A-2.6}$$

in which $a^\dagger(\boldsymbol{k}, t)$ is to be regarded as the variable canonically conjugate to $a(\boldsymbol{k}, t)$. The required expressions for the normal modes are now found to be

$$\xi_r(\boldsymbol{k}, t) = (2\omega_P k^2)^{-\frac{1}{2}} k_r\{a(\boldsymbol{k}, t)\exp[i\omega_P t] - ia^\dagger(-\boldsymbol{k}, t)\exp[-i\omega_P t]\}\,, \tag{A-2.7}$$

and

$$\pi_r(\boldsymbol{k}, t) = \left(\frac{\omega_P}{2k^2}\right)^{\frac{1}{2}} k_r\{-ia(-\boldsymbol{k}, t)\exp[i\omega_P t] + a^\dagger(\boldsymbol{k}, t)\exp[-i\omega_P t]\}\,. \tag{A-.28}$$

Although, in the linear theory, a and $a^\dagger$ are time-independent, the above transformation will be used in nonlinear theory so that these variables will be time-dependent. We note, for later purposes, the obvious relation

$$a^\dagger(\boldsymbol{k}, t) = -ia^*(\boldsymbol{k}, t)\,. \tag{A-2.9}$$

Appendix III

Nonlinear analysis by the derivative-expansion technique.

In this Section, we compute the dominant effect of the nonlinear terms $H^{(3)}$ and $H^{(4)}$ by using the derivative-expansion technique of ref. [4].

We find from (A-1.13), (A-1.14), (A-1.18) and (A-2.7) that the cubic and quadratic terms of the Hamiltonian are expressible as

$$H^{(3)} = i \sum_{\boldsymbol{k}+\boldsymbol{k}'+\boldsymbol{k}''=0} A(\boldsymbol{k}, \boldsymbol{k}', \boldsymbol{k}'')\big(a(-\boldsymbol{k}) \exp[i\omega_{\mathrm{P}} t] - ia^\dagger(\boldsymbol{k}) \exp[-i\omega_{\mathrm{P}} t]\big)\cdot \\ \cdot\big(a(-\boldsymbol{k}') \exp[i\omega_{\mathrm{P}} t] - ia^\dagger(\boldsymbol{k}') \exp[-i\omega_{\mathrm{P}} t]\big)\cdot \\ \cdot\big(a(-\boldsymbol{k}'') \exp[i\omega_{\mathrm{P}} t] - ia^\dagger(\boldsymbol{k}'') \exp[-i\omega_{\mathrm{P}} t]\big)\,, \tag{A-3.1}$$

$$H^{(4)} = \sum_{\boldsymbol{k}+\boldsymbol{k}'+\boldsymbol{k}''+\boldsymbol{k}'''=0} B(\boldsymbol{k}, \boldsymbol{k}', \boldsymbol{k}'', \boldsymbol{k}''')\big(a(-\boldsymbol{k}) \exp[i\omega_{\mathrm{P}} t] - ia^\dagger(\boldsymbol{k}) \exp[-i\omega_{\mathrm{P}} t]\big)\cdot \\ \cdot\big(a(-\boldsymbol{k}') \exp[i\omega_{\mathrm{P}} t] - ia^\dagger(\boldsymbol{k}') \exp[-i\omega_{\mathrm{P}} t]\big)\cdot \\ \cdot\big(a(-\boldsymbol{k}'') \exp[i\omega_{\mathrm{P}} t] - ia^\dagger(\boldsymbol{k}'') \exp[-i\omega_{\mathrm{P}} t]\big)\cdot \\ \cdot\big(a(-\boldsymbol{k}''') \exp[i\omega_{\mathrm{P}} t] - ia^\dagger(\boldsymbol{k}''') \exp[-i\omega_{\mathrm{P}} t]\big)\,, \tag{A-3.2}$$

where

$$A(\boldsymbol{k}, \boldsymbol{k}', \boldsymbol{k}'') = \frac{1}{12}\left(\frac{\omega_{\mathrm{P}}}{2}\right)^{\frac{1}{2}}\left(\frac{1}{kk'k''}\right)(\boldsymbol{k}\cdot\boldsymbol{k}'\,\boldsymbol{k}\cdot\boldsymbol{k}'' + \boldsymbol{k}\cdot\boldsymbol{k}'\,\boldsymbol{k}'\cdot\boldsymbol{k}'' + \boldsymbol{k}\cdot\boldsymbol{k}''\,\boldsymbol{k}'\cdot\boldsymbol{k}'')\,, \tag{A-3.3}$$

$$B(\boldsymbol{k}, \boldsymbol{k}', \boldsymbol{k}'', \boldsymbol{k}''') = -\left(\frac{1}{96}\right)\left(\frac{1}{kk'k''k'''}\right)\cdot \\ \cdot(\boldsymbol{k}\cdot\boldsymbol{k}'\,\boldsymbol{k}\cdot\boldsymbol{k}''\,\boldsymbol{k}\cdot\boldsymbol{k}''' + \boldsymbol{k}\cdot\boldsymbol{k}'\,\boldsymbol{k}'\cdot\boldsymbol{k}''\,\boldsymbol{k}'\cdot\boldsymbol{k}''' + \boldsymbol{k}\cdot\boldsymbol{k}''\,\boldsymbol{k}'\cdot\boldsymbol{k}''\,\boldsymbol{k}''\cdot\boldsymbol{k}''' + \boldsymbol{k}\cdot\boldsymbol{k}'''\,\boldsymbol{k}'\cdot\boldsymbol{k}'''\,\boldsymbol{k}''\cdot\boldsymbol{k}''') + \\ + \frac{1}{192}\left(\frac{1}{kk'k''k'''}\right)\big[|\boldsymbol{k}+\boldsymbol{k}'|^{-2}(\boldsymbol{k}+\boldsymbol{k}')\cdot\boldsymbol{k}(\boldsymbol{k}+\boldsymbol{k}')\cdot\boldsymbol{k}'(\boldsymbol{k}+\boldsymbol{k}')\cdot\boldsymbol{k}''(\boldsymbol{k}+\boldsymbol{k}')\cdot\boldsymbol{k}''' + \\ + |\boldsymbol{k}+\boldsymbol{k}''|^{-2}(\boldsymbol{k}+\boldsymbol{k}'')\cdot\boldsymbol{k}(\boldsymbol{k}+\boldsymbol{k}'')\cdot\boldsymbol{k}'(\boldsymbol{k}+\boldsymbol{k}'')\cdot\boldsymbol{k}''(\boldsymbol{k}+\boldsymbol{k}'')\cdot\boldsymbol{k}''' + \\ + |\boldsymbol{k}+\boldsymbol{k}'''|^{-2}(\boldsymbol{k}+\boldsymbol{k}''')\cdot\boldsymbol{k}(\boldsymbol{k}+\boldsymbol{k}''')\cdot\boldsymbol{k}'(\boldsymbol{k}+\boldsymbol{k}''')\cdot\boldsymbol{k}''(\boldsymbol{k}+\boldsymbol{k}''')\cdot\boldsymbol{k}''' + \\ + |\boldsymbol{k}'+\boldsymbol{k}''|^{-2}(\boldsymbol{k}'+\boldsymbol{k}'')\cdot\boldsymbol{k}(\boldsymbol{k}'+\boldsymbol{k}'')\cdot\boldsymbol{k}'(\boldsymbol{k}'+\boldsymbol{k}'')\cdot\boldsymbol{k}''(\boldsymbol{k}'+\boldsymbol{k}'')\cdot\boldsymbol{k}''' + \\ + |\boldsymbol{k}'+\boldsymbol{k}'''|^{-2}(\boldsymbol{k}'+\boldsymbol{k}''')\cdot\boldsymbol{k}(\boldsymbol{k}'+\boldsymbol{k}''')\cdot\boldsymbol{k}'(\boldsymbol{k}'+\boldsymbol{k}''')\cdot\boldsymbol{k}''(\boldsymbol{k}'+\boldsymbol{k}''')\cdot\boldsymbol{k}''' + \\ + |\boldsymbol{k}''+\boldsymbol{k}'''|^{-2}(\boldsymbol{k}''+\boldsymbol{k}''')\cdot\boldsymbol{k}(\boldsymbol{k}''+\boldsymbol{k}''')\cdot\boldsymbol{k}'(\boldsymbol{k}''+\boldsymbol{k}''')\cdot\boldsymbol{k}''(\boldsymbol{k}''+\boldsymbol{k}''')\cdot\boldsymbol{k}'''\big]\,, \tag{A-3.4}$$

where we again drop the argument t from $a(\boldsymbol{k}, t)$ for brevity. We now find from the canonical equations

$$\frac{\mathrm{d}a(\boldsymbol{k})}{\mathrm{d}t} = \frac{\partial H}{\partial a^\dagger(\boldsymbol{k}, t)}, \qquad \frac{\mathrm{d}a^\dagger(\boldsymbol{k}, t)}{\mathrm{d}t} = -\frac{\partial H}{\partial a(\boldsymbol{k}, t)}, \tag{A-3.5}$$

that, taking account of only the cubic and quadratic terms, $a(\boldsymbol{k}, t)$ satisfies the following differential equation:

$$\begin{aligned}\frac{\mathrm{d}a(\boldsymbol{k}, t)}{\mathrm{d}t} &= 3\tilde{\omega} \sum_{\boldsymbol{k}+\boldsymbol{k}'+\boldsymbol{k}''=0} A(\boldsymbol{k}, \boldsymbol{k}', \boldsymbol{k}'') \exp[-i\omega_{\mathrm{P}} t] \cdot \\ &\cdot \big(a(-\boldsymbol{k}') \exp[i\omega_{\mathrm{P}} t] - a^*(\boldsymbol{k}') \exp[-i\omega_{\mathrm{P}} t]\big) \cdot \\ &\cdot \big(a(-\boldsymbol{k}'') \exp[i\omega_{\mathrm{P}} t] - a^*(\boldsymbol{k}'') \exp[-i\omega_{\mathrm{P}} t]\big) - \\ &- 4i\omega^2 \sum_{h+h'+h''+h'''=0} B(\boldsymbol{k}, \boldsymbol{k}', \boldsymbol{k}'', \boldsymbol{k}''') \exp[-i\omega_{\mathrm{P}} t] \big(a(-\boldsymbol{k}') \exp[i\omega_{\mathrm{P}} t] - a^*(\boldsymbol{k}') \exp[-i\omega_{\mathrm{P}} t]\big) \cdot \\ &\cdot \big(a(-\boldsymbol{k}'') \exp[i\omega_{\mathrm{P}} t] - a^*(\boldsymbol{k}'') \exp[-i\omega_{\mathrm{P}} t]\big) \cdot \\ &\cdot \big(a(-\boldsymbol{k}''') \exp[i\omega_{\mathrm{P}} t] - a^*(\boldsymbol{k}''') \exp[-i\omega_{\mathrm{P}} t]\big) .\end{aligned} \tag{A-3.6}$$

We have at this stage introduced for present convenience the parameter $\tilde{\omega}$ which will separate terms and effects according to the powers of their dependence upon amplitude. We may confirm this by noting that, if $\tilde{\omega}$ is initially absent from (A-3.6), it enters as indicated when $a(\boldsymbol{k}, t)$ is replaced by $\tilde{\omega}a(\boldsymbol{k}, t)$.

We now seek an approximate solution of eq. (A-3.6) as follows. The effect of nonlinearity upon $a(\boldsymbol{k}, t)$ is of two types: the introduction of harmonics, and slow time-variation of amplitude. We allow for these two effects by replacing a by the expansion

$$a \to a + \tilde{\omega} a^{\mathrm{I}} + \tilde{\omega}^2 a^{\mathrm{II}} + \dots, \tag{A-3.7}$$

and by a similar expansion of the contributions to the time-derivative of any quantity:

$$\frac{\mathrm{d}}{\mathrm{d}t} \to \frac{\mathrm{d}}{\mathrm{d}t} + \omega \frac{\mathrm{d}^{\mathrm{I}}}{\mathrm{d}t} + \omega^2 \frac{\mathrm{d}^{\mathrm{II}}}{\mathrm{d}t} + \dots, \tag{A-3.8}$$

We see that the right-hand side of (A-3.6) contains no zero-frequency term to order $\tilde{\omega}$ so that

$$\frac{\mathrm{d}^{\mathrm{I}} a}{\mathrm{d}t} = 0 . \tag{A-3.9}$$

On examining the frequency-dependence of terms of order $\tilde{\omega}$ in (A-3.6) we see that a^{I} may be analyzed as follows:

$$a^{\mathrm{I}} = a_1^{\mathrm{I}} \exp[i\omega_{\mathrm{P}} t] + a_{-1}^{\mathrm{I}} \exp[-i\omega_{\mathrm{P}} t] + a_{-3}^{\mathrm{I}} \exp[-3i\omega_{\mathrm{P}} t], \tag{A-3.10}$$

of which a_1^{I} and a_{-3}^{I} will contribute to the principal second harmonic content, and a_{-1}^{I} will contribute to the principal zero-frequency content of the electrostatic waves.

On making the substitutions (A-3.7), (A-3.8) and using (A-3.9), (A-3.10) in (A-3.6), and separating terms of order $\tilde{\omega}$, we find that

$$\text{(A-3.11)}\quad \left\{\begin{aligned} a_1^{\mathrm{I}}(\boldsymbol{k}) &= -3i\omega_{\mathrm{P}}^{-1}\sum_{\boldsymbol{k}+\boldsymbol{k}'+\boldsymbol{k}''=0} A(\boldsymbol{k},\boldsymbol{k}',\boldsymbol{k}'')a(-\boldsymbol{k}')a(-\boldsymbol{k}''),\\ a_{-1}^{\mathrm{I}}(\boldsymbol{k}) &= -6i\omega_{\mathrm{P}}^{-1}\sum_{\boldsymbol{k}+\boldsymbol{k}'+\boldsymbol{k}''=0} A(\boldsymbol{k},\boldsymbol{k}',\boldsymbol{k}'')a^*(\boldsymbol{k}')a(-\boldsymbol{k}''),\\ a_{-3}^{\mathrm{I}}(\boldsymbol{k}) &= i\omega_{\mathrm{P}}^{-1}\sum_{\boldsymbol{k}+\boldsymbol{k}'+\boldsymbol{k}''=0} A(\boldsymbol{k},\boldsymbol{k}',\boldsymbol{k}'')a^*(\boldsymbol{k}')a^*(\boldsymbol{k}'').\end{aligned}\right.$$

We now consider terms of (A-3.6) which are of order $\tilde{\omega}^2$, and remember that a is to be replaced by (A-3.7) on the left-hand side and also on the right-hand side. We are interested only in the zero-frequency contribution of order $\tilde{\omega}^2$ and this is found to be

$$\text{(A-3.12)}\quad \frac{\mathrm{d}^{\mathrm{II}}a(\boldsymbol{k})}{\mathrm{d}t} = 6\sum_{\boldsymbol{k}+\boldsymbol{k}'+\boldsymbol{k}''=0} A(\boldsymbol{k},\boldsymbol{k}',\boldsymbol{k}'')\cdot$$
$$\cdot\{a_{-1}^{\mathrm{I}}(-\boldsymbol{k}')a(-\boldsymbol{k}'') - a^*(\boldsymbol{k}')a_1^{\mathrm{I}}(-\boldsymbol{k}'') - a_{-1}^{\mathrm{I}*}(\boldsymbol{k}')a(-\boldsymbol{k}'') + a_{-3}^{\mathrm{I}*}(\boldsymbol{k}')a^*(\boldsymbol{k}'')\} +$$
$$+ 12i\sum_{\boldsymbol{k}+\boldsymbol{k}'+\boldsymbol{k}''+\boldsymbol{k}'''=0} B(\boldsymbol{k},\boldsymbol{k}',\boldsymbol{k}'',\boldsymbol{k}''')a^*(\boldsymbol{k}')a(-\boldsymbol{k}'')a(-\boldsymbol{k}''').$$

On using formulas (A-3.11), we find that (A-3.12) may be rewritten as

$$\text{(A-3.13)}\quad \frac{\mathrm{d}^{\mathrm{II}}a(\boldsymbol{k}_1)}{\mathrm{d}t} = i\sum_{\boldsymbol{k}_1+\boldsymbol{k}_2=\boldsymbol{k}_3+\boldsymbol{k}_4} C(\boldsymbol{k}_1,\boldsymbol{k}_2,\boldsymbol{k}_3,\boldsymbol{k}_4)a^*(k_2)a(k_3)a(k_4),$$

where

$$\text{(A-3.14)}\quad C(k_1,k_2,k_3,k_4) = 12\omega_{\mathrm{P}}^{-1}\{A(\boldsymbol{k}_1,\boldsymbol{k}_2,-\boldsymbol{k}_1-\boldsymbol{k}_2)A(\boldsymbol{k}_3,\boldsymbol{k}_4,-\boldsymbol{k}_3-\boldsymbol{k}_4) -$$
$$-3A(\boldsymbol{k}_1,-\boldsymbol{k}_4,\boldsymbol{k}_4-\boldsymbol{k}_1)A(\boldsymbol{k}_2,-\boldsymbol{k}_3,\boldsymbol{k}_3-\boldsymbol{k}_2) -$$
$$-3A(\boldsymbol{k}_1,-\boldsymbol{k}_3,\boldsymbol{k}_3-\boldsymbol{k}_1)A(\boldsymbol{k}_2,-\boldsymbol{k}_4,\boldsymbol{k}_4-\boldsymbol{k}_2) + B(\boldsymbol{k}_1,\boldsymbol{k}_2,-\boldsymbol{k}_3,-\boldsymbol{k}_4)\}.$$

Equation (A-3.13) represents the dominant energy-transfer process due to nonlinearity as a process which may be analysed into the interactions of groups of four waves. The energy $\varepsilon(\boldsymbol{k})$ to be attributed to each wave is proportional to $a^*(\boldsymbol{k})a(\boldsymbol{k})$. We now find, from the symmetry properties of the « kernel » $C(\boldsymbol{k}_1,\boldsymbol{k}_2,\boldsymbol{k}_3,\boldsymbol{k}_4)$,

$$\text{(A-3.15)}\quad C(\boldsymbol{k}_2,\boldsymbol{k}_1,\boldsymbol{k}_3,\boldsymbol{k}_4) = C(\boldsymbol{k}_1,\boldsymbol{k}_2,\boldsymbol{k}_4,\boldsymbol{k}_3) = C(\boldsymbol{k}_3,\boldsymbol{k}_4,\boldsymbol{k}_1,\boldsymbol{k}_2) = C(\boldsymbol{k}_1,\boldsymbol{k}_2,\boldsymbol{k}_3,\boldsymbol{k}_4),$$

that

$$\text{(A-3.16)}\quad \frac{\mathrm{d}\varepsilon(\boldsymbol{k}_1)}{\mathrm{d}t} = \frac{\mathrm{d}\varepsilon(\boldsymbol{k}_2)}{\mathrm{d}t} = -\frac{\mathrm{d}\varepsilon(\boldsymbol{k}_3)}{\mathrm{d}t} = -\frac{\mathrm{d}\varepsilon(\boldsymbol{k}_4)}{\mathrm{d}t}.$$

Hence energy is drawn in equal amounts from one pair of waves and distributed equally among the other pair of waves. Momentum is exchanged in the same way. Equation (A-3.16) agrees with the action-transfer relations for four waves of equal frequency, the wave numbers of which are related by

$$\mathbf{k}_1 + \mathbf{k}_2 = \mathbf{k}_3 + \mathbf{k}_4 \,. \tag{A-3.17}$$

On using (A-3.3) and (A-3.4) we find that

$$\begin{aligned}
C(\mathbf{k}_1, \mathbf{k}_2, \mathbf{k}_3, \mathbf{k}_4) = &-\tfrac{1}{8}(k_1 k_2 k_3 k_4)^{-1} \cdot \\
&\cdot [\mathbf{k}_1\cdot\mathbf{k}_2\,\mathbf{k}_1\cdot\mathbf{k}_3\,\mathbf{k}_1\cdot\mathbf{k}_4 + \mathbf{k}_1\cdot\mathbf{k}_2\,\mathbf{k}_2\cdot\mathbf{k}_3\,\mathbf{k}_2\cdot\mathbf{k}_4 + \mathbf{k}_1\cdot\mathbf{k}_3\,\mathbf{k}_2\cdot\mathbf{k}_3\,\mathbf{k}_3\cdot\mathbf{k}_4 + \mathbf{k}_1\cdot\mathbf{k}_4\,\mathbf{k}_2\cdot\mathbf{k}_4\,\mathbf{k}_3\cdot\mathbf{k}_4] + \\
&+ \tfrac{1}{24}(k_1k_2k_3k_4)^{-1}|\mathbf{k}_1+\mathbf{k}_2|^{-2}[3\mathbf{k}_1\cdot(\mathbf{k}_1+\mathbf{k}_2)\mathbf{k}_2\cdot(\mathbf{k}_1+\mathbf{k}_2)\mathbf{k}_3\cdot(\mathbf{k}_1+\mathbf{k}_2)\mathbf{k}_4\cdot(\mathbf{k}_1+\mathbf{k}_2) + \\
&+ \{k_1^2k_2^2 - (\mathbf{k}_1\cdot\mathbf{k}_2)^2\}\{k_1^2k_4^2 - (\mathbf{k}_3\cdot\mathbf{k}_4)^2\}] + \\
&+ \tfrac{1}{8}(k_1k_2k_3k_4)^{-1}|\mathbf{k}_1-\mathbf{k}_3|^{-2}[\mathbf{k}_1\cdot(\mathbf{k}_1-\mathbf{k}_3)\mathbf{k}_2\cdot(\mathbf{k}_1-\mathbf{k}_3)\mathbf{k}_3\cdot(\mathbf{k}_1-\mathbf{k}_3)\mathbf{k}_4(\mathbf{k}_1-\mathbf{k}_3) - \\
&- \{k_1^2k_3^2 - (\mathbf{k}_1\cdot\mathbf{k}_3)^2\}\{k_2^2k_4^2 - (\mathbf{k}_2\cdot\mathbf{k}_4)^2\}] + \\
&+ \tfrac{1}{8}(k_1k_2k_3k_4)^{-1}|\mathbf{k}_1-\mathbf{k}_4|^{-2}[\mathbf{k}_1\cdot(\mathbf{k}_1-\mathbf{k}_4)\mathbf{k}_2\cdot(\mathbf{k}_1-\mathbf{k}_4)\mathbf{k}_3\cdot(\mathbf{k}_1-\mathbf{k}_4)\mathbf{k}_4\cdot(\mathbf{k}_1-\mathbf{k}_4) - \\
&- \{k_1^2k_4^2 - (\mathbf{k}_1\cdot\mathbf{k}_4)^2\}\{k_2^2k_3^2 - (\mathbf{k}_2\cdot\mathbf{k}_3)^2\}] \,.
\end{aligned} \tag{A-3.18}$$

By further algebraic manipulations it is possible to write the formula in the form

$$\begin{aligned}
C = \tfrac{1}{4}(k_1k_2k_3k_4)^{-1}[&-\mathbf{k}_1\cdot\mathbf{k}_3\,\mathbf{k}_1\cdot\mathbf{k}_4\,\mathbf{k}_2\cdot(\mathbf{k}_3+\mathbf{k}_4) - \mathbf{k}_2\cdot\mathbf{k}_3\,\mathbf{k}_2\cdot\mathbf{k}_4\,\mathbf{k}_1\cdot(\mathbf{k}_3+\mathbf{k}_4)] + \\
&+ \tfrac{1}{12}(k_1k_2k_3k_4)^{-1}|\mathbf{k}_1+\mathbf{k}_2|^{-2}\cdot \\
&\cdot[3\mathbf{k}_1\cdot(\mathbf{k}_1+\mathbf{k}_2)\mathbf{k}_2\cdot(\mathbf{k}_1+\mathbf{k}_2)\mathbf{k}_3\cdot(\mathbf{k}_1+\mathbf{k}_2)\mathbf{k}_4\cdot(\mathbf{k}_1+\mathbf{k}_2) - \\
&- \{k_1^2k_2^2 - (\mathbf{k}_1\cdot\mathbf{k}_2)^2\}\{k_3^2k_4^2 - (\mathbf{k}_3\cdot\mathbf{k}_4)^2\}] \,,
\end{aligned} \tag{A-3.19}$$

which shows explicitly that there is no singularity at $\mathbf{k}_1 - \mathbf{k}_3 = \mathbf{k}_4 - \mathbf{k}_2 = 0$ or at $\mathbf{k}_1 - \mathbf{k}_4 = \mathbf{k}_3 - \mathbf{k}_2 = 0$. One may also verify that there is no singularity at $\mathbf{k}_1 + \mathbf{k}_2 = \mathbf{k}_3 + \mathbf{k}_4 = 0$.

APPENDIX IV

Canonical transformation for reduction of perturbation Hamiltonian.

Consider a dynamical system described by dynamical variables p_r, q_r, and a Hamiltonian $h(p_r, q_s, t)$ which is given as a perturbation expansion in a parameter $\tilde{\omega}$ of the form

$$h = \tilde{\omega} h^{\mathrm{I}} + \tilde{\omega}^2 h^{\mathrm{II}} + \dots \,. \tag{A-4.1}$$

Since there is no zero-order term, p_r, q_r are constants of the motion in zero-order theory.

We consider a canonical transformation of the system to new variables P_r, Q_r and a new Hamiltonian $H(P_r, Q_r, t)$. We consider the transformation or variables to be expressible in the form

$$\text{(A-4.2)} \qquad \begin{cases} p_r = P_r + \varpi P_r^{\mathrm{I}} + \varpi^2 P_r^{\mathrm{II}} + \ldots \\ q_r = Q_r + \varpi Q_r^{\mathrm{I}} + \varpi^2 Q_r^{\mathrm{II}} + \ldots \end{cases}$$

wherein P_r^{I} etc. are to be expressed as functions of P_r, Q_r, t. The transformation is then the identity transformation in zero-order theory. The new Hamiltonian will in general be expressible as

$$\text{(A-4.3)} \qquad H = \varpi H^{\mathrm{I}} + \varpi^2 H^{\mathrm{II}} + \ldots .$$

We wish to find a canonical transformation of the above form for which $H^{\mathrm{I}} = 0$.

The essential properties of the original variables and Hamiltonian are summarized by the relation

$$\text{(A-4.4)} \qquad \delta h = \dot{q}_r\, \delta p_r - \dot{p}_r\, \delta q_r .$$

The term h may be replaced by the expression (A-4.1) and the argument of h^{I} etc. replaced by the expressions (A-4.2). The right-hand side of (A-4.4) may also be expressed in terms of P_r, Q_r by means of (A-4.2). If the resulting relation reduces to

$$\text{(A-4.5)} \qquad \delta H = \dot{Q}_r\, \delta P_t - \dot{P}_r\, \delta Q_r ,$$

then the transformation represented by (A-4.2) is canonical.

On making the above substitutions, (A-4.4) takes the form

$$\text{(A-4.6)} \qquad \varpi \left[\frac{\partial h^{\mathrm{I}}}{\partial P_r} \delta P_r + \frac{\partial h^{\mathrm{I}}}{\partial Q_r} \delta Q_r \right] + \varpi^2 \left[\delta P_r^{\mathrm{I}} \frac{\partial h^{\mathrm{I}}}{\partial P_r} + Q_r^{\mathrm{I}} \frac{\partial h^{\mathrm{I}}}{\partial Q_r} \right] +$$
$$+ \varpi^2 \left[\frac{\partial h^{\mathrm{II}}}{\partial P_r} \delta P_r + \frac{\partial h^{\mathrm{II}}}{\partial Q_r} \delta Q_r \right] + \ldots = \left[\dot{Q}_r + \varpi \left(\frac{\partial Q_r^{\mathrm{I}}}{\partial P_s} \dot{P}_s + \frac{\partial Q_r^{\mathrm{I}}}{\partial Q_s} \dot{Q}_s + \frac{\partial Q_r^{\mathrm{I}}}{\partial t} \right) + \right.$$
$$\left. + \varpi^2 \left(\frac{\partial Q_r^{\mathrm{II}}}{\partial P_s} \dot{P}_s + \frac{\partial Q_r^{\mathrm{II}}}{\partial Q_s} \dot{Q}_s + \frac{\partial Q_r^{\mathrm{II}}}{\partial t} \right) + \ldots \right] \cdot \left[\delta P_r + \varpi \left(\frac{\partial P_r^{\mathrm{I}}}{\partial P_t} \delta P_t + \frac{\partial P_r^{\mathrm{I}}}{\partial Q_t} \delta Q_t \right) + \right.$$
$$\left. + \varpi^2 \left(\frac{\partial P_r^{\mathrm{II}}}{\partial P_t} \delta P_t + \frac{\partial P_r^{\mathrm{II}}}{\partial Q.} \delta Q_t \right) + \ldots \right] - \left[-\dot{P}_r + \varpi \left(\frac{\partial P_r^{\mathrm{I}}}{\partial P_s} \dot{P}_s + \frac{\partial P_j^{\mathrm{I}}}{\partial Q_s} \dot{Q}_s + \frac{\partial P_r^{\mathrm{I}}}{\partial t} \right) + \right.$$
$$\left. + \varpi^2 \left(\frac{\partial P_r^{\mathrm{II}}}{\partial P_s} \dot{P}_s + \frac{\partial P_r^{\mathrm{II}}}{\partial Q_s} \dot{Q}_s + \frac{\partial P_r^{\mathrm{II}}}{\partial t} \right) + \ldots \right] \cdot \left[\delta Q_r + \varpi \left(\frac{\partial Q_r^{\mathrm{I}}}{\partial P_t} \delta P_t + \frac{\partial Q_r^{\mathrm{I}}}{\partial Q_t} \delta Q_t \right) + \right.$$
$$\left. + \varpi^2 \left(\frac{\partial Q_r^{\mathrm{II}}}{\partial P_t} \delta P_t + \frac{\partial Q_r^{\mathrm{II}}}{\partial Q_t} \delta Q_t \right) + \ldots \right] .$$

We now replace $\dot{Q}_r \delta P_r - \dot{P}_r \delta Q_r$ on the right-hand side of (A-4.6) by the variation of the expression (A-4.3) expressing this variation in terms of δP_r, δQ_r. If we now separate terms of first order in $\tilde{\omega}$, we find that (A-4.6) is satisfied to this order and that we may satisfy the requirement

$$H^{\mathrm{I}} = 0, \tag{A-4.7}$$

by introducing a generating function $U^{\mathrm{I}}(P_r, Q_s, t)$ which is to satisfy

$$\frac{\partial U^{\mathrm{I}}}{\partial t} = - h^{\mathrm{I}}, \tag{A-4.8}$$

and from which P_r^{I}, Q_r^{I} are obtained by means of the equations

$$P_r^{\mathrm{I}} = \frac{\partial U^{\mathrm{I}}}{\partial Q_r}, \qquad Q_r^{\mathrm{I}} = - \frac{\partial U^{\mathrm{I}}}{\partial P_r}. \tag{A-4.9}$$

We now wish to find second-order terms of (A-4.2) and (A-4.3) which are consistent with (A-4.7). We first note from (A-4.8) and (A-4.9) that

$$\frac{\partial h^{\mathrm{I}}}{\partial P_r} = \frac{\partial Q_r^{\mathrm{I}}}{\partial t}, \quad \frac{\partial h^{\mathrm{I}}}{\partial Q_r} = - \frac{\partial P_r^{\mathrm{I}}}{\partial t}, \tag{A-4.10}$$

and so express the second term of the left-hand side of (A-4.6) in terms of δP_r, δQ_r. If we now separate terms of (A-4.6) of order $\tilde{\omega}^2$, we find that (A-4.6) is satisfied to this order if P^{II} and Q^{II} are given by

$$\left\{ \begin{aligned} P_r^{\mathrm{II}} &= \left(-\frac{1}{2} + \theta\right) P_s^{\mathrm{I}} \frac{\partial Q_s^{\mathrm{I}}}{\partial Q_r} + \left(\frac{1}{2} + \theta\right) Q_s^{\mathrm{I}} \frac{\partial P_s^{\mathrm{I}}}{\partial Q_r}, \\ Q_s^{\mathrm{II}} &= \left(\frac{1}{2} - \theta\right) P_s^{\mathrm{I}} \frac{\partial Q_s^{\mathrm{I}}}{\partial P_r} + \left(-\frac{1}{2} - \theta\right) Q_s^{\mathrm{I}} \frac{\partial P_s^{\mathrm{I}}}{\partial P_r}, \end{aligned} \right. \tag{A-4.11}$$

and H^{II} is given by

$$H^{\mathrm{II}} = h^{\mathrm{II}} + \left(\frac{1}{2} + \theta\right) P_r^{\mathrm{I}} \frac{\partial Q_r^{\mathrm{I}}}{\partial t} + \left(-\frac{1}{2} + \theta\right) Q_r^{\mathrm{I}} \frac{\partial P_r^{\mathrm{I}}}{\partial t}, \tag{A-4.12}$$

where θ may be chosen arbitrarly.

APPENDIX V

Nonlinear analysis by canonical transformation.

In this Appendix we use a technique alternative to that of Section 4 for calculating the dominant wave-interaction process due to nonlinearity of the equation describing electrostatic waves. If $H^{(3)}$ contains a term of zero-frequency, this term would represent the dominant wave interaction. Since $H^{(4)}$ contains a zero-frequency term, this certainly contributes to the dominant interaction, but one must also consider the possibility that $H^{(3)}$ gives rise to a wave-interaction process comparable in magnitude with that due to $H^{(4)}$. The processes to be considered are characterized by the « Feynmann diagrams » shown in Fig. 2.

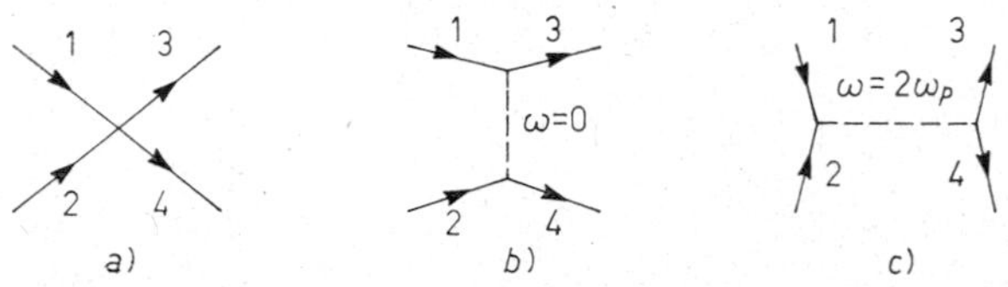

Fig. 2.

Figure 2*a* represents the direct interaction of four waves associated with the zero-frequency part of $H^{(4)}$. Figures 2*b* and 2*c* represent competing processes associated with $H^{(3)}$, which involve the excitation of « virtual waves » of frequencies zero and $2\omega_P$, respectively.

Since we wish to calculate the wave-interaction process only to order $H^{(4)}$, we may proceed as follows. We carry out a canonical transformation of the dynamical variables which eliminates the cubic contribution to the Hamiltonian, and then pick out the zero-frequency contribution to the resulting quartic part of the Hamiltonian. The technique for finding this transformation is given in Appendix IV.

We apply the method of Appendix IV to the present problem by identifying the initial set of variables with the complex variables $a^\dagger(\boldsymbol{k})$, $a(\boldsymbol{k})$ and identifying h^{I}, h^{II} with $H^{(3)}$, $H^{(4)}$ as given by (A-3.1) and (A-3.2). We write the transformed variables as $b^\dagger(\boldsymbol{k})$, $b(\boldsymbol{k})$ and the transformed Hamiltonian as $\widetilde{H}$. Then we see from Appendix I that the generating function $U^{\mathrm{I}}\big(b^\dagger(\boldsymbol{k}), (\boldsymbol{k}), t\big)$ is defined by the equation

$$\text{(A-5.1)} \qquad \frac{\mathrm{d}U^{\mathrm{I}}}{\mathrm{d}t} = -H^{(3)}\,,$$

in which the arguments $a^\dagger$, a of $H^{(3)}$ are replaced by $b^\dagger$, b. We find that

$$\text{(A-5.2)} \qquad U^{\mathrm{I}} = \omega_{\mathrm{P}}^{-1} \sum_{\boldsymbol{k}+\boldsymbol{k}'+\boldsymbol{k}''=0} A(\boldsymbol{k}, \boldsymbol{k}', \boldsymbol{k}'')\cdot$$

$$\cdot\big[-\tfrac{1}{3}\, b(-\boldsymbol{k})\, b(-\boldsymbol{k}')\, b(-\boldsymbol{k}'') \exp[3i\omega_{\mathrm{P}}t] + 3ib^\dagger(\boldsymbol{k})\, b(-\boldsymbol{k}')\, b(-\boldsymbol{k}'') \exp[i\omega_{\mathrm{P}}t] -$$

$$- 3b^\dagger(\boldsymbol{k})\, b^\dagger(\boldsymbol{k}')\, b(-\boldsymbol{k}'') \exp[-i\omega_{P}t] + \tfrac{1}{3}\, ib^\dagger(\boldsymbol{k})\, b^\dagger(\boldsymbol{k}')\, b^\dagger(\boldsymbol{k}'') \exp[-3i\omega_{\mathrm{P}}t]\big]\,.$$

On noting that

$$\text{(A-5.3)} \qquad A(-\boldsymbol{k}, -\boldsymbol{k}', -\boldsymbol{k}'') = A(\boldsymbol{k}, \boldsymbol{k}', \boldsymbol{k}'')\,,$$

we find from (A-4.9) that

(A-5.4) $$b^{\dagger\mathrm{I}}(\boldsymbol{k}) = \omega_{\mathrm{P}}^{-1} \sum_{\boldsymbol{k}+\boldsymbol{k}'+\boldsymbol{k}''=0} A(\boldsymbol{k}, \boldsymbol{k}', \boldsymbol{k}'')[-b(\boldsymbol{k}')b(\boldsymbol{k}'')\exp[3i\omega_{\mathrm{P}}t] + \\ + 6ib^{\dagger}(-\boldsymbol{k}')b(\boldsymbol{k}'')\exp[i\omega_{\mathrm{P}}t] - 3b^{\dagger}(-\boldsymbol{k}')b^{\dagger}(-\boldsymbol{k}'')\exp[-i\omega_{\mathrm{P}}t]],$$

(A-5.5) $$b^{\mathrm{I}}(\boldsymbol{k}) = \omega_{\mathrm{P}}^{-1} \sum_{\boldsymbol{k}+\boldsymbol{k}'+\boldsymbol{k}''=0} A(\boldsymbol{k}, \boldsymbol{k}', \boldsymbol{k}'')[-3ib(-\boldsymbol{k}')b(-\boldsymbol{k}'')\exp[i\omega_{\mathrm{P}}t] + \\ + 6b^{\dagger}(\boldsymbol{k}')b(-\boldsymbol{k}'')\exp[-i\omega_{\mathrm{P}}t] - ib^{\dagger}(\boldsymbol{k}')b^{\dagger}(\boldsymbol{k}'')\exp[-3i\omega_{\mathrm{P}}t]],$$

We may now find the function $\widetilde{H}^{(4)}$ by eq. (A-4.12). However, on noting that we shall need only the zero-frequency part of this function, we see that the term involving the arbitrary coefficient θ drops out of the expression, leaving

(A-5.6) $$\widetilde{H}^{(4)}_{\mathrm{ZF}} = H^{(4)}_{\mathrm{ZF}} + \frac{1}{2}\sum_{\boldsymbol{k}}\left[b^{\dagger\mathrm{I}}(\boldsymbol{k})\frac{\partial b^{\mathrm{I}}(\boldsymbol{k})}{\partial t} - b^{\mathrm{I}}(\boldsymbol{k})\frac{\partial b^{\dagger\mathrm{I}}(\boldsymbol{k})}{\partial t}\right]_{\mathrm{ZF}},$$

where the subscript ZF refers to the zero-frequency part of a function. Evaluating (A-5.6) by means of (A-3.3), (A-3.4), (A-5.4) and (A-5.5), we find that

(A-5.7) $$\widetilde{H}^{(4)}_{\mathrm{ZF}} = -\tfrac{1}{2}\sum_{\boldsymbol{k}_1+\boldsymbol{k}_2=\boldsymbol{k}_3+\boldsymbol{k}_4} C(\boldsymbol{k}_1, \boldsymbol{k}_2, \boldsymbol{k}_3, \boldsymbol{k}_4)b^{\dagger}(\boldsymbol{k}_1)b^{\dagger}(\boldsymbol{k}_2)b(\boldsymbol{k}_3)b(\boldsymbol{k}_4),$$

where $C(\boldsymbol{k}_1, \boldsymbol{k}_2, \boldsymbol{k}_3, \boldsymbol{k}_4)$ is identical with the function (A-3.14). We see from (A-3.5) that the canonical equation derivable from (A-5.7) is

(A-5.8) $$\frac{\mathrm{d}b(\boldsymbol{k}_1)}{\mathrm{d}t} = i\sum_{\boldsymbol{k}_1+\boldsymbol{k}_2=\boldsymbol{k}_3+\boldsymbol{k}_4} C(\boldsymbol{k}_1, \boldsymbol{k}_2, \boldsymbol{k}_3, \boldsymbol{k}_4)b^{*}(\boldsymbol{k}_2)b(\boldsymbol{k}_3)b(\boldsymbol{k}_4),$$

which is identical in form with (A-3.13). Hence the derivative-expansion technique and the canonical-transformation technique lead to the same formula for the dominant nonlinear wave interaction.

Appendix VI

Coupling of electromagnetic and electrostatic waves in plasmas.

If it is assumed that the plasma is of zero temperature, the disturbance of the electrons may be conveniently represented by the displacement vector $\xi_r(x, t)$. If it is assumed that there is no d.c. electric or magnetic field, the electromagnetic field may be conveniently represented by the vector potential $A_r(x, t)$.

The system may then be represented by the Lagrange density

$$\mathscr{L} = \frac{1}{2}\,mn\left(\frac{\partial \xi_r}{\partial t}\right)^2 + \frac{1}{8\pi}\left[\frac{1}{c^2}\left(\frac{\partial A_r}{\partial t}\right)^2 - \left(\frac{\partial A_r}{\partial x_s}\right)^2 + \frac{\partial A_r}{\partial x_s}\frac{\partial A_s}{\partial x_r}\right] - \frac{e}{c}\,n\frac{\partial \xi_r}{\partial t}\,A_r(x+\xi, t)\,, \tag{A-6.1}$$

where each function is evaluated with arguments x, t unless otherwise indicated.

The Lagrange density (A-6.1) may be expanded as

$$\mathscr{L} = \mathscr{L}^{(2)} + \mathscr{L}^{(3)} + \dots\,, \tag{A-6.2}$$

where

$$\mathscr{L}^{(2)} = \frac{1}{2}\,mn\left(\frac{\partial \xi_r}{\partial t}\right)^2 + \frac{1}{8\pi}\left[\frac{1}{c^2}\left(\frac{\partial A_r}{\partial t}\right)^2 - \left(\frac{\partial A_r}{\partial x_s}\right)^2 + \frac{\partial A_r}{\partial x_s}\frac{\partial A_s}{\partial x_r}\right] - \frac{e}{c}\,n\frac{\partial \xi_r}{\partial t}\,A_r\,, \tag{A-6.3}$$

and

$$\mathscr{L}^{(3)} = -\frac{e}{c}\,n\frac{\partial \xi_r}{\partial t}\,\xi_s\frac{\partial A_r}{\partial x_s}\,. \tag{A-6.4}$$

The Euler-Lagrange equations derivable from (A-6.2) are

$$\frac{\partial^2 \xi_r}{\partial t^2} - \frac{e}{mc}\frac{\partial A_r}{\partial t} = \frac{e}{mc}\,\xi_s\frac{\partial^2 A_r}{\partial t\,\partial x_s} + \frac{e}{mc}\frac{\partial \xi_s}{\partial t}\left(\frac{\partial A_r}{\partial x_s} - \frac{\partial A_s}{\partial x_r}\right). \tag{A-6.5}$$

$$\frac{1}{c^2}\frac{\partial^2 A_r}{\partial t^2} - \frac{\partial^2 A_r}{\partial x_s^2} + \frac{\partial^2 A_s}{\partial x_r\,\partial x_s} + \frac{4\pi n e}{c}\frac{\partial \xi_r}{\partial t} = \frac{4\pi n e}{c}\frac{\partial}{\partial x_s}\left(\frac{\partial \xi_r}{\partial t}\,\xi_s\right), \tag{A-6.6}$$

where we have taken account only of the terms $\mathscr{L}^{(2)}$ and $\mathscr{L}^{(3)}$ and grouped the linear contribution of the equations on the left-hand side, and the non-linear contributions on the right-hand side.

The solution of the linearized form of eqs. (A-6.5) and (A-6.6) may be written as the sum of longitudinal waves (plasma oscillations) and transverse waves (electromagnetic waves), as follows

$$A_r = \sum\left[l_r a_{\mathrm{L}} \exp\left[i(\boldsymbol{k}\cdot\boldsymbol{x} - \omega_{\mathrm{P}} t)\right] + l_r a_{\mathrm{L}}^* \exp\left[-i(\boldsymbol{k}\cdot\boldsymbol{x} - \omega_{\mathrm{P}} t)\right]\right] + \sum\left[e_r a_{\mathrm{T}} \exp\left[i(\boldsymbol{k}\cdot\boldsymbol{x} - \omega_{\mathrm{T}} t)\right] + e_r a_{\mathrm{T}}^* \exp\left[-i(\boldsymbol{k}\cdot\boldsymbol{x} - \omega_{\mathrm{P}} t)\right]\right], \tag{A-6.7}$$

and

$$\xi_r = \frac{ie}{mc\omega_{\mathrm{P}}}\sum\left[l_r a_{\mathrm{L}} \exp\left[i(\boldsymbol{k}\cdot\boldsymbol{x} - \omega_{\mathrm{P}} t)\right] + l_r a_{\mathrm{L}}^* \exp\left[-i(\boldsymbol{k}\cdot\boldsymbol{x} - \omega_{\mathrm{P}} t)\right] + \frac{ie}{mc}\sum \omega_{\mathrm{T}}^{-1}\left[e_r a_{\mathrm{T}} \exp\left[i(\boldsymbol{k}\cdot\boldsymbol{x} - \omega_{\mathrm{T}} t)\right] - e_r a_{\mathrm{T}}^* \exp\left[-i(\boldsymbol{k}\cdot\boldsymbol{x} - \omega_{\mathrm{T}} t)\right]\right], \tag{A-6.8}$$

where l_r is the unit vector parallel to k_r, e_r is the unit polarization vector normal to k_r, ω_{P} is the plasma frequency and ω_{T} satisfies the dispersion relation

(A-6.9) $$\omega_{\mathrm{T}}^2 = \omega_{\mathrm{P}}^2 + c^2 \boldsymbol{k}^2 .$$

The summation is over all vectors $\boldsymbol{k}$ permitted by the geometry of the model and all pairs of values of the polarization vector $\boldsymbol{e}$.

We now anticipate that, to order $\mathscr{L}^{(3)}$, the effect of nonlinearity of the equations of motion may be represented as the interaction of electrostatic and electromagnetic waves in groups of three. We consider the effect of interactions between two longitudinal waves and one transverse wave. We therefore consider such a group, writing the wave vectors of the longitudinal waves as $\boldsymbol{k}_{\mathrm{L1}}$, $\boldsymbol{k}_{\mathrm{L2}}$, and the wave vector of the transverse wave as $\boldsymbol{k}_{\mathrm{T}}$. If there is to be nonzero interaction, frequencies and wave vectors must satisfy the selection rules

(A-6.10) $$\boldsymbol{k}_{\mathrm{T}} = \boldsymbol{k}_{\mathrm{L1}} + \boldsymbol{k}_{\mathrm{L2}} ,$$

and

(A-6.11) $$\omega_{\mathrm{T}} = 2\omega_{\mathrm{P}} .$$

It follows from (A-6.9) that

(A-6.12) $$k_{\mathrm{T}}^2 = 3\,\frac{\omega_{\mathrm{P}}^2}{c^2} .$$

We mow assume that solutions of the eqs. (A-6.5) and (A-6.6) may be written in the forms (A-6.7) and (A-6.8) if small terms A_r^{I}, ξ_r^{I} are added to these formulas and if the amplitudes are allowed to vary slowly in space and time. We separate the rapid temporal and spatial variation of quantities, due to the wave nature of the linearized solutions, from the slow variation due to nonlinearity by writing formally

(A-6.13) $$\frac{\partial}{\partial t} \to \frac{\partial}{\partial t} + \frac{\partial^{\mathrm{I}}}{\partial t} , \qquad \frac{\partial}{\partial x_r} \to \frac{\partial}{\partial x_r} + \frac{\partial^{\mathrm{I}}}{\partial x_r} .$$

The terms A_r^{I}, ξ_r^{I} may be expanded as

(A-6.14) $$A_r^{\mathrm{I}} = A_r^{\mathrm{I}(-2)} \exp 2i\omega_{\mathrm{P}} t] + A_r^{\mathrm{I}(-1)} \exp[i\omega_{\mathrm{P}} t] + \\ + A_r^{\mathrm{I}(0)} + A_r^{\mathrm{I}(1)} \exp[-i\omega_{\mathrm{P}} t] + A_r^{\mathrm{I}(4)} \exp[-2i\omega_{\mathrm{P}} t] , \text{ etc.} .$$

If these substitutions are made, the eqs. (A-6.5) and (A-6.6) may be separated into a series of equations involving the slow derivatives of amplitudes and the « correction » terms of the type (A-6.14).

If, following the lines indicated, we pick out the terms in $\exp[i\omega_{\mathrm{P}} t]$ of (A-6.5), we obtain

(A-6.15) $$k_{\mathrm{L1}}^{-1} k_{\mathrm{L1}r} \frac{\partial^{\mathrm{I}} a_{\mathrm{L1}}}{\partial t} - \frac{mc\omega_{\mathrm{P}}^2}{e} \xi_{r\mathrm{L1}}^{\mathrm{I}(1)} + i\omega_{\mathrm{P}} A_{r\mathrm{L1}}^{\mathrm{I}(1)} = \\ = \frac{ie}{mc} k_{\mathrm{L2}}^{-1} \cdot [-\boldsymbol{k}_{\mathrm{T}} \cdot \boldsymbol{k}_{\mathrm{L2}} \boldsymbol{e}_r + \tfrac{1}{2}\, \boldsymbol{e} \cdot \boldsymbol{k}_{\mathrm{L2}} \boldsymbol{k}_{r\mathrm{L2}} - \boldsymbol{e} \cdot \boldsymbol{k}_{\mathrm{L2}} \boldsymbol{k}_{r\mathrm{T}}]\, a_{\mathrm{T}} a_{\mathrm{L2}}^* .$$

If we pick out the contribution to (A-6.6) involving $\exp[-i\omega_P t]$, we obtain

$$\text{(A-6.16)}\qquad \frac{-i\omega_P}{c^2}\, k_{L1}^{-1} k_{rL1} \frac{\partial^I a_{L1}}{\partial t} - i k_{L1}^{-1} k_{rL1} k_{sL1} \frac{\partial^I a_{L1}}{\partial x_s} + i k_{L1} \frac{\partial^I a_{L1}}{\partial x_r} -$$

$$- i\omega_P k_{L1}^{-1} k_{rL1} \frac{\partial^I a_{L1}}{\partial t} - i c^2 k_{L1}^{-1} k_{rL1} \frac{\partial^I a_{L1}}{\partial x_s} + i c^2 k_{L1} \frac{\partial^I a_{L1}}{\partial x_r} +$$

$$+ (-\omega_P^2 + c^2 k_{L1}^2) A_{rL1}^{I(1)} - c^2 k_{rL1} k_{sL1} A_{sL1}^{I(1)} - i \frac{mc}{e}\, \omega_P^3 \xi_{rL1}^{I(1)} =$$

$$= \frac{e}{mc}\, \omega_P k_{L2}^{-1} \left[\boldsymbol{k}_{L1}\cdot \boldsymbol{k}_{L2} \boldsymbol{e}_r - \tfrac{1}{2}\, \boldsymbol{e}\cdot \boldsymbol{k}_{L1} \boldsymbol{k}_{rL2}\right] a_T a_{L2}^* .$$

It is possible to eliminate $A_{L1}^{I(1)}$ and $\xi_{L1}^{I(1)}$ from (A-6.15) and (A-6.16) to obtain

$$\text{(A-6.17)}\qquad \frac{\partial^I a_{L1}}{\partial t} = \frac{1}{2}\, i \frac{e}{mc}\, k_{L1}^{-1} k_{L2}^{-1} (k_{L1}^2 - k_{L2}^2)\, \boldsymbol{e}\cdot \boldsymbol{k}_{L1} a_T a_{L2}^* .$$

We may obtain a companion equation for the slow rate of change of the transverse amplitude by picking out the terms of (A-6.5) involving $\exp[-2i\omega_P t]$,

$$\text{(A-6.18)}\qquad e_r \frac{\partial^I a_T}{\partial t} - 4 \frac{mc}{e}\, \omega_P^2 \xi_{rT}^{I(2)} + 2i\omega_P A_{rT}^{I(2)} = i \frac{e}{mc}\, k_{L1}^{-1} k_{L2}^{-1} \boldsymbol{k}_{L1}\cdot \boldsymbol{k}_{L2} k_{Tr} a_{L1} a_{L2} ,$$

and the terms of (A-6.6) involving $\exp[-2i\omega_P t]$,

$$\text{(A-6.19)}\qquad -\frac{1}{2}\, i\omega_P e_r \frac{\partial^I a_T}{\partial t} - 2ic^2 e_r k_{sT} \frac{\partial^I a_T}{\partial x_s} + i c^2 k_{rT} e_s \frac{\partial^I a_T}{\partial x_s} - \omega_P^2 A_{rT}^{I(2)} - c^2 k_{rT} k_{sT} A_{sT}^{I(2)} -$$

$$- 2i \frac{mc}{e}\, \omega_P^3 \xi_{rT}^{I(2)} = -\frac{e}{mc}\, \omega_P k_{L1}^{-1} k_{L2}^{-1} (\boldsymbol{k}_T\cdot \boldsymbol{k}_{L2} k_{rL1} + \boldsymbol{k}_T\cdot \boldsymbol{k}_{L1} k_{rL2}) a_{L1} a_{L2} .$$

By suitable combination of (A-6.18) and (A-6.19), we obtain

$$\text{(A-6.20)}\qquad \frac{\partial^I a_T}{\partial t} + \frac{1}{2}\, c^2 \omega_P^{-1} k_{rT} \frac{\partial^I a_T}{\partial x_r} = \frac{1}{4}\, i \frac{e}{mc}\, k_{L1}^{-1} k_{L2}^{-1} (k_{L1}^2 - k_{L2}^2) \boldsymbol{e}\cdot \boldsymbol{k}_{L1} a_{L1} a_{L2} .$$

REFERENCES

[1] A. A. Vedenov, E. P. Velikhov and R. Z. Sadgeev: *Nucl. Fusion*, **1**, 82 (1961).
[2] J. M. Dawson: *Phys. Rev.*, **113**, 383 (1959).
[3] I. B. Bernstein, J. M. Greene and M. D. Kruskal: *Phys. Rev.*, **108**, 546 (1957).
[4] P. A. Sturrock: *Proc. Roy. Soc.*, A **242**, 277 (1957).
[5] D. Bohm and E. P. Gross: *Phys. Rev.*, **75**, 1851 (1949).

[6] P. A. Sturrock: *Journ. Nucl. En.*, Part C. **2**, 158 (1961).
[7] D. A. Tidman and G. H. Weiss: *Phys. Fluids*, **4**, 866 (1961).
[8] P. A. Sturrock: *Ann. Phys.*, **9**, 422 (1960).
[9] P. A. Sturrock: *Ann. Phys.*, **15**, 250 (1961).
[10] P. Penfield: *Frequency-Power Formulas* (New York, 1960).
[11] J. M. Manley and H. E. Rowe: *Proc. Inst. Radio Engr.*, **44**, 904 (1956).
[12] W. E. Drummond and D. Pines: Gen. Atomics Report GA-2386 (1961).
[13] P. A. Sturrock: *Journ. Math. Phys.*, **1**, 405 (1960).
[14] L. Landau: *Journ. Phys. U.S.S.R.*, **10**, 25 (1946).
[15] P. A. Sturrock: *Phys. Rev.*, **121**, 18 (1961).
[16] P. A. Sturrock: *Proc. Sixth Lockheed Symposium on Magnetohydrodynamics* (Stanford, 1962).
[17] P. A. Sturrock: *Journ. Appl. Phys.*, **31**, 2052 (1960).
[18] H. Lamb: *Hydrodynamics* (Cambridge, 1930), p. 399.
[19] J. H. Adlam and J. E. Allen: *Phil. Mag.*, **3**, 448 (1958).
[20] J. P. Dougherty and D. T. Farley: *Proc. Roy. Soc.*, A **259**, 79 (1960); A **263**, 238 (1961).
[21] F. J. Fishman, A. R. Kantrowitz and H. E. Petschek: *Rev. Mod. Phys.*, **32**, 959 (1960).

Equilibre et stabilité des systèmes toroïdaux en magnétohydrodynamique au voisinage d'un axe magnétique.

C. Mercier

Groupe de Recherches de l'Association Euratom
C. E. A. sur la Fusion - Fontenay-aux-Roses - (Seine)

1. – Introduction.

La magnétohydrodynamique ne peut évidemment pas expliquer tous les aspects des plasmas. Cependant, envisagé comme une première approximation elle permet, grâce à sa simplicité relative, d'en étudier certains aspects microscopiques; dans ces deux conférences, nous allons étudier des plasmas de forme toroïdale quelconque et leurs propriétés, en particulier celles liées à leur topologie.

Cependant l'étude complète des solutions toroïdales M.H.D. est encore trop difficile. Les champs magnétiques et les courants sont tangents à des surfaces S, surfaces magnétiques, emboitées les unes dans les autres, autour d'un « axe magnétique ». Quand l'axe magnétique est dans le plasma, il est très intéressant d'étudier celui-ci au voisinage de cet axe car déjà apparaissent des propriétés globales liées à des singularités dans les solutions.

On peut par l'étude de ces équilibres au voisinage de l'axe prévoir des domaines de stabilité (changement de stabilité aux limites d'équilibre des solutions dépendant d'un paramètre).

Ceci nous amène à préciser ces domaines de stabilité à l'aide du principe d'énergie [1], ou plutôt d'un critère necessaire et suffisant pour les déplacements localisés que j'ai exposé à la conférence de Salzsburg (1961) [2, 3].

Ce critère envisage des déplacements localisés seulement perpendiculairement à une surface magnétique, ces déplacements sont quelconques par aillleurs. Ainsi les propriétés topologiques des systèmes envisagés peuvent être prises en compte.

Remarquons d'ailleurs que cette localisation est d'autant moins nécessaire pour obtenir le critère que les quantités physiques à l'équilibre varient peu au voisinage d'une surface magnétique.

Ce critère s'applique séparément sur chaque ligne magnétique fermée. Nous le développerons au voisinage de l'axe magnétique sous la forme un peu moins générale d'une intégrale de surface.

2. - Etude des champs à divergence nulle.

Un champ à divergence nulle peut s'écrire:

$$\boldsymbol{B} = \operatorname{grad} \chi \wedge \operatorname{grad} \psi \tag{1}$$

et l'élement de flux $d\varphi$ s'écrit

$$d\varphi = d\psi \cdot d\chi \,. \tag{2}$$

Si le champ est toroïdal, on peut définir la fonction ψ regulière dans tout l'espace en la choisissant fonction de la surface toroïdale S sur laquelle les lignes magnétiques sont tracées.

Définissons d'abord un système de coordonnées au moyen des deux fonctions η et θ. η est multiforme continue partout sauf sur l'axe magnétique, augmentant d'une unité sur toutes les courbes qui tournent une fois autour de l'axe magnétique. Elles définissent sur S des lignes $C_{1,0}$ que nous dirons de classe $C_{1,0}$ (un grand tour, pas de petit tour).

θ est une fonction multiforme, continue partout et augmentant d'une unité quand on parcourt l'axe magnétique.

Elles définissent sur les surfaces S des lignes $C_{0,1}$, de la classe (0, 1) (un petit tour, pas de grand tour).

Par exemple en géométrie de révolution, les parallèles des tores sont les lignes $C_{1,0}$ et les méridiens des lignes $C_{0,1}$.

Nous préciserons ψ comme suit:

$$\psi = \int_V d\tau \, \boldsymbol{B} \cdot \operatorname{grad} \eta \,, \tag{3}$$

l'intégration étant étendue sur le volume intérieur à un tore S.

Cette fonction est une fonction uniforme et constante sur les surfaces S et elle est égale, à une constante près, au flux à travers toute surface dont la limite est formée par une courbe de la classe (1, 0).

Ainsi $d\psi$ représente le flux à travers deux surfaces voisines et par conséquent d'après (2)

$$d\psi = \int_{C_{10}} d\chi \cdot d\psi \,,$$

ce qui fournit une première condition sur la fonction χ

$$\int_{C_{10}} d\chi = 1 \,. \tag{4}$$

Définissons maintenant la fonction

$$\bar{\psi} = \int_V d\tau \, \bar{B} \cdot \operatorname{grad} \bar{\theta} \,, \tag{5}$$

qui représente donc le flux à travers une surface limitée par une courbe C_{01}. $d\bar{\psi}$ représente alors le flux à travers la surface $\theta = \text{const}$ entre deux surfaces voisines et par conséquent d'après (2)

$$d\bar{\psi} = -\int_{C_{01}} d\chi \cdot d\psi \,,$$

ce qui fournit une deuxième condition sur χ.

$$\int_{C_{01}} d\chi = -\frac{d\bar{\psi}}{d\psi} = -\frac{\iota}{2\pi} = -\frac{2\pi}{\iota_c} \,. \tag{6}$$

ι ou ι_c comme les fonctions ψ et $\bar{\psi}$ est une fonction de surface, c'est-à-dire que l'on peut écrire $\iota(\psi)$. Nous appellerons ι et ι_c angles de transformation rotationnelle.

La fonction χ est une fonction de η, $\bar{\theta}$ et ψ; ψ caractérisant la surface magnétique S. Elle doit satisfaire à (4) et (6) qui s'écrivent:

$$\chi(\eta, \bar{\theta} + 1) = \chi(\eta, \bar{\theta}) + 1 \,,$$

$$\chi(\eta + 1, \bar{\theta}) = \chi(\eta, \bar{\theta}) - \frac{\iota}{2\pi} \,.$$

Elle aura donc la forme:

$$\chi = \bar{\theta} - \frac{\iota}{2\pi}\eta + \hat{\chi}(\bar{\theta}, \eta, \psi) \,, \tag{7}$$

$\hat{\chi}$ étant une fonction doublement périodique de période 1 en η et $\bar{\theta}$. Sur une ligne magnétique, χ est constante par définition. Si nous suivons une ligne magnétique de $A(\theta_A, \eta_A)$ à $B(\theta_B, \eta_A + m)$, c'est-à-dire jusqu'au point où la ligne recoupe C_{10} passant par A pour la m-ème fois, on peut écrire:

$$\chi_A = \bar{\theta}_A - \frac{\iota}{2\pi}\eta_A + \hat{\chi}(\bar{\theta}_A \eta_A \psi)\,,$$

$$\chi_B = \bar{\theta}_B - \frac{\iota}{2\pi}(\eta_A + m) + \hat{\chi}(\bar{\theta}_B, \eta_A, \psi) = \chi_A\,,$$

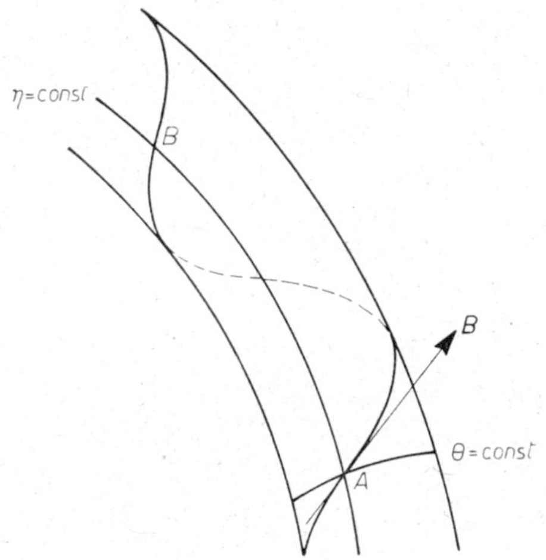

Fig. 1.

c'est-à-dire:

$$\frac{\iota}{2\pi} = \frac{(\bar{\theta}_B + \hat{\chi}(\theta_B, \eta_A, \psi)) - (\bar{\theta}_A + \hat{\chi}(\theta_A, \eta_A, \psi))}{m}\,.$$

Si la ligne magnétique se renferme, après k grands tours

$$\bar{\theta}_B - \bar{\theta}_A = k \to \hat{\chi}(\theta_B, \eta_A, \psi) = \hat{\chi}(\theta_A, \eta_A, \psi)\,,$$

d'où

$$\frac{\iota}{2\pi} = \frac{k}{m}\,. \tag{8}$$

Ainsi si $\iota/2\pi$ est rationnel, les lignes magnétiques sur la surface S défini par ψ, sont fermées; sinon, les lignes magnétiques remplissent toutes les surfaces: elles sont ergodiques. $\iota/2\pi$ caractèrise donc l'entortillement des lignes magnétiques sur une surface: c'est une propriété topologique.

3. – Transformation rotationnelle au voisinage de l'axe magnétique.

On définit l'axe magnétique, Γ par ses équations intrinsèques: $R(s)$ et $T(s)$, R et T sont les rayons de courbure et de torsion de la courbe, s l'abscisse curviligne. On sait que ces équations suffisent à déterminer une courbe gauche à un déplacement près.

L'axe magnétique est une courbe fermée, aussi $R(s)$ et $T(s)$ doivent être des fonctions périodiques de s, de période L, la longeur de la courbe. Les conditions générales que doivent satisfaire R et T pour que la courbe soit fermée ne sont pas connues malheureusement.

Un point P dans l'espace sera repéré par ses coordonnées ϱ, θ_0, s: ϱ, θ_0 coordonnées polaires dans le plan normal à Γ passant par P, s abscisse curviligne mesurée à partir d'un point O origine) (Fig. 2). Si θ est l'angle de $\boldsymbol{n}$ avec $\boldsymbol{MP}$

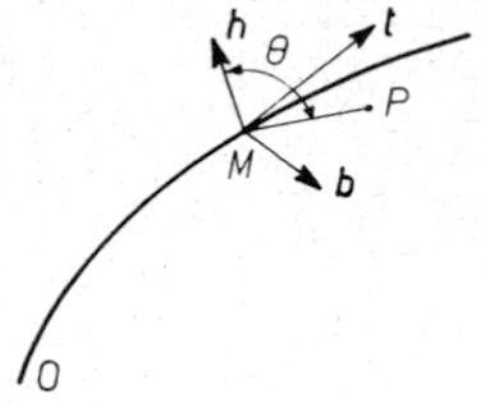

Fig. 2.

$$\theta_0 = \theta + \int_0^s \frac{\mathrm{d}s}{T(s)} . \tag{9}$$

Ce système de coordonnées est un système triple orthogonal

ϱ = const représente une surface « canal »,

θ_0 = const représente une surface reglée développable

s = const représente les plans normaux à Γ

(sur les surfaces $\varrho = \text{const}$ les courbes définies par $\theta_0 = \text{const}$ et $s = \text{const}$ forment les deux systèmes de lignes de courbure)

$$\boldsymbol{P} = \boldsymbol{M} + \varrho \cos\theta \boldsymbol{n} + \varrho \sin\theta \boldsymbol{b} .$$

En utilisant les formules de Seret et Frenet

$$\mathrm{d}\boldsymbol{p} = \boldsymbol{t}\left(1 - \frac{\varrho}{R}\cos\theta\right)\mathrm{d}s + \boldsymbol{n}(\cos\theta\,\mathrm{d}\varrho - \varrho\sin\theta\,\mathrm{d}\theta_0) + \boldsymbol{b}(\sin\theta\,\mathrm{d}\varphi + \varrho\cos\theta\,\mathrm{d}\theta),$$

d'où

$$\mathrm{d}S^2 = \mathrm{d}\varrho^2 + \varrho^2\mathrm{d}\theta_0^2 + \left(1 - \varrho\frac{\cos\theta}{R}\right)^2 \mathrm{d}s^2 . \tag{10}$$

A l'aide de (10) nous pourrons tirer les expressions du gradient, de la divergence et du rotationnel. Les fonctions physiques écrites en θ et s doivent être périodiques en θ (période 2π) et s (période L, L longueur de Γ).

Développons alors les fonctions ψ et χ, le champ magnétique $\boldsymbol{B}$, la densité de courant $\boldsymbol{J}$ sous la forme suivante

$$\boldsymbol{B} = \begin{cases} B_\varrho = a_1\varrho + a_2\varrho^2 + \ldots \\ B_\theta = b_1\varrho + b_2\varrho^2 + \ldots \\ B_s = c_0 + c_1\varrho + c_2\varrho^2 + \ldots \end{cases}$$

$$\boldsymbol{j} = \operatorname{rot} \boldsymbol{B} \begin{cases} J_\varrho = \dfrac{\partial}{\partial \theta_0}\left(c_1 - c_0 \dfrac{\cos\theta}{R}\right) + \dots \\ J_\theta = -\left(c_1 - c_0 \dfrac{\cos\theta}{R}\right) + \dots \\ J_s = 2b_1 - \dfrac{\partial a_1}{\partial \theta_0}, \end{cases}$$

$$\psi = \psi_0 + \varrho^2 \psi_2 + \varrho^3 \psi_3$$

$$\chi = \chi_0 + \varrho \chi_1 + \dots$$

Chercher des solutions qui sont analytiques près de l'axe magnétique nous oblige à choisir

$$c_0 \text{ et } 2b_1 - \frac{\partial a_1}{\partial \theta_0} = j_{s_0} \text{ fonctions de } s \text{ seul,}$$

et

$$\varphi_2 = \frac{\psi_2}{c_0} = a + b \cos 2u\,,$$

$$u = \theta + \frac{d}{2}\,,$$

$a(s)$ et $b(s)$ fonctions périodiques (periode L) et $\mathrm{d}(s)$ est telle que

$$\mathrm{d}(s + L) = \mathrm{d}(s) + 4K\pi\,.$$

Après quelques calculs, on tire au premier ordre:

$$(11) \qquad \begin{cases} \dfrac{\partial \chi_0}{\partial s} = K - \dfrac{d'}{4\varphi_2} + \dfrac{b' \sin 2u}{4a\varphi_2}\,, \\ \dfrac{\partial \alpha_0}{\partial \theta} = -\dfrac{1}{2\varphi_2}\,, \end{cases} \qquad K = \frac{1}{2a}\left(\frac{j_{s_0}}{c_0} + d' - \frac{2}{T}\right).$$

La relation de compatiblité conduit immédiatement à

$$aa' - bb' = \text{const} \qquad \text{soit} \qquad a^2 - b^2 = \text{const}\,.$$

On obtient alors

$$\chi_0 = -\frac{1}{2\sqrt{a^2 - b^2}} \operatorname{arctg}\left(\frac{a - b}{\sqrt{a^2 - b^2}} \operatorname{tg} u\right) + \int_0^s K\,\mathrm{d}s + \text{const}\,.$$

Pour finir le calcul nous devons choisir les lignes C_{10} et C_{01}. Nous pouvons prendre les courbes à $s = \text{const}$ comme courbe C_{01}. Il est difficile de trouver une courbe C_{10}. Cependant si nous choisissons une courbe fermée faisant un grand tour et un nombre k de petits tours, nous trouverons $-(2\pi/\iota)+k$, c'est-à-dire que nous aurons déterminé $2\pi/\iota=\iota_c/2\pi$ à un nombre entier près, ce qui ne change pas les résultats sur l'équilibre et la stabilité. Aussi appellerons nous encore $\iota/2\pi$ la quantité obtenue en prenant à la place d'une courbe C_{10} une courbe $\theta=\text{const}$ qui fait un grand tour et un nombre non precisé de petits tours. Remarquons que dans le cas d'une courbe plane, ces courbes font effectivement 0 petits tours.

Alors

$$\int_{C_{10}} d\chi = \oint K\, ds = 1 \qquad (\theta = \text{const},\ s \text{ variable}),$$

$$\int_{C_{01}} d\chi = -\frac{2\pi}{2\sqrt{a^2-b^2}} = -\frac{\iota}{2\pi} \qquad (s = \text{const},\ \theta \text{ variable}),$$

φ_2 n'est définie qu'à une constante près, donc a et b. Ainsi:

$$\left(\frac{\iota}{2\pi}\right)_{\text{axe}} = \frac{2\mu}{2\sqrt{a^2-b^2}\oint K\, ds} = \frac{2\pi}{\iota_c}$$

$$\iota_{c_{\text{axe}}} = \frac{1}{2}\oint \frac{\sqrt{a^2-b^2}}{a}\left(\frac{Js_0}{Bs_0} + d' - \frac{2}{T}\right) ds\,.$$

4. - Equilibre en MHD.

Jusqu'à présent nous n'avons pas fait appel à d'autre équation de champ que $\operatorname{div} B = 0$.

Si l'on pose $\boldsymbol{j} = \operatorname{grad} K \wedge \operatorname{grad} \psi$, l'équation d'équilibre $\boldsymbol{j}\wedge\boldsymbol{B} = \operatorname{grad} P$ peut s'écrire:

$$(\operatorname{grad} K \wedge \operatorname{grad}\cdot\chi)\operatorname{grad}\psi = \frac{dp}{d\Psi}\,.$$

car p une fonction de Ψ.

Cette équation impose à tous les ordres des conditions supplémentaires.

Tout d'abord, sur l'axe magnétique, la densité de courant $\boldsymbol{j}$ n'a qu'une composante j_s non nulle.

Par conséquent $j_\theta = 0$, c'est-à-dire:

$$c_1 - c_0\frac{\cos\theta}{R} = 0\,.$$

Le développement de $(\operatorname{grad} K \wedge \operatorname{grad} \psi)\cdot \operatorname{grad} \chi = -p'$ donne au premier ordre

$$2\psi_2 \frac{\partial K_0}{\partial s}\frac{\partial \chi_0}{\partial \theta_0} - 2\psi_2 \frac{\partial K_0}{\partial \theta_0}\frac{\partial \chi_0}{\partial s} = -p_0' . \tag{12}$$

Ainsi $K_0 = f(\chi_0) +$ solution particulière.

En tenant compte de

$$\frac{\partial \chi_0}{\partial \theta} = -\frac{1}{2\varrho_2},$$

une solution particulière est facilement trouvée:

$$p_0' \int_0^s \frac{\mathrm{d}s}{c_0(s)},$$

et comme K_0 de même que χ_0 doit être linéaire en θ et s

$$K_0 = p_0' \int_0^s \frac{\mathrm{d}s}{c_0(s)} + \frac{\lambda}{2}\chi_0 .$$

Or d'après $\boldsymbol{j} = \operatorname{grad} K \wedge \operatorname{grad} \psi$

$$j_s(s) = -2\psi_2 \frac{\partial K_0}{\partial \theta_0} = \frac{\lambda}{2} c_0(s) .$$

Ainsi

$$\frac{j_s}{c_0(s)} = \frac{j_{s_0}}{B_{s_0}} = \frac{\lambda}{2} . \tag{13}$$

Le rapport sur l'axe de la densité de courant sur le champ magnétique est *constant*.

Les expressions de $\partial\chi_0/\partial s$, $\partial\chi_0/\partial\theta_0$, $(\iota/2\pi)_0$ et φ_2 sont les mêmes que précédemment avec cette condition en plus.

Tels sont les résultats de l'équilibre au premier ordre.

Au second ordre les calculs se compliquent. Limitons nous ici à quelques formules importantes.

De $(\operatorname{grad} K \wedge \operatorname{grad} \psi)\cdot \operatorname{grad} \chi = -p'$ on tire au second ordre

$$2\psi_2^{\frac{3}{2}} \left[\frac{\partial k_1}{\partial s}\frac{\partial \chi_0}{\partial \theta_0} - \frac{\partial k_1}{\partial \theta_0}\frac{\partial \chi_0}{\partial s} + \frac{\partial K_0}{\partial s}\frac{\partial g_1}{\partial \theta_0} - \frac{\partial K_0}{\partial \theta_0}\frac{\partial g_1}{\partial s}\right] + \\ + \left[\frac{\partial K_0}{\partial s}\frac{\partial \chi_0}{\partial \theta_0} - \frac{\partial K_0}{\partial \theta_0}\frac{\partial \chi_0}{\partial s}\right]\left[3\psi_3 + 2\psi_2 \frac{\cos\theta}{R}\right] = 0 ,$$

en posant

$$\chi_1 = g_1 \psi_2^{\frac{1}{2}},$$

$$K_1 = k_1 \psi_2^{\frac{1}{2}},$$

qui après quelques manipulations devient

$$\frac{\partial \chi_0}{\partial \theta_0} \frac{\partial}{\partial s}(2k_1 - \lambda g_1) - \frac{\partial \chi_0}{\partial s} \frac{\partial}{\partial \theta_0}(2k_1 - \lambda g_1) = \frac{2p_0' \cos\theta}{R\psi_2^{\frac{3}{2}}}. \tag{14}$$

Mais remarquons que χ ou K peuvent s'écrire:

$$a(\psi)\theta + b(\psi)s + \dots \text{ fonction périodique},$$

$$a(\psi) = a_0 + a_0' \psi_2 \varrho^2 + \dots,$$

$$b(\psi) = b_0 + b_0' \psi_2 \varrho^2 + \dots.$$

Par conséquent

$$\chi_0 \text{ ou } K_0 = a_0\theta + b_0 s + \text{fonction périodique},$$

$$\chi_1 \text{ ou } K_1 = \text{fonction périodique},$$

$$\chi_2 \text{ ou } K_2 = (a_0'\theta + b_0' s)\psi_2 + \text{fonction périodique}.$$

De façon précise, on trouve:

$$\left|\begin{aligned} &\chi_0 = \frac{s}{L} - \left(\frac{i}{2\pi}\right)_0 \frac{\theta}{2\pi} + \text{fonction périodique}, \\ &\chi_1 \text{ ou } g_1 = \text{fonction périodique}, \\ &\frac{\chi_2}{\Psi_2} \text{ ou } g_2 = -\left(\frac{i}{2\pi}\right)_0' \frac{\theta}{2\pi} + \text{fonction périodique}, \end{aligned}\right.$$

et

$$\left|\begin{aligned} &K_0 = \left(\frac{\mathrm{d}T}{\mathrm{d}\psi}\right)_0 \frac{s}{L} - \left(\frac{\mathrm{d}T}{\mathrm{d}\psi}\right)_0 \frac{\theta}{2\pi} + \text{fonction périodique}, \\ &K_1 \text{ ou } k_1 = \text{fonction périodique}, \\ &\frac{K_2}{\Psi_2} \text{ ou } k_2 = \left(\frac{\mathrm{d}^2 T}{\mathrm{d}\psi^2}\right)_0 \frac{s}{L} - \left(\frac{\mathrm{d}^2 \overline{T}}{\mathrm{d}\psi^2}\right)_0 \frac{\theta}{2\pi} + \text{fonction périodique}, \end{aligned}\right.$$

où

$$\left|\begin{aligned} &T = \int_V \boldsymbol{J} \cdot \operatorname{grad} \eta \, \mathrm{d}\tau, \\ &\overline{T} = \int_V \boldsymbol{J} \cdot \operatorname{grad} \bar{\theta} \, \mathrm{d}\tau. \end{aligned}\right.$$

Nous devons donc chercher la solution périodique de (14) soit U

(15)
$$\frac{\partial \chi_0}{\partial \theta_0} \frac{\partial U}{\partial s} - \frac{\partial \chi_0}{\partial s} \frac{\partial U}{\partial \theta_0} = -\frac{2 p_0' \cos \theta}{R \psi_2^{\frac{3}{2}}} \; .$$

Pour des raison d'analyticité, u à la forme:

$$U = \frac{\alpha \cos u + \beta \sin u}{\varphi_2^{\frac{1}{2}}} \, , \qquad u = \theta + \frac{d}{2} \, ,$$

on doit alors chercher une solution périodique en s de

(16)
$$z' - \frac{i\mu}{2a} \left[\frac{\lambda}{2} + d' - \frac{2}{T} \right] z = \frac{4 p_0'}{R c_0^{\frac{3}{2}}} \left[\cos \frac{d}{2} \exp[-\eta/2] + i \sin \frac{d}{2} \exp[\eta/2] \right] ,$$

où

$$
\begin{array}{l|ll}
z = x + iy & \varphi_2 = a + b \cos 2u \, , & \\
\alpha = x \exp[\eta/2] \, , & a = \mu \cosh \eta & \mu = \mathrm{const}. \\
\beta = y \exp[-\eta/2] \, , & b = \mu \sinh \eta &
\end{array}
$$

Cette équation rentre dans le type suivant

$$\frac{\partial Z}{\partial s} + i a(s) Z + b(s) = 0 \, ,$$

a et b fonctions périodiques de période L.

Cette équation a une seule solution périodique

$$Z = \frac{1}{1 - \exp\left[i \oint a(x)\, \mathrm{d}x \right]} \int_s^{s+L} b(x) \exp \left[i \int_s^x a(u)\, \mathrm{d}u \right] \mathrm{d}x \, .$$

Cette solution n'existe que si

$$\oint a(x)\, \mathrm{d}x \neq 2 K \pi \, ,$$

à moins que pour

$$\oint a(x)\, \mathrm{d}x = 2 K \pi \, ,$$

$$\oint b(x) \exp \left[i \int_s^x a(u)\, \mathrm{d}u \right] \mathrm{d}x = 0 \, .$$

Nous allons mettre Z sous une autre forme où les pôles seront plus visibles. Posons

$$\int_{s_0}^{x} a(u)\,\mathrm{d}u = A_0(x)\,.$$

Comme

$$\int_{s_0}^{x+L} a(u)\,\mathrm{d}u = \int_{s_0}^{x} a(u)\,\mathrm{d}u + \oint a(u)\,\mathrm{d}u\,,$$

A_0 peut s'écrire

$$A_0(x) = \widetilde{A}_0(x) + \oint \cdot \frac{x}{L}\,.$$

$\widetilde{A}_0(x)$ est périodique (période L) et $\oint = \oint a(u)\,\mathrm{d}u$

$$Z = \frac{\exp[-i\widetilde{A}_0(s)]\exp[-i\oint(s/L)\,\mathrm{d}s]}{1-\exp[i\oint a(x)\,\mathrm{d}x]}\int_{s}^{s+L} b(x)\exp[i\widetilde{A}_0(x)]\cdot\exp\left[i\oint\frac{x}{L}\,\mathrm{d}x\right].$$

$b(x)\,\exp[i\widetilde{A}_0(x)]$ est périodique. Donc

$$b(x)\exp[i\widetilde{A}_0(x)] = \sum_K a_K \exp\left[-2Ki\pi\frac{x}{L}\right].$$

En intégrant il vient

$$Z = iL\exp[-i\widetilde{A}_0(s)]\sum_K\frac{a_K}{\oint - 2K\pi}\exp\left[-2Ki\pi\frac{s}{L}\right],$$

la fonction périodique Z est donc discontinue aux points où $\oint = 2K\pi$ à moins que $a_K = 0$.

Revenons à l'éq. (16)

$$z = \frac{-Lp_0'}{1-\exp[-i\iota_c]}\int_{s}^{s+L}\frac{\mathrm{d}x}{R_{\iota_0}^{3/2}}\left(\cos\frac{\theta}{2}\exp[-\eta/2] + i\sin\frac{d}{2}\exp[\eta/2]\right)\cdot$$
$$\cdot\exp\left[-i\left[\int_0^x\frac{\mu}{2a}\left(\frac{j_s}{B_s}+d'-\frac{2}{T}\right)\mathrm{d}u\right]\right].$$

En général les solutions d'équilibre n'existeront pas quand $\iota_c = 2K\pi$.

5. - Cas particulier.

Choisissons comme axe magnétique un cercle $T=\infty$, $R=\text{const}$. Prenons a et b constant, mais $\mathrm{d}(s)=2Ks/R$ c'est-à-dire que les surfaces $\psi=\text{const}$ au voisinage de l'axe ont pour méridienne des ellipses qui tournent régulièrement autour de l'axe magnétique K fois quand s augmente.

Dans ce cas on peut intégrer et on trouve:

$$z=\frac{-4ip_0'}{R\iota_0^{3/2}}\left[\frac{\cosh(\eta/2)\exp[iKs/R]}{(K/R)(1-\mu/a)-\mu_0\,js/2aB_s}+\frac{\sinh(\eta/2)\exp-[iKs/R]}{(K/R)(1+\mu/a)+\mu j_s/2aB_s}\right].$$

On aura ainsi 2 pôles à z pour

$$\frac{Rj_s}{2B_s}=K\left(\frac{a}{\sqrt{a^2-b^2}}-1\right)\qquad\text{et}\qquad\frac{Rj_s}{2B_s}=-K\left(\frac{a}{\sqrt{a^2-b^2}}+1\right).$$

soit $\iota_c=\pm 2\pi K$.

La symétrie de révolution s'obtient en faisant $K=0$

$$z=4ip_0'\frac{a}{\sqrt{a^2-b^2}}\left(\frac{2B_s}{Rj_s}\right)\exp[-\eta/2].$$

Il n'y a pas de pôle sauf pour $j_s=0$ ($B_s\neq 0$). C'est le cas bien connu: il n'y aura pas d'équilibre en symétrie de révolution sans courant avec champ longitudinal seul.

En général pas d'équilibre pour:

$$i_c=\frac{1}{2}\oint\left(1-\frac{b^2}{a^2}\right)^{\frac{1}{2}}\left(\frac{j_s}{B_s}-d'-\frac{2}{T}\right)\mathrm{d}s=2K\pi.$$

Cette condition permet de classer diverses configurations proposées pour confiner les plasmas:

1) *Le pinch*: c'est une configuration avec courant. En général $d_0'=0$ et $1/T=0$ (symétrie de révolution). Quand le courant augmente (toutes choses égales par ailleurs), et que $(Rj_s/2B_s)(1-(b^2/a^2))=K$ on passera par des états singuliers, limite d'équilibre. Ces états sont des limites de stabilité. Il est alors à prévoir que dans certaines conditions on pourra constater expérimentalement des accidents sur les différents signaux mesurés quand $\iota_c=2K\pi$. Voir en particulier les constatations faites sur Zéta.

2) *Configuration sans courant.* Alors il faut soit:

a) $1/T \neq 0$ on agit sur la forme du plasma: cas du stellerator en 8

$$\iota_c = \oint \frac{\mathrm{d}s}{T}\,.$$

b) Soit $d' \neq 0$, rotation de la configuration en fonction de s. Si $1/T = 0$ on aura que $i_c \neq 2K\pi$ si les surfaces à pression constante au voisinage de l'axe ne sont pas des cercles, car

$$d' = \frac{4k\pi}{L} + \hat{d}'\,.$$

$\hat{d}$ périodique

$$\iota_c = \frac{1}{2}\oint \sqrt{1 - \frac{b^2}{a^2}}\left(\frac{4k\pi}{L} + \hat{d}'\right)\mathrm{d}s\,.$$

et pour $b = 0$

$$\iota_c = 2k\pi\,,$$

c'est le cas des stellarators droits (stellarator C).

3) On peut évidemment combiner ces différents moyens. Par exemple sur certaines stellarators en 8, $1/T \neq 0$ ainsi que d'.

6. – Stabilité.

Comme nous l'avons dit dans l'introduction nous n'étudierons que le critère de stabilité vis-à-vis des déplacements localisés.

Ce sont des déplacements centrés autour d'une surface magnétique S_0 ($\psi = \psi_0$) tels que la composante ξ_1 du déplacement $\boldsymbol{\xi}$ perpendiculairement à S_0 n'est différente de zéro que sur les surfaces ψ comprises entre les surfaces $\psi_0 \pm \varepsilon$.

On montre alors que l'expression δW du principe d'énergie est toujours positive (donc stabilité) pour ε assez petit sauf si la surface S_0 est une surface sur laquelle les lignes magnétiques sont fermées ($\iota/2\pi$ rationnel).

Le critère alors s'écrit comme une condition sur toutes lignes magnétiques fermées

$$\varphi = \frac{1}{(4/m_0 \oint (B_2/|\operatorname{grad}\psi|^2)(\mathrm{d}l/B)}\left[\frac{\mathrm{d}}{\mathrm{d}\psi}\left(\frac{\iota}{2\pi}\right) + \frac{2}{m_0}\oint \frac{\boldsymbol{j}\cdot\boldsymbol{B}}{|\operatorname{grad}\psi|^2}\frac{\mathrm{d}l}{B}\right]^2 - \frac{2}{m_0}\oint \frac{(\boldsymbol{j}\wedge\boldsymbol{n})(\boldsymbol{B}\cdot\nabla\boldsymbol{n})(\mathrm{d}l/B)}{|\operatorname{grad}\psi|^2} > 0\,,$$

m_0 = nombre de petits tours avant que la ligne magnétique ne se referme.

Le critère limite pour obtenir m_0 tendant vers l'infini s'écrit

$$\bar{\varphi} = \frac{1}{4\int_S B^2\,\mathrm{d}S/|\mathrm{grad}\,\psi|^3}\left[\frac{\mathrm{d}}{\mathrm{d}\psi}\left(\frac{\iota}{2\pi}\right) + 2\int_S \boldsymbol{j}\cdot\boldsymbol{B}\,\frac{\mathrm{d}S}{|\mathrm{grad}\,\psi|^3}\right]^2 -$$

$$- 2\int_S (\boldsymbol{j}\wedge\boldsymbol{n})(\boldsymbol{B}\,\nabla\boldsymbol{n})\,\frac{\mathrm{d}S}{|\mathrm{grad}\,\psi|^3} > 0\,.$$

Or au voisinage d'une surface S quelconque, on peut trouver des lignes magnétiques qui se ferment après un nombre arbitrairement grand de tours et par conséquent une valeur de φ arbitrairement voisine de $\bar{\varphi}$, donc $\bar{\varphi}$ est un critère necessaire. Dans le cas particulier de la symétrie de révolution $\bar{\varphi} = \varphi$ et on retrouve le critère publié dans mon article de *Nuclear Fusion* [4]. Enfin en symétrie cylindrique, ce critère est identique au critère de Suydam [5].

7. – Applications.

Les critères φ ou $\bar{\varphi}$ ne sont pas facilement exploitables. On peut:

a) soit étudier sur toutes lignes ou surfaces magnétiques un équilibre connu analytiquement, par exemple;

b) soit les développer au voisinage de l'axe magnétique ce qui fournit des critères plus simples et peut suggérer les équilibres à réaliser pour obtenir, peut-être, un plasma stable.

En symétrie de révolution, on a calculé le critère pour différentes configurations avec pression maximum sur l'axe magnétique et dependant de plusieurs paramètres.

Dans ce cas, il a été trouvé que c'était le voisinage de l'axe magnétique qui était le plus difficile à stabiliser pour ces déplacements localisés; ce qui donne un nouvel intérêt au développement au voisinage de l'axe si de nombreuses configurations sont principalement instables à son voisinage.

Pour obtenir le critère au voisinage de l'axe, il est nécessaire de connaître les solutions, d'équilibre jusqu'au 3° ordre. Les calculs sont compliqués dans le cas général.

Notons les principaux résultats obtenus dans cette voie jusqu'à maintenant.

1) En symétrie de révolution dans le cas où les courbes $\psi = \text{const}$ au voisinage de l'axe sont des cercles ($b = 0$).

Le critère s'écrit alors très simplement:

$$\left(\frac{\mathrm{d}^2 p}{\mathrm{d}\varrho^2}\right)_0 \left[-1 + \frac{R^2\,j_{s_0}^2}{4B_{s_0}^2}\right] > 0\,.$$

Ainsi si la pression est maximum sur l'axe, (Fig. 3) le critère s'écrit:

$$\frac{Rj_{s_0}}{2B_{s_0}} < 1 \qquad \text{ou} \qquad \iota_c < 2\pi \,,$$

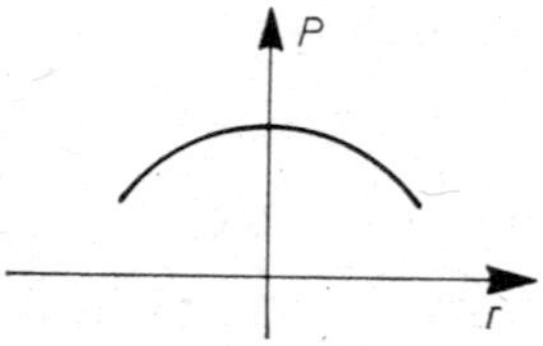

Fig. 3.

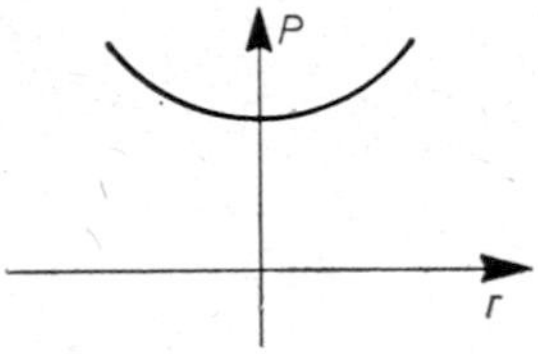

Fig. 4.

si la pression est minimum sur l'axe:

$$\frac{Rj_{s_0}}{2B_{s_0}} > 1 \qquad \text{ou} \qquad \iota_c > 2\pi \,,$$

R étant le rayon du cercle axe magnétique.

Remarquons que si on s'approche de la symétrie cylindrique ($R \to \infty$) le premier cas devient instable, le second devient stable, ce qui montre l'influence du rayon de courbure de l'axe magnétique.

8. – Remarque.

Pour le cas $p'' < 0$, c'est-à-dire pression maximum sur l'axe magnétique, la condition s'écrit

$$\frac{Rj_{s_0}}{2B_{s_0}} = \frac{RI}{2\pi r_0^2 B_i}\left(\frac{B_i}{B_0}\frac{j_{s_0}}{\bar{j}}\right) < 1 \,,$$

r_0: rayon du pinch.
$\bar{j}$ = densité de courant moyenne $= I/\pi r_0^2$.
B_i: champ longitudinal initial.

La quantité $(B_i/B_0)(js_0/\bar{j})$ est en général voisine de 1 pour les courants faibles. (Il en est ainsi dans Zéta et TA 2 000).

Ainsi ce critère est dans ce cas équivalent à la limite de Kruskal:

$$\frac{RI}{2\pi r_{0i}^2 B} < 1 \,.$$

2) Cas général [5] des surfaces magnétiques de forme surface « canal » au voisinage de l'axe magnétique: $b = 0$, $c_0(s) = \text{const}$.

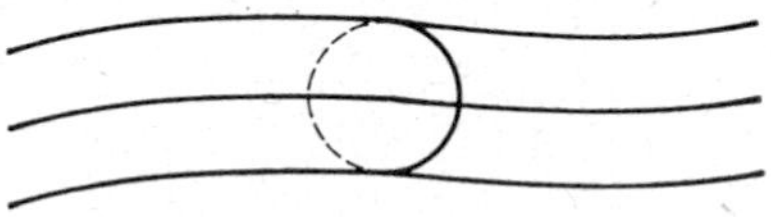

Fig. 5.

Dans ce cas le critère est plus compliqué:

$$\varrho'' \left[1 + \frac{j_s}{B_s L}\, \bar{R}^2 \oint \frac{V}{R}\, \mathrm{d}s + \frac{j_s^2 \bar{R}^2}{4B_s^2}\right] > 0 \,,$$

où

$$V = \mathrm{Re}\left[\frac{-i}{1 - \exp[-i\oint (j_s/2B_s - 1/T)]\,\mathrm{d}s} \int_s^{s+L} \frac{\mathrm{d}x}{R} \exp\left[-i\int_s^x \left(\frac{j_s}{2B_s} - \frac{1}{T}\right) \mathrm{d}u\right]\right],$$

et

$$\frac{1}{\bar{R}^2} = \frac{1}{L} \oint \frac{\mathrm{d}s}{R^2} .$$

L = longueur de l'axe magnétique.

Dans le cas particulier de la symétrie de révolution, le deuxième terme devient -2 et on retrouve le résultat précédent.

Il est commode de répresenter ce résultat dans un graphique avec $X = j_s L/4\pi B_s$ en abscisse et $Y = (1/2\pi)\oint \mathrm{d}s/T$ en ordonnée.

Quand $i_c = \oint((j_s/2B_s) - (1/T))\,\mathrm{d}s = 2K\pi$ on sait que, en général, il n'y a pas d'équilibre possible: se fait se traduit dans le critère par un changement de stabilité le long des droites: $X - Y = K$.

Les autres limites de stabilité dépendent encore de l'équilibre choisi, mais ont approximativement les formes indiquées sur la figure.

On peut en particulier faire plusieurs remarques sur ces zones de stabilité (on a choisi le cas $p'' < 0$, cas d'un maximum de pression sur l'axe).

Si l'axe n'est pas plan (example du stellarator en 8) il y aura instabilité pour des densités de courants faibles.

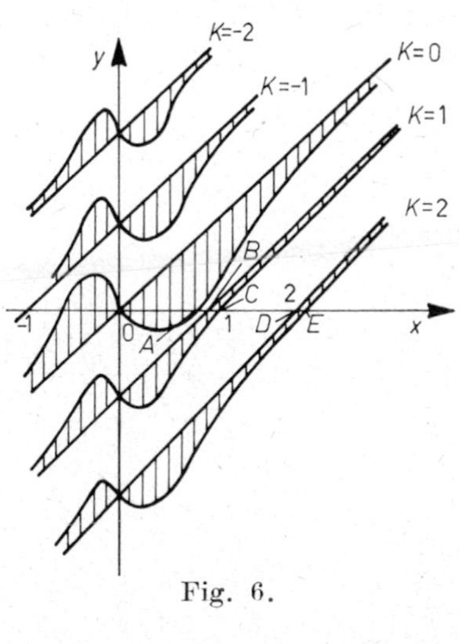

Fig. 6.

D'une manière générale pour un $\oint \mathrm{d}sT/$ donné, on trouve des domaines de stabilité successifs possibles.

Dans le cas où l'axe magnétique est plan $\oint \mathrm{d}s/T = 0$ on constate la même chose, sauf dans le cas très particulier de la symétrie de révolution parfaite; alors A et B sont confondus en $X=1$ et les domaines BC, DE etc., se réduisent à un point; ils disparaissent. On retrouve le cas déjà traité. Comme en realité la symétrie de révolution n'est jamais physiquement réalisée, on doit s'attendre à des particularités dans les plasmas en ces points.

Pour préciser ces courbes dans un cas particulier, un exemple numérique a été traité sur machine à calculer:

Les équations de l'axe magnétique s'écrivent (cas simple du type stellarator en 8):

$$x = \sin\theta\,,$$

$$y = \alpha \sin 2\theta\,,$$

$$z = \beta\cos\theta\,.$$

Si on choisit $\alpha^2 = (1-\beta^2)^2/32(1+\beta^2)^2$ les calculs sont très simples.

Pour $\beta = 1$ l'axe magnétique est un cercle.

Pour $\beta = 0$ c'est une courbe plane, avec un point double, du genre 8.

Pour des β intermédiaires nous avons une courbe de forme ∞ du genre de l'axe magnétique d'un stellarator en 8

$$\oint \frac{\mathrm{d}s}{T} = -2\pi\left(1 - \frac{4\beta}{\sqrt{3+10\beta^2+3\beta^4}}\right) \qquad \left| \begin{array}{l} \beta = 1 \Rightarrow 0\,, \\ \beta = 0 \Rightarrow -2\pi. \end{array}\right.$$

Remarquons que la courbe pour $\beta = 0$ est tracée dans un plan mais a cependant un $\oint \mathrm{d}s/T \neq 0$ provenant du point double.

$$\frac{1}{\overline{R}^2} = \frac{8(1+\beta^2)^2(15+2\beta^2+15\beta^4)}{(3+10\beta^2+3\beta^4)^{\frac{5}{2}}}\,, \qquad L = \frac{+2\pi}{\sqrt{2}}(1+\beta^2)^{\frac{1}{2}}\,.$$

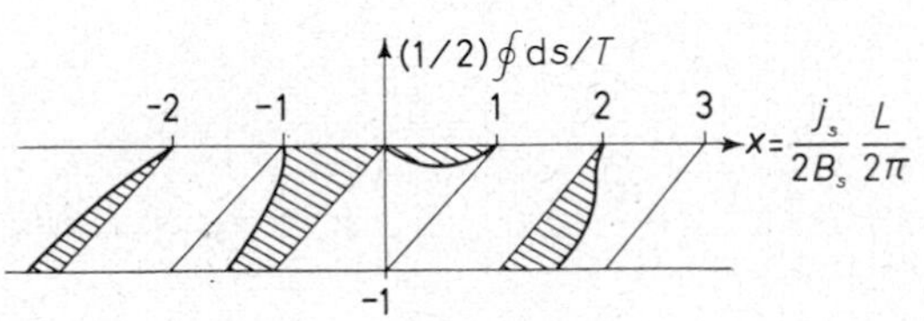

Fig. 7.

Les calculs numériques confirment les formes des domaines de stabilité ci-dessus.

9. – Application à un axe magnétique-plan.

Dans ce cas

$$1/T(s) = 0\,.$$

Le critère s'écrit d'une façon simple en développant $1/R(s)$, en série de Fourier:

$$\frac{1}{R(s)} = \sum_k a_k \exp\left[2\pi i k \frac{s}{L}\right].$$

On obtient:

$$1 + \frac{j_{s_0}}{B_{s_0}} \frac{4\bar{R}^2}{2\pi} \sum_k \frac{|ak|^2}{k - (\iota/2\pi)_c} + \frac{j_s^2 \bar{R}^2}{4B_{s_0}^2} < 0\,.$$

Sous cette forme il est visible qu'en symétrie de revolution $(R = \bar{R})$

$$a_k = 0 \qquad \text{sauf} \qquad a_0 = \frac{1}{R}\,.$$

On retrouve

$$-1 + \left(\frac{\iota_c}{2\pi}\right)^2 < 0\,,$$

soit

$$\left|\frac{\iota_c}{2\pi}\right|_0 < 1\,.$$

Il n'y a stabilité que si

$$0 < \left|\frac{\iota_c}{2\pi}\right| < 1\,.$$

Mais si on abandonne la symétrie de révolution d'autres plages de stabilité apparaissent dont la grandeur est liée à la valeur du coefficient de Fourier de $1/R$.

Si l'on prenait $1/R(s) = (1/R_0)\cos 2\pi K_0(s/L)$, on obtiendrait une plage au voisinage de $(\iota_c/2\pi) = K_0$ seulement. Malheureusement cette courbe n'est pas convenable (forme de 8 parcourue K_0 fois) mais on peut s'en inspirer pour choisir $R(s)$ de façon à agrandir la plage de stabilité K_0.

Ainsi pour avoir des plages de stabilité importantes pour $K = 2$, on cherchera des courbes du type 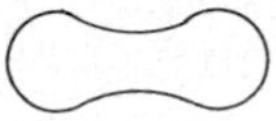 et pour $K_0 = 3$, des courbes du type 

Il est évident que d'autre plages que celles à K_0 indiquées apparaissent avec ces formes.

Des calculs ont été faits en approximant ces formes avec des arcs de cercles tangents [8].

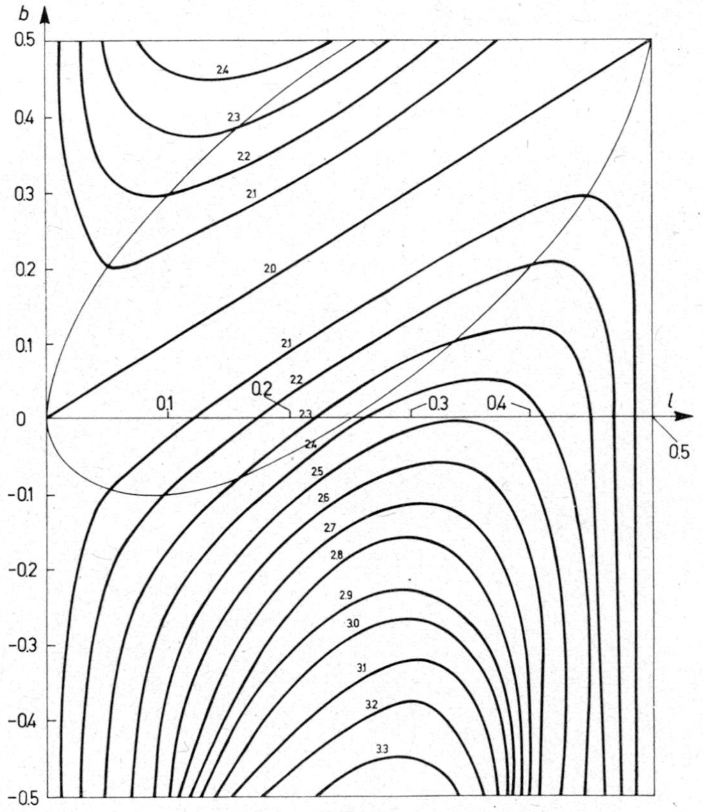

Fig. 8.

Par example pour le cas $K_0 = 2$, on obtient les résultats indiqués dans la Fig. 8 où les deux paramètres b et l sont indiqués

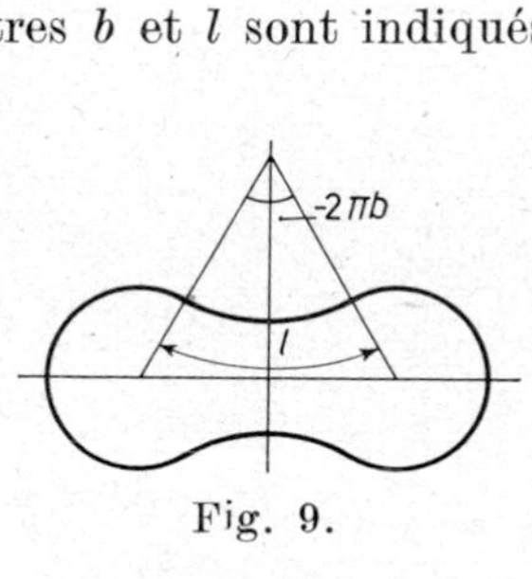

Fig. 9.

la longueur L de l'axe magnétique est égale à 1.

Pour $b=0$, la forme est celle d'un hippodrome.

On peut obtenir dans ce cas des largeurs de plage de stabilité de l'ordre de 0.5 en unités $\iota_c/2\pi$.

Pour le stellarator C, qui a cette forme ($b=0$, $l=0.22$) on obtient une première plage

$$0 < \frac{\iota_c}{2\pi} < 0.40 ,$$

et une deuxième plage

$$2 < \frac{\iota_c}{2\pi} < 2.35 .$$

Pour Zéta, qui possède aussi des parties droites ($b=0$, $l=0.06$) la plage à $K_0=2$ est de l'ordre de 0.02.

En résumé ces resultats théoriques expliquent très bien les résultats expérimentaux obtenus sur Zéta; accidents sur les courbes de courant et d'impédance correspondants à $\iota_c/2\pi$ entiers.

Si la théorie est exacte, les plages de stabilité qui doivent exister au délà sont très réduites (système presque en symétrie de révolution) d'où difficulté pour stabiliser par court-circuit le système sur une de ces plages. Mais ceci étant obtenu, des durées de l'ordre de plusieurs millisecondes ont été observées.

Enfin le résultat théorique qu'il n'y a pas en général d'équilibre pour $\iota_c/2\pi$ entier se traduit par une variation brusque du volume du plasma au voisinage de ces valeurs ce qui peut expliquer des pertes de plasma à la paroi et une variation brusque de l'impédance et du courant.

REFERENCES

[1] I. B. Bernstein, E. A. Frieman, M. D. Kruskal et R. Kulsrud: *Proc. Roy. Soc. (London)*, A **244**, 17 (1958).
[2] C. Mercier: *Compt. Rend.*, **252**, 1577 (1961).
[3] C. Mercier: *Proc. of the Conference on Plasma Physics and Controlled Nuclear Fusion Research* (Salzburg, 1961); *Fusion Nucléaire*, Supplement 1962, (Partie II).
[4] C. Mercier: *Nucl. Fusion*, **1**, 47 (1960).
[5] B. R. Suydam: *Proc. of Second U.N. International Conference on the Peaceful Uses of Atomic Energy*, **31**, 157 (1958).
[6] C. Mercier et M. Cotsaftis: *Compt. Rend.*, **252**, 2203 (1961).
[7] Conférences sur la magnéto-hydrodynamique, INSTN (Saclay, 1962).
[8] D. Voslamber: *Rapport Interne*, EUR.C.E.A.F.C.-196.
[9] C. Mercier: *Nucl. Fusion*, **3**, No. 2 (1963).

Boundary Layer Problems in Plasma Physics.

B. Bertotti

Laboratorio Gas Ionizzati (EURATOM - C.N.E.N.) - Frascati

1. – Boundary layer problems.

The purpose of these lectures is to point out a general method to solve a wide class of problems which often occur in plasma theory, as well as in many other branches of physics; an introduction will be given to the mathematical theory and a number of examples will be presented.

Two simple differential equations will show the kind of questions one has to deal with. The solution of

$$\varepsilon y'' + y = 0\,, \tag{1}$$

with the boundary conditions

$$y(0) = 0\,, \qquad y(1) = 1\,, \tag{2}$$

is

$$y(x, \varepsilon) = \frac{\sinh x\varepsilon^{-\frac{1}{2}}}{\sinh \varepsilon^{-\frac{1}{2}}}\;; \tag{3}$$

for $\varepsilon \to 0$ it exhibits a rather queer behaviour: for all values of $x \neq 1$, $y(x) \to \to y_0 \equiv 0$, which is the solution of (1) when one takes $\varepsilon = 0$; but its derivative becomes very large at $x = 1$. One could say, the limiting process $\varepsilon \to 0$ « drags » the solution toward $y = y_0$ as much as possible, but not completely, since the function $y(x)$ is « hanged » to the fixed right-boundary condition. A small region of rapid change (boundary layer) is thus generated which, eventually, becomes a discontinuity. Mathematically, we encounter a phenomenon of *nonuniform convergence* of $y(x, \varepsilon)$ to zero. It is clear from (3) that the thickness $\delta(\varepsilon)$ of the boundary layer is of $O(\sqrt{\varepsilon})$; more precisely, one can define δ

by the relationship

$$\delta = \int_0^1 dx\, y(x, \varepsilon) = \sqrt{\varepsilon}\, \text{tgh}\, \frac{1}{2\sqrt{\varepsilon}}\,. \tag{4}$$

δ is the base of a rectangle whose area is the area between $y(x, \varepsilon)$ and $y_0 = 0$ and whose height is $1 = y(1)$.

Notice also that, in general, eq. (1) shows the appearance of *two* boundary layers, corresponding to the two boundary conditions ($e.g.$ $y(-1) = y(1) = 1$). In our example one of them is absent, having chosen the left boundary condition to fit the fundamental solution $y_0 = 0$.

An entirely different behaviour occurs instead for the differential equation

$$\varepsilon y'' + y' + y = 0\,, \tag{5}$$

which is different from (1) in that, when $\varepsilon \to 0$, it reduces to a *differential* equation

$$y_0' + y_0 = 0\,.$$

We can then use the freedom left to us to fulfill the right boundary condition $y(1) = 1$ and have

$$y_0(x) = e \exp[-x]\,.$$

The actual solution of (5) and (2) for small ε

$$y(x, \varepsilon) = \frac{\exp[\alpha_+ x] - \exp[\alpha_- x]}{\exp[\alpha_+] - \exp[\alpha_-]} \qquad \left(\alpha = -\frac{1}{2\varepsilon}\left[1 \pm \sqrt{1 - 4\varepsilon}\right]\right) \tag{6}$$

does tend to $y_0(x)$ for all $x \neq 0$ and shows a single boundary layer at the origin. One can easily see that now the thickness is smaller:

$$\delta = \int_0^1 dx\, [y_0(x) - y(x)] = e\varepsilon + \ldots\,. \tag{7}$$

The critical feature of these kinds of problems is the fact that the differential equation changes its nature in the limiting process $\varepsilon \to 0$; indeed, boundary layers can be classified according to the kind of « catastrophe » they show. More specifically, the class of ∞^2 functions solutions of (1), for example, collapses to the single function $y_0 \equiv 0$; one can say, every boundary layer phenomenon corresponds to a freezing of degrees of freedom.

It is clear that the classical perturbation theory is not applicable to these cases. Such a theory is based on the fundamental theorem by Poincarè concerning ordinary differential equations containing a small parameter ε *analytically*; whereby it is meant that, having reduced the differential equations to first order for a vector x

$$\frac{\mathrm{d}x}{\mathrm{d}t} = F(x, t, \varepsilon)\,, \tag{7}$$

the vector F is analytic around $\varepsilon = 0$. The theorem says then that a solution $x(t, \varepsilon)$ of (7) for fixed initial conditions converges uniformly for $\varepsilon \to 0$ to a solution of

$$\frac{\mathrm{d}x}{\mathrm{d}t} = F(x, t, 0) \tag{8}$$

and, moreover, can be computed as a power series in ε by a trivial expansion procedure (*).

The main problem we are facing is, how to devise an expansion procedure in ε to calculate approximately the solution of a differential (or even integral) equation where the analyticity condition is lost and nonuniform convergence appears.

2. – Mathematical details.

Boundary layer problems have received surprisingly little attention by mathematicians, perhaps because of the fact that their variety and complexity arises in general from particular physical situations [1]. Many papers in the literature are concerned with the problem of the existence of some uniformly convergent solutions, which in practice correspond to « good » boundary conditions; but the physicist is, of course, interested in situations where the boundary layer does exist. As an indication to a rigorous theory, we quote here the three main theorems proved by WASOW [2] for the rather general differential equation

$$\varepsilon y'' = F_1(x, y, \varepsilon)\, y' + F_2(x, y, \varepsilon)\,, \tag{8}$$

with the boundary conditions:

$$y(\alpha) = l_1\,, \qquad y(\beta) = l_2 > l_1\,, \qquad (\alpha < \beta). \tag{9}$$

(*) See, *e.g.*, E. A. CODDINGTON and N. LEVINSON: *Theory of Ordinary Differential Equations* (New York, 1955).

The functions F_1 and F_2 are assumed to be regular and, in particular, analytic with respect to ε; moreover, they are such that the reduced differential equation

$$F_1(x, y, 0)y' + F_2(x, y, 0) = 0 \tag{10}$$

has a solution $y = y_0(x)$ which fulfils the second boundary condition.

The first theorem proves that there exist solutions $y = y_0(x, \varepsilon)$ of (8) which fulfil

$$y_0(\beta, \varepsilon) = l_2 ,$$

but not, in general, the other boundary condition, and converge uniformly to $y_0(x)$ in all the interval $l_1 \leqslant x \leqslant l_2$ as $\varepsilon \to 0$. One can say, therefore, the appearance of a boundary layer is to be put at the door of a « violent », though very probable, choice of a boundary condition. One could wonder then, wether it is possible to solve (8) and (9) approximatively when

$$\mu = l_1 - y_0(\alpha, \varepsilon) \tag{11}$$

is small. The second theorem answers affirmatively and indicats a way to construct a convergent expansion of $y(x, \varepsilon)$ in a power series with respect to μ:

$$y(x, \varepsilon) = \sum_{r=0}^{\infty} \mu^r y_r(x, \varepsilon) . \tag{12}$$

The functions $y_r(x, \varepsilon)$ are of the form

$$y_r(x, \varepsilon) = \exp\left\{-\frac{1}{\varepsilon}\int_{\alpha}^{x} dx' F_1(x', y_0(x'), 0)\right\} \times \begin{matrix}\text{bounded function}\\ \text{of } x \text{ and } \varepsilon > 0 ,\end{matrix} \tag{13}$$

which shows how the nonuniform convergence occurs in this problem. For small ε, y_r $(r > 0)$ is small for all values of x not very near to α.

We are interested, however, in an arbitrary value of μ, and in the detailed structure of the boundary layer thus arising. The trick is to consider a limiting procedure where not only y, but also the independent variable x depends on ε and approaches to left point α as $\varepsilon \to 0$. In other words, one considers the function

$$y(\alpha + \varepsilon z, \varepsilon) \equiv w(z, \varepsilon) \tag{14}$$

which, when z is fixed and finite, describes only points infinitesimally near to α. One can say, since the boundary layer is very small, we have to use a

« magnifying glass » to look at it, which « stretches » it to a finite width. The substitution (14) makes ε disappear as a coefficient of y'' and, in the limit, transforms (8) into the simpler equation

$$\frac{d^2w}{dz^2} = F_1(\alpha, w, 0)\frac{dw}{dz}\,. \tag{15}$$

This is the boundary layer equation, to be solved with the boundary conditions

$$w(0) = l_1\,, \qquad w(\infty) = y_0(\alpha)\,. \tag{16}$$

While the first is pretty obvious (it « hangs » the curve to the prescribed point), the second calls for at least an intuitive explanation. One can say, it is a matching condition consistent with the limiting procedure itself. If ε is very small, a large value of z corresponds to values of x still very near to α, for which $y_0(x)$ is practically equal to $y_0(\alpha)$; in other words, the boundary layer completes its decay to its asymptotic value before the fundamental solution $y_0(x)$ has appreciably moved off its boundary value.

The actual solution $y(x, \varepsilon)$ is then approximated by *two* different functions $y_0(x)$ and $w(z)$ in the two regions; both functions, however, are defined and can be computed over the complete domains $\alpha \leqslant x \leqslant \beta$ and $0 \leqslant z < \infty$, respectively. For a given ε, the approximation is the better, the farther or the nearer

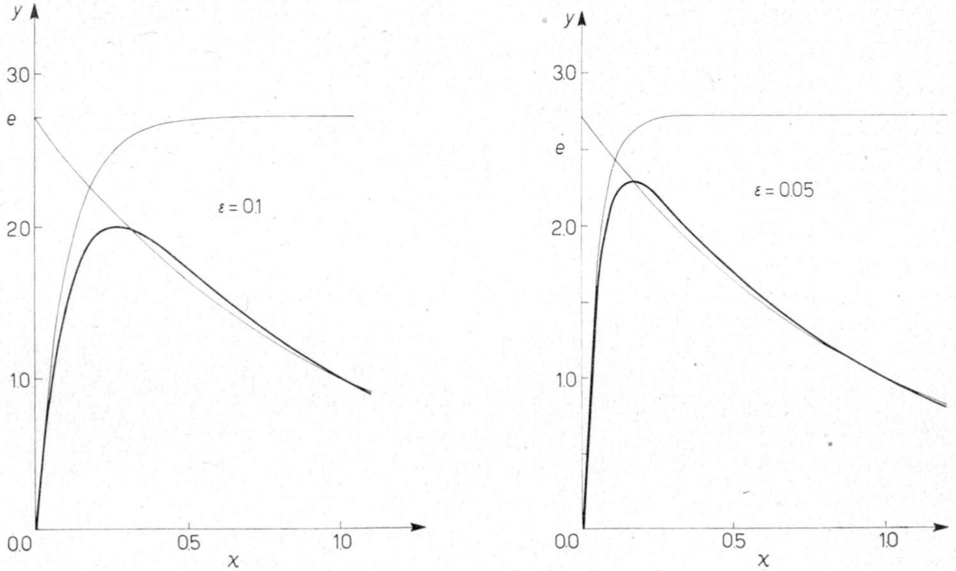

Fig. 1. – Solution of $\varepsilon y'' + y' + y = 0$, $y(0) = 0$, $y(1) = 1$ with the inner and outer approximations for $\varepsilon = 0.1$ and $\varepsilon = 0.05$.

to $\alpha\, x$ is, respectively. The problem is then reduced to the solution of two differential eqs. (10) and (15), each of which is much simpler than original one (8). The last theorem by WASOW provides for a rigorous justification of this method. A more accurate approximation in the boundary layer could be provided by an expansion of $w(x, \varepsilon)$ in ε; for more details and a far reaching generalization of this method, see [3].

In the case of eqs. (5) and (2) the boundary layer eq. (11) reads

$$\frac{d^2w}{dz^2} + \frac{dw}{dz} = 0 \qquad (x = \varepsilon z),$$

with

$$w(0) = 0\,, \qquad w(\infty) = e\,.$$

Hence

$$w(z) = e(1 - e^{-z})\,.$$

Figure 1 clearly indicates in what sense the two functions $y_0(x)$ and $w(z)$ approximate the rigorous solution (6).

3. – Boundary layer problems in physics.

The smallness of viscosity in ordinary fluids provides for the best known application of this theory; regions of sudden change generated in the limit of very large Reynolds number may be realized near walls or in shocks; the latter may, in a way, be termed free boundary layers between two uniform states of the gas. The difficulty of these cases, however, is confined to the boundary layer itself, because the analogue of the function $y_0(x)$ is a constant and the problem of the exterior boundary conditoin for $w(z)$ does not arise.

In plasma physics every charge sheath problem features two characteristic lengths, whose ratio is very small: the Debye length and the mean free path for collisions or ionization. It can be expected, therefore, that the transition from vacuum or from a conductor to a free plasma occurs in two stages, the first of which corresponds to a space-charge region, while in the second collisions slowly delete any trace of the presence of the boundary. The theory of an electrostatic probe, in particular, was extensively studied by a number of people, starting from BOHM [4-6]; their discussion, however, has a number of unsatisfactory points. They usually start from the concept of sheath thickness, while the main task of the theory should be to arrive at its definition and its explicit calculation: and assume a discontinuous transition between the sheath and the rest of the plasma. The asymptotic analysis of the problem shows instead that such a point does not exist, but a meaningful sheath concept can instead be defined and conveniently used in the calculation of physical quantities. More about this problem later.

Aside from this, a wide range of other problems is still open to the application of this method. The most interesting one, perhaps, is the transition between a plasma and magnetic field—a fundamental problem in the theory of confinement. One considers a situation where every quantity depends only on x and there is a magnetic field $\boldsymbol{B}(x)=\boldsymbol{e}_z B(x)$, with

$$B(-\infty)=0\,,\qquad B(\infty)=B_0\,. \tag{17}$$

While the purely magnetic transition is known [7] (which occurs rigorously when the masses of two species are equal), one would like to know the structure of the electrostatic potential $\varphi(x)$ generated by the charge separation. Qualitatively, the electrons, whose Larmor radius is smaller, should turn back before the ions, so that, as one proceeds from $-\infty$ to ∞, he encounters regions of negative and positive charge in succession. There are two basic questions to answer is concerned: what is the size of the two regions? Given the total potential jump $\varphi(\infty)-\varphi(-\infty)$, is it possible to construct a solution where all particles start from and go back to $x=-\infty$? Two characteristic lengths occur in this problem, the Debye radius λ_D and the thickness of the magnetic transition

$$\delta=\left(\frac{m_e c^2}{4\pi n_0 e^2}\right)^{\frac{1}{2}}. \tag{18}$$

In a nonrelativistic situation

$$\lambda_D \ll \delta$$

and the asymptotic analysis should apply (*).

I can point out at least two other examples of irregular perturbation theory in plasma physics, which, albeit they do not show the clear cut features of these examples, still present the typical « freezing » of degrees of freedom. The theory of orbits in a strong magnetic field is based on a formal expansion in powers of m/e [8], the coefficient of the acceleration. The drift motion is then determined only by four initial data (position and longitudinal velocity) instead of six. Even the relaxation from Liouville to the Fokker-Planck equation and then to thermal equilibrium constitutes a « contraction » in the description of the fluid from a function of $6N$ variables to one of 6 (the particle distribution in phase space) and next to five functions of position (the hydrodynamical density, temperature and velocity). These two relaxation processes occur in two different time scales (the duration of a collision and the mean free time)

(*) *Note added in proof.* - See *Ann. Phys.*, **25**, 271 (1963).

and, as it was done by FRIEMAN and SANDRI [3], can be discussed by expansion methods where also the time scale is involved.

One cannot exaggerate in saying that phenomena of nonuniform convergence play a prominent role in the mathematical schematization of most physical processes. For a general abstract theory of the differential equations with nearly periodic solutions, which is closely related to our topic, see KRUSKAL [8].

4. – Electrostatic probe in a strong magnetic field.

We will consider now an example which I have worked out in detail. It concerns the theory of charge-collection by an electrode whose size is larger than the Larmor radius of the collected particles, with special reference to an electrostatic probe, whose characteristic (voltage-current relation) is widely used as a diagnostic tool. In order to stress the application of the boundary layer technique to the problem, we confine the presentation of the physical aspects of the theory to the minimum necessary for understanding it and refer for full details to the two papers of the author [9].

We assume that the magnetic field is so strong as to impair collective transversal drifts; the region wherefrom changes are collected has then the shape of a long tube or force, whose section is determined by the probe's cross-section (which we take to be circular). A diffusion process—whose nature we need not specify here—exchanges continuously particles between the interior of the tube and the rest of the plasma; this introduces in the model a phenomenological parameter, the probability s that a particle is scattered outside per unit time. It can be shown that it is related to the transversal diffusion coefficient D by the order of magnitude relation

$$s \sim \frac{2D}{aR}, \tag{19}$$

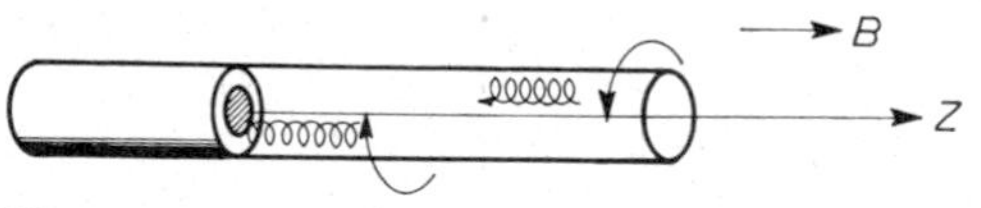

Fig. 2. – The geometry of the collection of charges by a small electrode in a very strong magnetic field.

where a is the Larmor radius of the collected particles and R the radius of the tube. Once a particle is inside the tube, it is assumed to fall freely in the electrostatic field of the probe, unless it is scattered outside again (see Fig. 2). Notice also that only particles of one kind are collected, so that the other ones have Maxwellian distribution.

As one can easily guess, this model may be described by Poisson's equation in one variable only, the longitudinal distance from the probe z; it is just this feature that makes it solvable. For the sake of simplicity, we assume also that the plasma is composed of two equal streams moving with velocities v_0 and $-v_0$ along $\boldsymbol{B}$; and that the diffusion process does not change the longitudinal velocity. Whether this fundamentally simple model is correct, only the experiments can give the final answer.

We now proceed to write down Poisson's equation in terms of the dimensionless variables

$$(20)\qquad \begin{cases} \varphi(z) = \dfrac{e}{kT} \times \text{electrostatic potential} \\[2ex] z = \left(\dfrac{4\pi e^2 n_0}{kT}\right)^{\frac{1}{2}} \times \text{distance from the probe.} \end{cases}$$

T is the temperature of the collected particles and e their (signed) charge. We consider the particular case

$$(21)\qquad 2\alpha\varphi = \frac{2kT}{mv_0^2}\,\varphi \ll 1\,,$$

which means essentially that the potential is so small that the collected particles (of mass m) do not feel appreciably its effect. (21) can be fulfilled in the case of electron collection in a gas of cold ions. It can be shown then that $\varphi(z)$ fulfills

$$(22)\qquad \frac{\mathrm{d}^2\varphi}{\mathrm{d}z^2} + \exp[-\varphi] - 1 + \frac{1}{2}\exp[-sz/v_0] = 0\,,$$

which shows that the collected particles' density is affected by the geometrical presence of the probe even a long distance away (the dimensionless parameter v_0/s measures the diffusion mean free path in units of Debye distance and is usually very large).

We have to solve (22) with the boundary conditions

$$(23)\qquad \varphi(0) = \varphi_P\,, \qquad \varphi(\infty) = 0\,.$$

The boundary layer character of this problem becomes apparent when the new variable

$$(24)\qquad z^* = sz/v_0$$

and the function

$$(25)\qquad \varphi^*(z^*) = \varphi\left(\frac{v_0}{s}\,z^*\right)$$

are introduced:

$$(22')\qquad \frac{s^2}{v_0^2}\frac{\mathrm{d}^2\varphi^*}{\mathrm{d}s^2} + \exp[-\varphi^*] - 1 + \frac{1}{2}\exp[-z^*] = 0\,.$$

One proceeds then in the following way.

a) Check that the reduced equation

$$\exp[-\varphi_\infty^*] - 1 + \tfrac{1}{2}\exp[-z^*] = 0\,, \tag{26}$$

i.e.

$$\varphi_\infty^*(z^*) = -\ln\left(1 - \tfrac{1}{2}\exp[-z^*]\right)\,, \tag{26'}$$

does not fulfil both boundary conditions (23). If $\varphi_P \neq \ln 2$, hence, a boundary layer exists. The solution (26′) is valid only for values of z^* not near the origin, where the boundary layer creating condition has been imposed; (26′), in fact, corresponds to a chargeless region, which extends far away into the plasma.

b) The potential profile near the origin must be investigated with the « squeezed » variable z and eq. (22) in the limit $s/v_0 \to 0$:

$$\frac{d^2\varphi_0}{dz^2} + \exp[-\varphi_0] - \frac{1}{2} = 0\,. \tag{27}$$

We are looking for a solution such that

$$\varphi_0(0) = \varphi_P\,, \qquad \varphi_0(\infty) = \varphi_\infty^*(0) = \ln 2 = b\,. \tag{28}$$

The second is the crucial matching condition and corresponds to the circumstance that φ_0 and φ_∞^* are practically constant outside and inside the sheath respectively. Notice that (27) offers a check of the method, since because of (28),

$$\lim_{z\to\infty} \frac{d^2\varphi_0}{dz^2} = 0\,,$$

as it should be.

The second of (28) shows that the collected particles at the edge of the sheath (where $\varphi \sim \ln 2$) have an energy

$$\tfrac{1}{2}mv_0^2 + kT\ln 2\,,$$

a fact already noticed first by BOHM for the case $B = 0$: the long potential tail outside the sheath accelerates the collected particles and gives them an extra energy of the order of the thermal energy of the other species. Bohm's criterion appears now in an entirely new light, which makes it rigorous and completely understandable. Its intuitive justification is always the same and consists in the equilibrium condition of the sheath edge. With cold ions the pressure exerted on it from the plasma is about $n_0 kT_e$, since the electron density is very low inside; this has to be balanced by the momentum transferred by the accelerated ions, which is about mv^2 per ion. One could say, the plasma is kept outside the sheath by the reaction of the ions.

The integration of (27) and (28) is easily performed with one numerical quadrature.

c) The last step is to compute the measurable quantities, the main of which in our case is the collected current. The role played by the two approximation to φ becomes then apparent.

One can show that the collected current density, for an arbitrary value of the parameter α (see (21)) is

$$j = \frac{1}{2} s n_0 e \int_0^\infty d\xi \exp[-st(\xi)] , \tag{29}$$

where the $t(\xi)$ is the time that a particle entering the tube at $z = \xi$ takes to reach the origin. In the lowest approximation in α

$$t(\xi) = \frac{\xi}{v_0}$$

and the current has the thermal value

$$j_{th} = \tfrac{1}{2} n_0 e v_0 . \tag{30}$$

Because of energy conservation, in an expansion in powers of α, we have

$$t(\xi) = \frac{1}{v_0}\int_0^\xi dz\,[1 + 2\alpha\varphi(z) - 2\alpha\varphi(\xi)]^{-\frac{1}{2}} = \frac{\xi}{v_0}[1 + \alpha\varphi(\xi)] - \frac{\alpha}{v_0}\int_0^\xi dz\varphi(z); \tag{31}$$

(29) then yields, after an integration by parts

$$j = j_{th} - \frac{1}{2} s n_0 e \alpha \int_0^\infty d\xi\,\xi\varphi'(\xi) \exp[-s\xi/v_0] . \tag{32}$$

The increase in j is due to the particles' acceleration, which makes them spend less time within the tube, so that fewer of them are lost. Because of the small coefficient α in front of the integral, we can use the approximations already computed for $\alpha = 0$.

To evaluate (32) we introduce provisionally the sheath thickness δ and use in each region of integration the suitable approximation and variable of integration:

$$j = j_{th} - \frac{1}{2} s n_0 e\alpha \int_0^\delta d\xi \, \xi \varphi_0'(\xi) \exp[-s\xi/v_0] - \frac{1}{2} n_0 e v_0 \alpha \int_{s\delta/v_0}^\infty d\xi^* \xi^* \varphi_\infty^{*\prime}(\xi^*) \exp[-\xi^*] .$$

In the limit $s/v_0 \to 0$ the second integral gives a constant, hence uninteresting, contribution to j; and the upper limit can be replaced with ∞, since for $\xi > \delta \varphi_0(\xi)$ is practically constant. The final result, after another integration by parts, reads

$$(33) \qquad j = j_{th} - \frac{1}{2} n_0 e v_0 \alpha \int_0^\infty d\xi^* \xi^* \varphi_\infty^{*\prime}(\xi^*) \exp[-\xi^*] + \frac{1}{2} s n_0 e\alpha \int_0^\infty d\xi [\varphi_0(\xi) - \ln 2] ,$$

which clearly exhibits the way how j varies with φ_P. Remembering the definition of sheath thickness given in Section **1**, we can write

$$(33') \qquad j = \text{const} + \frac{1}{2} s n_0 e\alpha\delta(\varphi_P) , \qquad \delta = \int_0^\infty d\xi [\varphi_0(\xi) - \ln 2] .$$

The general case for arbitrary values of α has been partially discussed. It shows two noteworthy complications. The plasma eq. (26) is replaced by an integral equation; moreover, the plasma solution appears explicitly in the sheath eq. (27), as evidence of the fact that the particles enter the sheath with a velocity determined by φ^*.

5. – Ionization near a wall.

CAVALIERE and CARUSO [10] have recently analysed the equilibrium situation which is produced near a plane wall by an ionizing agent; the second characteristic length is now the ionization mean free path. Collision of ions with neutrals or other ions are neglected, so that they, once they are created, undergo a free fall towards the wall under the electrostatic force. One assumes also that only ions are collected, the electrons being in thermal equilibrium. We will make no particular assumption about the rate $S(x)$ at which ions are generated in the unit volume; the theory may apply to a wide variety of situations. The most interesting one is a plane model of a low pressure mercury discharge, where the current flows between the electrodes. If the neutrals' density n_n is uniform, we should take

$$S(x) = \langle v\sigma \rangle_e n_n n_0 \exp\left(\frac{eV}{kT_e}\right)$$

($\langle v\sigma \rangle_e$ is the ionization cross-section, averaged over the electron distribution). If the discharge is operated near limiting current conditions, the spatial variation of n_n should be explicitely taken into account by discussing the neutrals' dynamics as they are emitted from the wall and diffuse into the plasma. In any case, the free fall assumption for the ions means that

$$\frac{1}{n_n \sigma} \gg \text{radius of the tube} = R\,,$$

where σ is either the collision or charge-exchange cross-section. Notice that, because of Bohm's criterion, the longitudinal distance an ion goes while moving from the center of the discharge to the wall is about

$$R\sqrt{\frac{T_i}{T_e}} < R.$$

It is easy to see that, if the initial energy of the ions is neglected (which corresponds to the limit $\alpha \to \infty$ of the preceding section), Poisson's equation reads

$$\text{(34)} \qquad \frac{d^2\varphi}{dx^2} = e^{\varphi} - \frac{1}{n_0}\left(\frac{m_i}{2kT_e}\right)^{\frac{1}{2}} \int_x^R d\xi \frac{S(\xi)}{\sqrt{\varphi(\xi) - \varphi(x)}} \qquad \left(\varphi = \frac{|e|V}{kT} < 0\right).$$

The Debye distance is unity. The abscissa $x = R$ corresponds to the center of the discharge and must be kept finite, lest the ion current be infinite. Notice also that the neglect of collisions prevents one from considering nonmonotonic solutions: the ions would in fact accumulate in a potential well and produce a nonstationary situation. Introducing S_0, the value of the production rate S in the free plasma, we see that the problem is characterized by the «ionization length»

$$\text{(35)} \qquad L = \frac{n_0}{S_0}\left(\frac{2kT_e}{m_i}\right)^{\frac{1}{2}} \gg 1\,.$$

Then (34) reads

$$\text{(34')} \qquad \frac{d^2\varphi}{dx^2} = e^{\varphi} - \frac{1}{L}\int_x^R d\xi \frac{\sigma(\xi)}{\sqrt{\varphi(\xi) - \varphi(x)}} \qquad \left(\sigma(x) = \frac{S(x)}{S_0}\right).$$

The plasma solution is obtained by going over to the plasma variable

$$x^* = x/L \tag{36}$$

and then taking the limit $L \to \infty$; which boils down to the charge neutrality condition

$$\exp[\varphi^*_\infty(x^*)] = \int_{x^*}^{R^*} d\xi^* \frac{\sigma^*(\xi^*)}{\sqrt{\varphi^*_\infty(\xi^*) - \varphi^*_\infty(x^*)}} \tag{37}$$

($\sigma^*(x^*) = \sigma(Lx^*)$; $R^* = R/L$ is assumed to stay finite in the limit). The trick to tackle the problem is to take $\varphi^*_\infty = \chi$ as the independent variable in (37):

$$e^\chi = \int_\chi^0 d\chi_1 \frac{\Sigma_\infty(\chi_1) x^{*\prime}_\infty(\chi_1)}{\sqrt{\chi_1 - \chi}} \qquad \big(\Sigma_\infty(x) \equiv \sigma^*(x^*_\infty(\chi))\big)\,. \tag{38}$$

$x^*_\infty(\chi)$ is the inverse of the monotone function $\chi = \varphi^*_\infty(x^*)$. Abel's theorem (*) allows an easy inversion of (38):

$$\Sigma_\infty(\chi) x^{*\prime}_\infty(\chi) = C(\chi) \equiv \frac{1}{\pi}\frac{d}{d\chi}\int_0^\chi d\chi_1 \frac{\exp[\chi_1]}{\sqrt{\chi_1 - \chi}}\,. \tag{39}$$

The function $C(\chi)$ (see Fig. 3), which plays an important role in our theory behaves like $(-\chi)^{-\frac{1}{2}}$ when $\chi \to -0$ and has a zero at $\chi = -0.853\,9$. For large $-\chi$ it is approximated by

$$C(\chi) \sim -\frac{1}{2\pi}(-\chi)^{-\frac{3}{2}}\,.$$

If $\Sigma_\infty(\chi)$ is known, like in the example mentioned, another quadrature yields the required solution $x^*_\infty(\chi)$. The plasma solution cannot go below the zero of $C(\chi)$, lest the single-valuedness of $\varphi^*_\infty(x^*)$ is lost. The major question at stake is to find Bohm's constant $b = \varphi^*_\infty(0)$; so far we know only that $b \geqslant -0{,}8539$.

The sheath equation is obtained by taking the limit $L \to \infty$ in (34′).

(*) See F. T. WHITTAKER and G. N. WATSON: *Modern Analysis* (Cambridge, 1958), p. 229.

Splitting the integration into the intervals belonging to the sheath and the plasma and using the appropriate variable and the appropriate approximation,

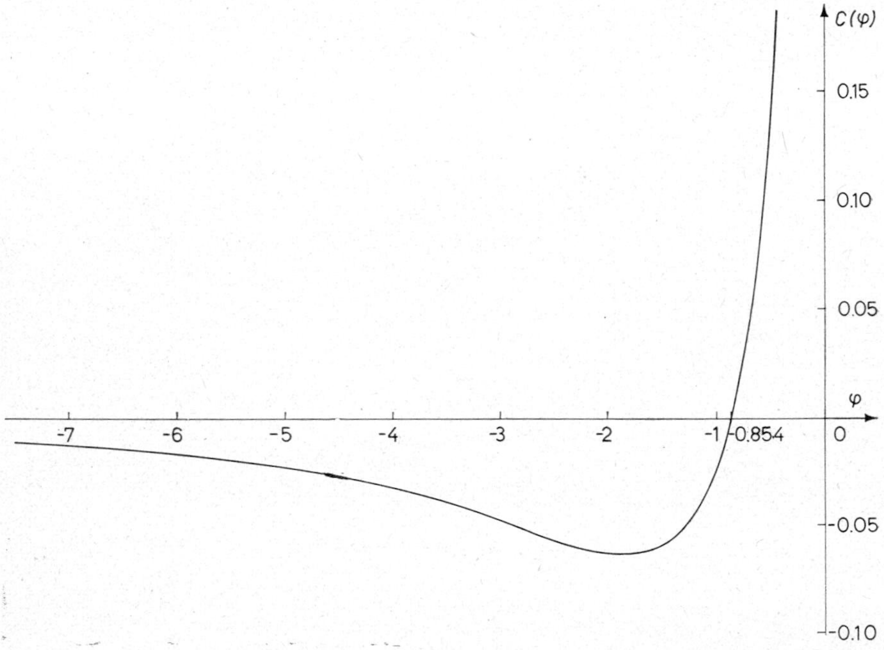

Fig. 3. – The function $C(\chi) = \dfrac{1}{\pi}\dfrac{\mathrm{d}}{\mathrm{d}\chi}\displaystyle\int_0^{\chi}\mathrm{d}\chi_1\,\frac{\exp[\chi_1]}{\sqrt{\chi_1 - \chi}}$.

it is easy to see that the first does not contribute, while the second can be extended right to the origin:

$$\frac{\mathrm{d}^2\varphi_0}{\mathrm{d}x^2} = e^{\varphi_0} - \int_0^{\infty}\mathrm{d}\xi^*\,\frac{\sigma^*(\xi^*)}{\sqrt{\varphi^*_\infty(\xi^*) - \varphi_0(x)}}\,. \tag{40}$$

That is to say, the sheath is so short that the ions generated within it do not contribute appreciably to its charge density. Using (39), (40) can be written in a form completely independent of the generation mechanism:

$$\frac{\mathrm{d}^2\varphi_0}{\mathrm{d}x^2} = e^{\varphi_0} - \int_b^{0}\mathrm{d}\chi\,\frac{C(\chi)}{\sqrt{\chi - \varphi_0(x)}}\,. \tag{41}$$

It can be proved that

$$e^{\varphi_0} = \int_{\varphi_0}^{0} \mathrm{d}\chi \frac{C(\chi)}{\sqrt{\chi - \varphi_0}}, \tag{42}$$

so that (41) reads, finally,

$$\frac{\mathrm{d}^2\varphi_0}{\mathrm{d}x^2} = \int_{\varphi_0}^{b} \mathrm{d}\chi \frac{C(\chi)}{\sqrt{\chi - \varphi_0(x)}} . \tag{43}$$

Because of reality requirements the sheath's solution must extend to the left of b; on the other hand the monotone character of the solution requires that $\mathrm{d}^2\varphi_0/\mathrm{d}x^2$ be nonpositive. Therefore, the interval of integration (φ_0, b) in (43) must cover only the region where $C(\chi) \leqslant 0$, thence

$$b \leqslant -0.8539 ;$$

with the previous inequality this fixes exactly the value of Bohm's constant:

$$b = -0.8539 . \tag{44}$$

The problem is now completely formulated and brought to analytical solution. The sheath eq. (43) with the boundary conditions (28) can be solved with a numerical quadrature of its first integral, which we write down for future reference:

$$\left(\frac{\mathrm{d}\varphi_0}{\mathrm{d}x}\right)^2 = -4\int_{\varphi_0}^{b} \mathrm{d}\chi\, G(\chi)\sqrt{\chi - \varphi_0(x)} . \tag{45}$$

6. – The collected current.

If a biased plane probe is inserted parallely to the wall in the discharge, its characteristic may give indications about the properties of the plasma. We leave aside here the classical theory of an electrostatic probe as developed by Bohm and discuss only the calculation of the current in our model and, in particular, the saturation effect.

Because of conservation of mass the collected ionic current density is

$$j = \int_{0}^{R} \mathrm{d}x\, S(x) \tag{46}$$

or, in dimensionless units (see (35) and (36))

$$\iota = \frac{j}{n_0}\left(\frac{m_i}{2kT_e}\right)^{\frac{1}{2}} = \int_0^{R^*} dx^* \sigma^*(x) \,. \tag{46'}$$

Just like in the previous example, in the limit $L \to \infty$ the current is a constant:

$$\iota_0 = \int_b^0 d\chi\, C(\chi) = \pi \int_0^b d\chi \frac{e^\chi}{\sqrt{\chi - b}} \,. \tag{47}$$

We have used here (39) and χ as the variable of integration. The collected current differs from L_0 by two terms: the contribution of the sheath

$$\frac{1}{L}\int_0^\delta dx\, \sigma(x) = \frac{1}{L}\int_{\varphi_P}^b d\chi \Sigma_0(\chi) x_0'(\chi)$$

$\big(x_0(\chi)$ is the inverse of the function $\chi = \varphi_0(x)$ and $\Sigma_0(\chi) \equiv \sigma(x_0(x))\big)$ and the term that we have to subtract from (47) because it should not cover the interval $(0, \delta)$:

$$\int_0^\delta dx^* \sigma^*(x^*) = \delta \Sigma_\infty^*(b) = \delta \Sigma_0(b) \,.$$

Combining together we find, finally

$$\iota = \iota_0 + \frac{1}{L}\int_{\varphi_P}^b d\chi \{\Sigma_0(\chi) - \Sigma_0(b)\}\, x_0'(\chi) \,. \tag{48}$$

Using (45) this reads, finally,

$$\iota = \iota_0 + \frac{1}{L}\int_{\varphi_P}^b d\chi \{\Sigma_0(\chi) - \Sigma_0(b)\}\left\{-\int_\chi^b d\chi_1\, C(\chi_1)\sqrt{\chi_1 - \chi}\right\}^{-\frac{1}{2}} \tag{49}$$

When $|\varphi_P| \gg 1$ the relevant contribution to (49) comes from $-\chi \gg 1$, for which

$$-\int_\chi^b d\chi_1\, C(\chi_1)\sqrt{\chi_1 - \chi} \sim \frac{1}{\pi}\sqrt{-\chi} \,.$$

If also $\Sigma_0(\chi) \to 0$ as

$$\iota \sim \iota_0 + \frac{4}{3}\sqrt{\pi}\,\frac{\Sigma_0(b)}{L}(-\varphi)^{\frac{3}{4}}. \tag{50}$$

We see that the saturation effect is very slight, in the sense that $d\iota/d\varphi_P$ goes to zero rather slowly as $-\varphi_P \to \infty$. Notice also that the sheath thickness δ goes with the same power of $-\varphi_P$. In any actual device it is possible that the increase in δ is actually limited by the size of the machine, thus giving rise to a much stronger saturation effect; moreover, when δ is very large the method of approximation may break down, in that ι may have an essentially nonlinear dependence from L^{-1}.

REFERENCES

[1] H. S. Tsien: *Adv. in Appl. Mech.*, Vol. **4** (New York, 1956).

[2] W. Wasow: *Comm. Pure and Appl. Math.*, **9**, 93 (1950); *Journ. Math. Phys.*, **23**, 75 (1944).

[3] G. Sandri: *The Foundations of Nonequilibrium Statistical Mechanics* (Rutgers University, Lecture Notes, 1962); a paper by G. Sandri and another by E. Frieman on this subject have to been submitted for publication to the *Journ. of Math. Phys.*

[4] D. Bohm, E. H. S. Burhop and H. S. W. Massey: Ch. 2 of the book *The Characteristics of Electrical Discharges in Magnetic Fields*, edited by A. Guthrie and R. K. Wakerling (New York, 1949).

[5] I. Tonks and I. Langmuir: *Phys. Rev.*, **33**, 1070 (1929); **34** 876 (1929).

[6] P. Auer: *Nuovo Cimento*, **22**, 548 (1961).

[7] H. Grad: *Phys. of Fluids*, **4**, 1366 (1961).

[8] M. Kruskal: *Journ. Math. Phys.*, **3**, 806 (1962); *Rendiconti del III Congresso Internazionale sui Fenomeni di Ionizzazione nei Gas* (Venezia, 1957) (this paper is reprinted in the present volume).

[9] B. Bertotti: *Phys. of Fluids*, **4**, 1047 (1961); **5**, 1010 (1962).

[10] A. Caruso and A. Cavaliere: *Nuovo Cimento*, **26**, 1389 (1962).

Introduction for Papers on Adiabatic Invariance.

R. M. KULSRUD

Project Matterhorn, Princeton University - Princeton, N. J.

One important invariant for a particle moving in a magnetic field is its magnetic moment $W_\perp/B$, where $W_\perp = v_\perp^2/2$ is the energy perpendicular to $\boldsymbol{B}$, the magnetic field. This invariant is not strictly a constant of the motion but only an approximate constant, the approximation getting better as the percentage change in B experienced by a spiralling particle during one gyration period approaches zero. Thus, if the relative change in B and E over one period is of order 0.001, one can show that the relative change in $W_\perp/B$ is of order 10^{-6} over one period or at most 0.001 in 1000 periods. This result follows from the lowest-order orbit theory. (From the simplest consideration one would have expected a relative change of 0.001 in one period.)

The constancy of $W_\perp/B$ represented by this result is not sufficient for most purposes since it is often the case that the particle makes millions or even billions of gyrations, and the constancy would soon be lost if it were only valid to this accuracy. An important example is the confinement of particles by magnetic mirrors for some reasonable length of time. This confinement depends directly on the constancy of the invariant for a large number of periods. Therefore, it is necessary to establish the constancy to higher order and, in fact, the goal was to show that the change of $W_\perp/B$ over many gyration periods went to zero faster than any power of the relative change in B over one gyration.

Preliminary to the actual solution of this problem by M. KRUSKAL, SPITZER suggested the consideration of the simpler problem of a simple harmonic oscillator with varying spring constant. This problem has a similar invariant whose constancy to higher order was a valid question. KULSRUD was able to show that the invariant of the harmonic oscillator was more constant than any power of the relative change of the spring constant over one period. The confidence generated by the existence of at least one such adiabatic invariant to all orders inspired KRUSKAL to attempt the more difficult problem, which he very beautifully resolved in the affirmative.

These two papers establishing the constancy of the adiabatic invariant of the oscillator and the spiralling particle follow. Both these papers are included as trivial cases in a much more general theory in a paper (*Journal of Mathematical Physics*, **3**, 806 (1962)).

Adiabatic Invariant of the Harmonic Oscillator (*).

R. M. Kulsrud

Project Matterhorn, Princeton University - Princeton, N. J.

1. – Introduction.

There are many problems in physics in which there exist quantities which change so slowly that they may be taken as constants of the motion to a high degree of accuracy. Any such quantity whose change approaches zero asymptotically as some physical parameter approaches zero or infinity is an adiabatic invariant. For instance, in Fermi's theory [1] for the acceleration of cosmic rays, it is assumed that the magnetic moment of a spiraling particle in a varying magnetic field remains constant. Combined with the conservation of energy this enables one to show that a magnetic field can reflect such a spiraling particle. The magnetic moment of this patricle is not really a rigorous constant but is nearly so if the relative space change in the magnetic field over the Larmor radius of the particle is small, or when the field is changing in time if its relative change during a Larmor period is also small. These conditions are satisfied to a high degree in many astrophysical situations [2].

The constancy of the magnetic moment to first order in these parameters of smallness was first derived by Alfvén [3] and was later shown to be true in the next order by Helwig [4] for a general field. Later, Kruskal [5] showed that it was valid to all orders for the special case of a particle moving in a magnetic field in the z direction which varies only in the y direction and a constant electric field in the x direction. From these results it seemed possible that the adiabatic constancy of the magnetic moment to all orders was a result of general validity.

That the magnetic moment of the particle is a constant in all orders would imply that any change in it must vanish more rapidly than any power of the parameter of smallness, *i.e.*, the relative change of the field over the Larmor radius. This does not imply that it must be a rigorous constant. For instance, $\Delta c = \exp[-1/\lambda]$ has this behavior since at $\lambda = 0$ all derivatives of Δc vanish,

(*) Reprinted from *Phys. Rev.*, **106**, 205 (1957).

An example of an adiabatic invariant in quantum mechanics would be the distribution over energy states of a system as the Hamiltonian is changed by external means, such as changing the volume of the boundaries of the system without adding heat to the system.

In order to approach the problem of the constancy of adiabatic invariants to all orders, this paper treats another simpler problem in which an adiabatic invariant exists. Consider the classic one-dimensional problem of an oscillator whose spring constant is slowly varied by some external means, such as a varying temperature, which only affects the motion through its spring constant. The counterpart of this problem was first considered by EINSTEIN [6] at the Solvay Congress of 1911 on the old quantum theory. LORENTZ asked how the amplitude of a simple pendulum would vary if its period were slowly changed by shortening its string. Would the number of quanta of its motion change? EINSTEIN immediately gave the answer that the action, E/ω, where E is its energy and ω its frequency, would remain constant and thus the number of quanta would remain unchanged, if $(1/\omega)(\mathrm{d}\omega/\mathrm{d}t)$ were small enough. BIRKHOFF [7] showed for problems such as these, that one can write the displacement

$$x = W(t) \sin\left(\int_0^t \omega(t)\,\mathrm{d}t + \delta\right),$$

where $\omega(t)$ is the frequency and $W(t)$ can be developed in a series which converges asymptotically. However, he did not evaluate the higher-order terms and find the variation of E/ω. This is easier to do if one writes

$$x = W(t) \sin\left(\int_0^t S(t)\,\mathrm{d}t + \delta\right),$$

and develops both W and S in asymptotic series. The relation between W and S may be chosen to simplify the problem and it can be seen that E/ω is indeed invariant to all orders in $(1/\omega)(\mathrm{d}\omega/\mathrm{d}t)$.

By generalizing this device KRUSKAL [8] is able to express the motion of a spiraling particle in a general electromagnetic field in terms of two such independent variables. Specifying a relation between them to simplify the problem, he is able to demonstrate that the magnetic moment of the particle is also invariant to all orders as has been suspected. It might be that there are many such invariants which are known to be constant to lowest order but which are actually invariant to all orders.

In this paper the details of the solution of the simple harmonic oscillator are given. The solution of the particle in the electromagnetic field will be published in a separate paper be KRUSKAL [8].

2. – Calculation.

Since, when ω is changing E is changing also, we restrict ourselves to the case where the frequency ω as a function of the time t is at first infinitely flat then changes in some arbitrary way, and finally becomes infinitely flat again. Thus the energy is well defined and constant in the initial and final regions and we compare E/ω in these two regions. In this case it is found that if ω has N continuous derivatives throughout, than the final value of E/ω is the same as its initial value to $N+1$ orders in $(1/\omega)(\mathrm{d}\omega/\mathrm{d}t)$. The change in the $(N+2)$-nd order is calculated. If $N=\infty$, that is if ω has all derivatives continuous, then E/ω is the same to all orders, although, as already remarked, this does not imply it is rigorously constant, since the result is only an asymptotic one.

If the mass is one, the equation of motion is

$$\mathrm{d}^2x/\mathrm{d}t^2+\omega^2(t/T)x=0\,, \tag{1}$$

where x is the displacement, $\omega(t/T)$ is a function with the properties listed above, and large T corresponds to small $(1/\omega)(\mathrm{d}\omega/\mathrm{d}t)$. Let $\tau=t/T$, so that eq. (1) becomes

$$T^{-2}\,\mathrm{d}^2x/\mathrm{d}\tau^2+\omega^2(\tau)x=0\,, \tag{2}$$

which is to be solved for large T. Since ω is changing slowly in time, we anticipate that x may be written in the form

$$x(\tau)=W\sin\left(T\int^{\tau}S\,\mathrm{d}\tau\right), \tag{3}$$

where W and S are slowly varying functions of τ, between which we may specify an arbitrary relation. Then

$$x'(\tau)=W'\sin\left(T\int^{\tau}S\,\mathrm{d}\tau\right)+TSW\cos\left(T\int^{\tau}S\,\mathrm{d}\tau\right), \tag{4}$$

$$x''(\tau)=(W''-T^2S^2W)\sin\left(T\int^{\tau}S\,\mathrm{d}\tau\right)+(2TSW'+TS'W)\cos\left(T\int^{\tau}S\,\mathrm{d}\tau\right), \tag{5}$$

where primes represent derivatives with respect to τ. Thus eq. (2) becomes

$$\begin{aligned}(W''/T^2-WS^2+\omega^2(\tau)W)\sin\left(T\int^{\tau}S\,\mathrm{d}\tau\right)&+\\ +T^{-1}(2W'S+WS')\cos\left(T\int^{\tau}S\,\mathrm{d}\tau\right)&=0.\end{aligned} \tag{6}$$

We choose the relationship between S and W to make the two coefficients in eq. (6) vanish separately, so that W and S are given by

$$W''/T^2 - WS^2 + \omega^2(\tau)W = 0\,, \tag{7}$$

$$2W'S + WS' = 0\,. \tag{8}$$

Equation (8) may be integrated to give

$$W^2S = a\,, \tag{9}$$

where a is a constant.

Let us assume that the oscillator is started in the initial region with a displacement x_0 and zero velocity. The initial conditions on W and S are chosen such that W is constant throughout the initial region. From eqs. (3), (4), (7) and (9).

$$S = \pm\,\omega \qquad \text{(in the initial region)}, \tag{10}$$

$$W = x_0 \qquad \text{(in the initial region)}, \tag{11}$$

$$\pm\,x_0^2\omega = a \qquad \text{(in the initial region)}. \tag{12}$$

We may take the positive sign in eqs. (10) and (12). Note that if we choose x_0 independent of T, a is also independent of T.

The energy is computed in the initial and final regions by calculating the kinetic energy when $x = 0$. Thus, by eqs. (3) and (4),

$$E = \tfrac{1}{2}x'^2/T^2 = \tfrac{1}{2}W^2S^2 = \tfrac{1}{2}aS \qquad \text{(when } x = 0)\,. \tag{13}$$

In the initial region, by eq. (10),

$$E/\omega = \tfrac{1}{2}aS/\omega = \tfrac{1}{2}a\,, \tag{14}$$

and if $S = \omega$ in the final region, than the value of E/ω would be the same in the two regions for all T. However, the representation (3) does not guarantee $S = \omega$, since many other functions S and W may represent a sinusoidal oscillation. For instance, if W changes in time, then eq. (7) shows that $S \neq \omega$ even though ω is constant. In general, after ω has varied in an arbitrary way the simple relations (10) and (11) will not hold in the final region.

To investigate the value of S in the final region, we therefore solve for S as a function of τ by developing S and W as asymptotic series in $1/T$.

$$S = S_0 + S_1/T + S_2/T^2 + \dots\,, \tag{15}$$

$$W = W_0 + W_1/T + W_2/T^2 + \dots\,. \tag{16}$$

Substituting these series in eqs. (7) and (9), we have, to lowest order,

$$S_0 = \omega\,, \tag{17}$$

$$W_0 = (a/\omega)^{\frac{1}{2}}\,. \tag{18}$$

The n-th order then reads, for an even n,

$$W''_{n-2} - W_{n-2}(S^2)_2 - W_{n-4}(S^2)_4 - \ldots - W_0(S^2)_n = 0\,, \tag{19}$$

$$(W^2)_n S_0 + (W^2)_{n-2} S_2 + \ldots + W_0^2 S_n = 0\,, \tag{20}$$

where we have introduced the abbreviations

$$(S^2)_m = S_m S_0 + S_{m-2} S_2 + \ldots + S_0 S_m\,, \tag{21}$$

and

$$(W^2)_m = W_m W_0 + W_{m-2} W_2 + \ldots + W_0 W_m\,. \tag{22}$$

In deriving eq. (19) we have used eq. (17) to cancel the ω term and in eq. (20) we have used the fact that a is independent of T. Note from the equations corresponding to (19) and (20) that all odd orders vanish since S_1 and W_1 vanish.

Equation (19) with (21) expresses S_n in terms of lower orders S_m and W_m and their derivatives. Similarly eq. (20) with (22) expresses W_n in terms of S_n and lower orders. Since S_0 and W_0 are given in terms of ω, we have found asymptotic series in $1/T$ for S and W which formally satisfy eqs. (7) and (9). Notice that W_n and S_n depend on ω and its first n derivatives. Thus if the $(N+1)$-st derivative of ω has a discontinuity at some time then S_{N+2} and W_{N+2} do also. Since x and x' must be continuous in all orders in $1/T$, the series given by (21) and (22) cannot represent a solution to $(N+2)$-nd order. This and higher orders in S and W must be found by solving eqs. (7) and (9) in regions where $\mathrm{d}^{N+1}\omega/\mathrm{d}^{N+1}$ is continuous and matching x and x' in these orders.

Suppose that ω has N continuous derivatives and that these N derivatives are zero in the final region. Then by eqs. (17) and (18) the first N derivatives of S_0 and W_0 are continuous and vanish in the final region. By eqs. (19) and (20) with eqs. (21) and (22), S_2 and W_2 have $N-2$ continuous derivatives, and S_2 and W_2 vanish together with these derivatives in the final region. Proceeding in this manner we find that S_N and W_N are continuous and vanish in the final region. In particular,

$$S_n = 0\,, \qquad 0 < n \leqslant N\,. \tag{23}$$

Comparing eq. (23) with eqs. (13) and (17), we see that in the final region

$$E/\omega = \tfrac{1}{2}a + S_{N+2}/\omega T^{N+2}\,, \tag{24}$$

so that E/ω is constant to $N+1$ orders in $1/T$. If ω has a jump z_{N+1} in its $(N+1)$-st derivative as a function of t, the relative change in E/ω is given by

$$\frac{\Delta(E/\omega)}{(E/\omega)} = -\frac{2z_{N+1}}{(2\omega)^{N+2}}\sin\left(2\varphi - \tfrac{1}{2}\,N\pi\right). \tag{25}$$

Here φ is the phase of the oscillator when it reaches the discontinuity. N may be either even or odd.

REFERENCES

[1] E. Fermi: *Phys. Rev.*, **75**, 1169 (1949).
[2] L. Spitzer: *Astrophys. Journ.*, **116**, 299 (1952).
[3] H. Alfvén: *Cosmical Electrodynamics* (Oxford, 1950), p. 19.
[4] G. Helwig: *Zeits. Naturfors.*, **10** a, 508 (1955).
[5] M. Kruskal: (private communication).
[6] P. Langevin and M. De Broglie: *La Théorie du Rayonnement et les Quanta* (*Report on meeting at Institute Solvay*, Brussels, 1911) (Paris, 1912), p. 450.
[7] G. D. Birkhoff: *Trans. Am. Math. Soc.*, **9**, 219 (1908).
[8] M. Kruskal: *Terzo Congresso Internazionale sui Fenomeni di Ionizzazione nei Gas* (Venezia, 1957), p. 562. This paper is reprinted here.

The Gyration of a Charged Particle (*).

M. KRUSKAL

Plasma Physics Laboratory - Princeton, N. J.

Let the mass of the particle be m and its charge e. The guiding-center approximation may be formalized by considering the ratio $\varepsilon = m/e$ to be numerically small, but the velocity of the particle and the electromagnetic fields to be finite. We are, therefore, interested in analysing the solution of the equation of motion of the particle, namely

$$\varepsilon\ddot{\boldsymbol{r}} = \boldsymbol{E}(\boldsymbol{r}, t) + \dot{\boldsymbol{r}} \times \boldsymbol{B}(\boldsymbol{r}, t) , \tag{1}$$

in the limit as ε approaches zero.

Because the order of (1) is reduced by formally going to the limit $\varepsilon = 0$, it is clear that simply a representation of $\boldsymbol{r}$ by a power series in ε is inappropriate. Instead, an asymptotic analysis is called for. It turns out to be appropriate to write

$$\boldsymbol{r} = \sum_{n=-\infty}^{\infty} \varepsilon^{|n|} \boldsymbol{R}_n(t) \exp[n\, C(t)/\varepsilon] , \tag{2}$$

where each vector function $\boldsymbol{R}_n(t)$ is itself a power series in ε, starting with a term of zeroth order. The characteristic function $C(t)$ can be chosen independent of ε, but it may (as we shall see) be convenient to choose it to be a power series in ε. The functions $\boldsymbol{R}_n$ and C may be complex, but $\boldsymbol{r}$ must be real.

The conditions on the $\boldsymbol{R}_n$ and C are found by substituting (2) and its derivatives into (1), expanding $\boldsymbol{E}$ and $\boldsymbol{B}$ in Taylor series around $\boldsymbol{R}_0$ (note that $\boldsymbol{r} - \boldsymbol{R}_0 = O(\varepsilon)$), and separately equating the corresponding coefficients of each exponential $\exp[nC/\varepsilon]$. These coefficients are themselves power series in ε, of course.

(*) Reprinted from *Terzo Congresso Internazionale sui Fenomeni di Ionizzazione nei Gas* (Venezia, 1957), p. 562.

For $n = 0$ this leads to

$$\varepsilon \ddot{\boldsymbol{R}}_0 = \boldsymbol{E} + \dot{\boldsymbol{R}}_0 \times \boldsymbol{B} + \varepsilon \dot{C}[\boldsymbol{R}_1 \times (\boldsymbol{R}_{-1} \cdot \nabla \boldsymbol{B}) - \boldsymbol{R}_{-1} \times (\boldsymbol{R}_1 \cdot \nabla \boldsymbol{B}] + O(\varepsilon^2) \,, \tag{3}$$

where here and later through (24) $\boldsymbol{E}$ and $\boldsymbol{B}$ and their derivatives are understood to have the arguments $\boldsymbol{R}_0$, t. The physical content of (3) to lowest order is that the velocity of the guiding center $\boldsymbol{R}_0$ orthogonal to $\boldsymbol{B}$ is $(\boldsymbol{E} \times \boldsymbol{B})/\boldsymbol{B}^2$, which is well known.

From (3) trivially follows

$$\boldsymbol{E} \cdot \boldsymbol{B} = O(\varepsilon) \,, \tag{4}$$

which is a condition not so much on the trajectory as on the given fields. Since physically $\boldsymbol{E}$ and $\boldsymbol{B}$ seem independent of ε, this would appear to require $\boldsymbol{E} \cdot \boldsymbol{B} = 0$. However, we certainly do not wish to be so severely restricted. The way out is to let the original fields $\boldsymbol{E}(\boldsymbol{r}, t)$ and $\boldsymbol{B}(\boldsymbol{r}, t)$ be given as power series in ε. Mathematically this is an entirely natural and almost trivial generalization, but it permits us to treat physical situations where the field vectors are not strictly orthogonal (though they must be so approximately; this will often be the case in plasmas of high electrical conductivity, for instance). We assume that $\boldsymbol{B} \neq O(\varepsilon)$ at any point on the trajectory.

The physical significance of (4) is that $\boldsymbol{E}$ can have no component along $\boldsymbol{B}$ to lowest order, because if it did, it would give the particle a large acceleration and hence a large velocity (of order $1/\varepsilon$). This would contradict the condition that the fields seen by the particle be nearly constant during one gyration period.

Only the components of $\dot{\boldsymbol{R}}_0$ orthogonal to $\boldsymbol{B}$ are determined directly from (3); information on the parallel component is hidden. The information may be uncovered by dotting with $\boldsymbol{B}$ and dividing through by ε, which yields

$$\ddot{\boldsymbol{R}}_0 \cdot \boldsymbol{B} = \varepsilon^{-1} \boldsymbol{E} \cdot \boldsymbol{B} + \dot{C}[\boldsymbol{R}_1 \times (\boldsymbol{R}_{-1} \cdot \nabla \boldsymbol{B}) - \boldsymbol{R}_{-1} \times (\boldsymbol{R}_1 \cdot \nabla \boldsymbol{B})] \cdot \boldsymbol{B} + O(\varepsilon) \,. \tag{5}$$

This is the desired condition. It determines the parallel component of the acceleration, rather than of the velocity.

Returning to the substitution of (2) into (1), from the terms involving $\exp[C/\varepsilon]$ (*i.e.* $n = 1$) we obtain

$$\dot{C}^2 \boldsymbol{R}_1 = \dot{C} \boldsymbol{R}_1 \times \boldsymbol{B} + O(\varepsilon) \,. \tag{6}$$

Now we do not permit $\dot{C}$ to be $O(\varepsilon)$, as this would vitiate the whole point of the representation (2). Hence we may divide through by $\dot{C}$ or $\dot{C}^2$.

Taken to lowest order, (6) may be construed as a homogeneous linear equation for $\boldsymbol{R}_1$. Indeed, we are presented with an eigenvalue problem, for $\boldsymbol{R}_1$

is not generally $O(\varepsilon)$. Since

$$\boldsymbol{B}\cdot\boldsymbol{R}_1 = O(\varepsilon) \tag{7}$$

(as seen by dotting (6) with $\boldsymbol{B}$), by using (6) twice we obtain

$$\dot{C}^2\boldsymbol{R}_1 = (\boldsymbol{R}_1\times\boldsymbol{B})\times\boldsymbol{B} + O(\varepsilon) \;=\; -\boldsymbol{B}^2\boldsymbol{R}_1 + O(\varepsilon)\,. \tag{8}$$

Thus we have determined the eigenvalues $\dot{C} = \pm i|\boldsymbol{B}| + O(\varepsilon)$, and if $\dot{C}$ were not permitted to depend on ε this would determine it completely (except for the trivial choice of sign). It greatly simplifies the formalism, however, to choose

$$\dot{C} = i|\boldsymbol{B}|, \tag{9}$$

which is a series in ε not only because $\boldsymbol{B}(\boldsymbol{r}, t)$ is, but also because so is $\boldsymbol{R}_0$, the argument of $\boldsymbol{B}$ in (9).

We pause to point out that although from (6), by dotting with $\boldsymbol{R}_1$, we deduce that $\boldsymbol{R}_1^2 = O(\varepsilon)$, this does not imply $\boldsymbol{R}_1 = O(\varepsilon)$. For $\boldsymbol{R}_1$ may be (indeed must be) a complex vector.

A further point to notice is that since $\dot{C}$ is purely imaginary, the condition we must impose to insure that $\boldsymbol{r}$ as given by (2) be real is

$$\boldsymbol{R}_{-n} = \boldsymbol{R}_n^*, \tag{10}$$

where the asterisk indicates the complex conjugate. We may, therefore, restrict ourxelves to the determination of $\boldsymbol{R}_n$ for $n\geqslant 0$.

Returning to (6), it is easily seen that the eigenvector $\boldsymbol{R}_1$ to lowest order is uniquely determined up to an arbitrary multiplicative complex scalar. For by (7) its component along $\boldsymbol{B}$ vanishes, while its components in any two mutually orthogonal directions both orthogonal to $\boldsymbol{B}$ are determinable from each other by (6) itself.

Let $\varepsilon\boldsymbol{F}$ denote the $O(\varepsilon)$ terms in (6), so that

$$\dot{C}^2\boldsymbol{R}_1 - \dot{C}\,\boldsymbol{R}_1\times\boldsymbol{B} = \varepsilon\boldsymbol{F}\,. \tag{11}$$

Viewing this as a linear system of algebraic equations for $\boldsymbol{R}_1$, we have determined $\dot{C}$ so as to make the system degenerate (which ensures that the corresponding homogeneous system has a nontrivial solution). It follows that there is a linear condition on the nonhomogeneous terms $\varepsilon\boldsymbol{F}$ for (11) to have a solution $\boldsymbol{R}_1$. It is easily verified that the necessary and sufficient condition is

$$\dot{C}^2 \boldsymbol{F} + \dot{C} \boldsymbol{F} \times \boldsymbol{B} + \boldsymbol{F} \cdot \boldsymbol{B}\boldsymbol{B} = 0 . \tag{12}$$

Carrying out the derivation of (6) to the next order gives

$$\boldsymbol{F} = \boldsymbol{R}_1 \cdot \nabla \boldsymbol{E} + \dot{\boldsymbol{R}}_1 \times \boldsymbol{B} + \dot{\boldsymbol{R}}_0 \times (\boldsymbol{R}_1 \cdot \nabla \boldsymbol{B}) - \ddot{C} \boldsymbol{R}_1 - 2\dot{C} \dot{\boldsymbol{R}}_1 + O(\varepsilon) . \tag{13}$$

This together with (12) gives to lowest order a first-order differential equation for $\boldsymbol{R}_1$. Since $\boldsymbol{R}_1$ has already been determined up to a complex multiplier, this amounts to a first-order differential equation for the dependence of the multiplier on t.

Returning once again to the substitution of (2) into (1), from the terms involving $\exp[nC/\varepsilon]$ $(n \geqslant 2)$ we obtain

$$n^2 \dot{C}^2 \boldsymbol{R}_n = n\dot{C} \boldsymbol{R}_n \times \boldsymbol{B} + \boldsymbol{G}_n , \tag{14}$$

where to lowest order (zeroth) $\boldsymbol{G}_n$ is a polynomial in the $\boldsymbol{R}_p$ $(1 \leqslant p \leqslant n-1)$ with space derivatives of $\boldsymbol{B}$ as coefficients. Since $\dot{C}$ as previously determined is not an eigenvalue of the homogeneous equation for $\boldsymbol{R}_n$ obtained from (14) by deleting $\boldsymbol{G}_n$, it follows that $\boldsymbol{R}_n$ is determined algebraically by (14).

We now have a complete set of recursion relations determining the $\boldsymbol{R}_n$ to every order (with respect to expansion in powers of ε) in terms of the lower order $\boldsymbol{R}_n$. The original exact eq. (1), formally an ordinary vector differential equation of second order, constitutes a sixth-order system of ordinary scalar differential equations. In our recursion relations these orders reapper as follows: Two orders from the two components of $\dot{\boldsymbol{R}}_0$ (which is real) orthogonal to $\boldsymbol{B}$ as determined by (3). Another two orders from the single component of $\ddot{\boldsymbol{R}}_0$ in the direction of $\boldsymbol{B}$ as determined by (5). None from (6) which determines $\boldsymbol{R}_1$ up to a complex scalar multiplication function. Finally, two more from the first-order complex scalar differential equation for that function.

Although our procedure has been entirely formal, is has been proved by BERKOWITZ and GARDNER [4] that the series (2) which we finally obtain really is an asymptotic series to all orders in ε for the exact solution of (1).

Incidentally, it turns out that by using (6) and (10) it is possible to write (5) in the form

$$\ddot{\boldsymbol{R}}_0 \cdot \boldsymbol{B} = \varepsilon^{-1} \boldsymbol{E} \cdot \boldsymbol{B} - \tfrac{1}{2} |\boldsymbol{R}_1|^2 \boldsymbol{B} \cdot \nabla B^2 + O(\varepsilon) . \tag{15}$$

Since the only way $\boldsymbol{R}_1$ (with two degrees of freedom) enters this equation is through its absolute value (with one degree of freedom), it is suggested that we examine how $|\boldsymbol{R}_1|$ varies. From our differential equation for $\boldsymbol{R}_1$, *i.e.*, (12) with (13), we can compute $|\boldsymbol{R}_1|^\bullet$. With the help of the Maxwell equations

$$\nabla \times \boldsymbol{E} + \frac{\partial}{\partial t}\boldsymbol{B} = 0\,, \qquad \nabla \cdot \boldsymbol{B} = 0\,, \tag{16}$$

we can reduce the resulting expression to

$$|\boldsymbol{R}_1|^\bullet = -|\boldsymbol{R}_1|\,|\boldsymbol{B}|^\bullet/2\,|\boldsymbol{B}| + O(\varepsilon)\,. \tag{17}$$

In view of this, $\boldsymbol{R}_0$ and $|\boldsymbol{R}_1|$ satisfy at each step of the recursion a system of differential equations of fifth order. The one remaining degree of freedom in $\boldsymbol{R}_1$ then satisfies a first-order differential equation involving $\boldsymbol{R}_0$ and $|\boldsymbol{R}_1|$.

It may be noted that (17) can be integrated explicitly to lowest order yielding

$$|\boldsymbol{R}_1|^2\,|\boldsymbol{B}| = k + O(\varepsilon)\,, \tag{18}$$

where k is a constant. This exhibits the well-known lowest-order constancy of the « magnetic moment » of the particle [1], although in the form written it more obviously exhibits the equivalent lowest-order constancy of the magnetic flux enclosed by the gyration of the particle, as viewed from the Galilean frame of reference in which the guiding center is instantaneously at rest.

Having completed the expansion of the equations of motion, we now claim that there is an adiabatic invariant which is constant to all orders in ε, and which is given by (18) to lowest order. (This has already been shown to the next order by HELLWIG [2].) It must be emphasized that this result is only asymptotic, *i.e.*, that constancy to all orders does not mean exact constancy, but merely that the deviation from constancy goes to zero faster than any power of ε. The adiabatic invariant, furthermore, is in general not just $|\boldsymbol{R}_1|^2|\boldsymbol{B}|$, but an infinite series whose leading term is $|\boldsymbol{R}_1|^2|\boldsymbol{B}|$. However, if the particle is in a region of space-time where the electromagnetic field is constant (both spatially and temporally), then the higher order terms vanish and the invariant is just $|\boldsymbol{R}_1|^2|\boldsymbol{B}|$.

The method of proof was suggested by Kulsrud's proof of an analogous statement for the harmonic oscillator [3]. If in (2) we replace the bracketed exponent by $ni\theta$ and specify that

$$\dot{\theta} = |\boldsymbol{B}|/\varepsilon\,, \tag{19}$$

we have in view of (9) an equivalent representation, in which $\boldsymbol{r}$ appears as a series of (nonnegative) powers of ε with coefficients which are functions of θ and t, periodic in θ with period 2π. From here on the argument proceeds quite generally for any system describable by a Hamiltonian; it is well known that a particle moving in a given electromagnetic field constitutes such a system.

Let $\boldsymbol{q}$ and $\boldsymbol{p}$ be canonical variables and $H(\boldsymbol{q}, \boldsymbol{p}, t)$ the Hamiltonian. Suppose $\boldsymbol{q}$ and $\boldsymbol{p}$ can be written as functions of two independent variables t and θ periodic in θ with period 2π, in such a way that there is a similar function α of t and θ with the property that if θ is made any function of t satisfying $\dot{\theta} = \alpha$ (independently of the initial value of θ), then $\boldsymbol{q}$ and $\boldsymbol{p}$ constitute a trajectory. (In our particular application here α is a function of t only, given by (19).) Using subscripts to denote partial derivatives, we have Hamilton's equations of motion

$$\left\{\begin{aligned} H_{\boldsymbol{p}} &= \mathrm{d}\boldsymbol{q}/\mathrm{d}t = \boldsymbol{q}_t + \boldsymbol{q}_\theta \alpha\,, \\ -H_{\boldsymbol{q}} &= \mathrm{d}\boldsymbol{p}/\mathrm{d}t = \boldsymbol{p}_t + \boldsymbol{p}_\theta \alpha\,. \end{aligned}\right. \tag{20}$$

Taking the inner product of these with $\boldsymbol{p}_\theta$ and $\boldsymbol{q}_\theta$ respectively and subtracting gives

$$\boldsymbol{p}_\theta \cdot \boldsymbol{q}_t - \boldsymbol{q}_\theta \cdot \boldsymbol{p}_t = H_{\boldsymbol{p}} \cdot \boldsymbol{p}_\theta + H_{\boldsymbol{q}} \cdot \boldsymbol{q}_\theta\,. \tag{21}$$

Remembering that θ and t are independent variables, we integrate with respect to θ from 0 to 2π. The right-hand side then obviously vanishes, and if the first on the left-hand side is integrated by parts the result can be written

$$-\int_0^{2\pi} (\boldsymbol{p} \cdot \boldsymbol{q}_{\theta t} + \boldsymbol{q}_\theta \cdot \boldsymbol{p}_t)\,\mathrm{d}\theta = 0\,. \tag{22}$$

But this is a perfect derivative with respect to t, whence

$$K = \int_0^{2\pi} \boldsymbol{p} \cdot \boldsymbol{q}_\theta\,\mathrm{d}\theta = \oint \boldsymbol{p} \cdot \mathrm{d}\boldsymbol{q}\,. \tag{23}$$

Thus the usual «adiabatic invariant» or action integral $\oint \boldsymbol{p} \cdot \mathrm{d}\boldsymbol{q}$ is indeed a constant K whenever and to whatever extent one can introduce an auxiliary variable θ with the properties specified.

In the application to the gyrating particle, it may be noted that in regions where the fields $\boldsymbol{E}$ and $\boldsymbol{B}$ are constant (spatially and temporally) and are orthogonal to all orders, it can be easily proved that $\boldsymbol{R}_0$ is a linear function of t, $\boldsymbol{R}_1$ is independent of t, and $\boldsymbol{R}_2$, $\boldsymbol{R}_3$, ... all vanish. When $\oint \boldsymbol{p} \cdot \mathrm{d}\boldsymbol{q}$ is evaluated in such a region, it turns out that K is given there (except for a trivial numerical factor) by $|\boldsymbol{R}_1|^2 |\boldsymbol{B}|$. To compute the higher-order terms at a general point is in principal straightforward but quickly becomes rather lengthy. To the next order above the lowest, however, it is not yet by any means prohibitive.

Actually, it is more useful to write the invariant in terms of familiar dynamical variables, here $\boldsymbol{r}$ and $\boldsymbol{v}$, where

$$(24)\qquad \boldsymbol{v} \equiv \dot{\boldsymbol{r}} = \sum_{n=-\infty}^{\infty} \varepsilon^{|n|}(n\dot{C}\boldsymbol{R}_n/\varepsilon + \dot{\boldsymbol{R}}_n)\exp[nC/\varepsilon]\,.$$

It is convenient to introduce

$$(25)\qquad \boldsymbol{v}_\perp \equiv \boldsymbol{v} - (\boldsymbol{B}\cdot\boldsymbol{v}\boldsymbol{B} + E\times\boldsymbol{B})/\boldsymbol{B}^2\,;$$

in (25) and the following equations $\boldsymbol{E}$ and $\boldsymbol{B}$ and their derivatives are to be evaluated at $\boldsymbol{r}$, and not at $\boldsymbol{R}_0$ as in the preceding formulas. Note that $\boldsymbol{v}_\perp$ is the component of the velocity $\boldsymbol{v}$ perpendicular to $\boldsymbol{B}$ in a frame of reference in which $\boldsymbol{E}$ vanishes instantaneously at $\boldsymbol{r}$.

Without giving any details or indications of the method of calculation, we merely state that the constant of motion written explicitly to the first two orders turns out to be

$$(26)\qquad |\boldsymbol{B}|^{-1}\boldsymbol{v}_\perp^2 + \varepsilon|\boldsymbol{B}|^{-5}\{-\tfrac{1}{2}\boldsymbol{v}_\perp^2\boldsymbol{v}_\perp\cdot(\boldsymbol{B}\times\nabla\boldsymbol{B}^2) + (\tfrac{1}{2}\boldsymbol{B}^2\boldsymbol{v}_\perp\boldsymbol{v}_\perp + \tfrac{3}{2}\boldsymbol{v}_\perp\times\boldsymbol{B}\boldsymbol{v}_\perp\times\boldsymbol{B}):(\nabla\boldsymbol{E} - \nabla\boldsymbol{B}\times\dot{\boldsymbol{R}}_0) - 2\boldsymbol{B}^2\boldsymbol{v}_\perp\cdot[\boldsymbol{E}_t + \dot{\boldsymbol{R}}_0\cdot\nabla\boldsymbol{E} + \dot{\boldsymbol{R}}_0\times(\boldsymbol{B}_t + \dot{\boldsymbol{R}}_0\cdot\nabla\boldsymbol{B}]\} + O(\varepsilon^2)\,;$$

here

$$(27)\qquad \dot{\boldsymbol{R}}_0 = \boldsymbol{v} - \boldsymbol{v}_\perp + O(\varepsilon) = (\boldsymbol{B}\cdot\boldsymbol{v}\boldsymbol{B} + \boldsymbol{E}\times\boldsymbol{B})/\boldsymbol{B}^2 + O(\varepsilon)\,.$$

The first term of (26) is the well-known lowest-order invariant [1]. If $\boldsymbol{E}$ vanishes identically the constant of motion can be simplified considerably and written

$$(28)\qquad |\boldsymbol{B}|^{-1}\boldsymbol{v}_\perp^2 - \varepsilon|\boldsymbol{B}|^{-5}[(\boldsymbol{v}\times\boldsymbol{B})\cdot(\nabla\boldsymbol{B})\cdot(\boldsymbol{v}^2\boldsymbol{B} + \boldsymbol{B}\cdot\boldsymbol{v}\boldsymbol{v}_\perp) + \boldsymbol{B}\cdot\boldsymbol{v}(\nabla\times\boldsymbol{B})\cdot(\tfrac{1}{2}\boldsymbol{v}_\perp^2\boldsymbol{B} + 2\boldsymbol{B}\cdot\boldsymbol{v}\boldsymbol{v}_\perp)] + O(\varepsilon^2)\,.$$

Similar methods can be used to prove that the adiabatic invariant for the anharmonic oscillator is constant to all orders. The same result holds for a large number of systems (at least when only one frequency is involved).

REFERENCES

[1] H. ALFVÉN: *Cosmical Electrodynamics* (Oxford, 1950).

[2] G. HELLWIG: *Zeits. Naturfors.*, **10** a, 508 (1955).

[3] R. KULSRUD: *Phys. Rev.*, **106**, 205 (1957).

[4] J. BERKOWITS and C. S. GARDNER: *On the Asymptotic Series Expansion of the Motion of a Charged Particle in Slowly Varying Fields*, NYO-7975 (1957).

Tipografia Compositori - Bologna - Italy